Clostridia

BIOTECHNOLOGY HANDBOOKS

Series Editors: Tony Atkinson and Roger F. Sherwood

PHLS Centre for Applied Microbiology and Research
Division of Biotechnology
Salisbury, Wiltshire, England

Volume 1   *PENICILLIUM* AND *ACREMONIUM*
Edited by John F. Peberdy

Volume 2   *BACILLUS*
Edited by Colin R. Harwood

Volume 3   CLOSTRIDIA
Edited by Nigel P. Minton and David J. Clarke

A Continuation Order Plan is available for this series. A continuation order will bring delivery of each new volume immediately upon publication. Volumes are billed only upon actual shipment. For further information please contact the publisher.

Clostridia

Edited by
Nigel P. Minton
and David J. Clarke

PHLS Centre for Applied Microbiology and Research
Division of Biotechnology
Porton Down, Wiltshire, England

Plenum Press • New York and London

Library of Congress Cataloging in Publication Data

Clostridia / edited by Nigel P. Minton and David J. Clarke.
    p.   cm. — (Biotechnology handbooks; 3)
    Includes bibliographical references and index.
    ISBN 0-306-43261-7
    1. Clostridium. 2. Clostridium — Biotechnology. I. Minton, Nigel P. II.
Clarke, David J. III. Series: Biotechnology handbooks; v. 3.
QR82.B3C56   1989   89-16076
660′.63 — dc20   CIP

© 1989 Plenum Press, New York
A Division of Plenum Publishing Corporation
233 Spring Street, New York, N.Y. 10013

Printed in the United States of America

Contributors

Jan R. Andreesen ● Institut für Microbiologie, Georg-August-Universität, D-3400 Göttingen, Federal Republic of Germany

Hubert Bahl ● Institut für Microbiologie, Georg-August-Universität, D-3400 Göttingen, Federal Republic of Germany

†Elizabeth P. Cato ● Department of Anaerobic Microbiology, Virginia Polytechnic Institute and State University, Blacksburg, Virginia 24061, U.S.A.

Gerhard Gottschalk ● Institut für Microbiologie, Georg-August-Universität, D-3400 Göttingen, Federal Republic of Germany

Peter Hambleton ● Division of Biologics, Public Health Laboratory Service Centre for Applied Microbiology and Research, Porton Down, Salisbury, Wiltshire SP4 0JG, England

Jeroen Hugenholtz ● Center for Biological Resource Recovery, and Department of Biochemistry, University of Georgia, Athens, Georgia 30602, U.S.A.

David T. Jones ● Microbiology Department, University of Cape Town, Rondebosch 7700, Cape Town, South Africa. *Present address:* Department of Microbiology, University of Otago, Dunedin, New Zealand

Raphael Lamed ● Center for Biotechnology, George S. Wise Faculty of Life Sciences, Tel Aviv University, Ramat Aviv, Israel

Lars G. Ljungdahl ● Center for Biological Resource Recovery, and Department of Biochemistry, University of Georgia, Athens, Georgia 30602, U.S.A.

Nigel P. Minton ● Molecular Genetics Group, Division of Biotechnology, Public Health Laboratory Service Centre for Applied Microbiology and Research, Porton Down, Salisbury, Wiltshire SP4 0JG, England

†Deceased.

J. Gareth Morris ● Department of Biological Sciences, University College of Wales, Penglais, Aberystwyth, Dyfed SY23 3DA, Wales

Badal C. Saha ● Michigan Biotechnology Institute, Lansing, Michigan 48910, U.S.A.

Clifford C. Shone ● Division of Biologics, Public Health Laboratory Service Centre for Applied Microbiology and Research, Porton Down, Salisbury, Wiltshire SP4 0JG, England

Erko Stackebrandt ● Institute for General Microbiology, Christian Albrechts University–Kiel, D-2300 Kiel, Federal Republic of Germany

Walter L. Staudenbauer ● Botanic and Microbiology Institute, Department of Microbiology, Technical University of Munich, D-8000 Munich 2, Federal Republic of Germany

Juergen Wiegel ● Center for Biological Resource Recovery, and Department of Microbiology, University of Georgia, Athens, Georgia 30602, U.S.A.

David R. Woods ● Microbiology Department, University of Cape Town, Rondebosch 7700, Cape Town, South Africa

Michael Young ● Department of Biological Sciences, University College of Wales, Penglais, Aberystwyth, Dyfed SY23 3DA, Wales

J. Gregory Zeikus ● Michigan Biotechnology Institute, Lansing, Michigan 48910, U.S.A.

Preface

To the uninitiated, the genus *Clostridium* is likely more to be associated with disease than biotechnology. In this volume, we have sought to remedy this misconception by compiling a series of chapters which, together, provide a practically-oriented handbook of the biotechnologic potential of the genus.

Clostridium is a broad grouping of organisms that together undertake a myriad of biocatalytic reactions. In the first two chapters, the reader is introduced to this diversity, both taxonomically and physiologically. In the following chapter, the current state of genetic analysis of members of the genus is reviewed. The remaining chapters concentrate on specific, exploitable aspects of individual *Clostridium* species—highlighting their range of unique capabilities (of potential or recognized industrial value), particularly in the areas of biotransformation, enzymology, and the production of chemical fuels. Fittingly, the final chapter demonstrates that even the most toxic of the clostridia can be of therapeutic value.

The contributors to this volume reflect the transnational interest in *Clostridium,* and we are indebted to each of them for making this volume possible. We particularly wish to acknowledge the contributions, both to this volume and to microbiology in general, of Dr. Elizabeth Cato, who, sadly, died shortly before publication of this volume. Finally, we would like to join the authors in recommending closer and wider consideration of the attributes and capabilities of this genus.

Nigel P. Minton
David J. Clarke

Porton, United Kingdom

Contents

Chapter 3

Genetics of *Clostridium* 63

Michael Young, Walter L. Staudenbauer, and Nigel P. Minton

Chapter 4

Solvent Production 105

David T. Jones and David R. Woods

Chapter 5

Acetogenic and Acid-Producing Clostridia **145**

Lars G. Ljungdahl, Jeroen Hugenholtz, and Juergen Wiegel

Chapter 6

Chapter 8

Toxigenic Clostridia 265

Clifford C. Shone and Peter Hambleton

Taxonomy and Phylogeny 1

ELIZABETH P. CATO† and ERKO STACKEBRANDT

1. INTRODUCTION

Bacterial taxonomy is the science that makes logical and rational communication possible among all scientists, microbiologists, physicians, biochemists, and others; indeed, all people who need to know and use microbiological information. Combining, as it does, the arts of classification and identification with stringent rules of nomenclature, there is much that remains subjective in the selection of limits allowed in defining each taxon. However, with the explosion of information that is accumulating regarding the chemical and genetic composition of bacterial cells, it is now becoming possible to approach the definition of limits so as to include in a taxon, at the level of either genus or species, only organisms that are truly closely related.

To classify bacteria into coherent units, as many properties as possible of pure cultures of the organisms must be determined and described. By selection of those characteristics that are stable and common to a group of organisms but that distinguish that group from other well-defined units, separate taxa can be constructed. Properties of new isolates can then be compared with those of accepted taxa and the new isolates identified. Where the new isolates are sufficiently unlike any that have been previously accepted, a new taxon must be described and an appropriate name assigned according to the International Code of Nomenclature of Bacteria prepared by Lapage *et al.* (1975).

For classification, a great deal of information must be gathered about the strains studied: morphological, biochemical, chemical, physiological, metabolic, genetic, etc. Much of this may never be needed again, but it should be available in the literature so that it can be related to as new

†Deceased.

_____

ELIZABETH P. CATO ● Department of Anaerobic Microbiology, Virginia Polytechnic Institute and State University, Blacksburg, Virginia 24061. ERKO STACKEBRANDT ● Institute for General Microbiology, Christian Albrechts University–Kiel, D-2300 Kiel, Federal Republic of Germany.

information is developed. Where only one or a few strains are described, they should be tested several times under different conditions to determine any variations that might exist. For identification, as few properties as possible are selected that distinguish one organism from others that are closely related. These properties must be common to all strains, stable, and preferably easy to determine by simple, rapid, reliable, and inexpensive tests.

2. HISTORICAL REVIEW

The genus *Clostridium* was described originally by Prazmowski (1880). With the limited information available at that time, the genus was separated from the genus *Bacillus* to include those gram-positive rods that are obligately anaerobic in their metabolism with central or subterminal heat-resistant spores that swell the cell. As the numbers of presumably different species that fit that description proliferated, there have been numerous proposals, principally on morphological grounds (position of spores, motility, gram stain, possession of capsule), to split the genus into smaller units (Chester, 1901; Weinberg *et al.*, 1937; Handuroy *et al.*, 1937; Prévot, 1938, 1940). With current routine diagnostic techniques, there has been little or no advantage gained from splitting the genus into smaller genera for purposes of identification, although most recent genetic data indicate that several only distantly related groups are included in the genus as it is currently defined. The definition adopted by Bergey *et al.* (1923), a slightly expanded version of that of Prazmowski, is essentially that given today in *Bergey's Manual of Systematic Bacteriology*, Volume 2 (Cato *et al.*, 1986). For inclusion in the genus, there are few requirements: anaerobic or micro-aerophilic spore-forming rods that do not form spores in the presence of air, are usually gram-positive, and do not carry out a dissimilatory sulfate reduction.

During the first half of this century, many species of clostridia were named, often with a sketchy description of a single strain, which might differ in one or two reactions from species already described. Some inconsistencies were due to the use of impure substrates or less than optimal methods or growth conditions, and some to the particular interest, capability, or bias of the different investigators. Many original strains were lost; others became contaminated and results could not be confirmed. This situation was greatly alleviated by the publication of the 1980 "Approved Lists of Bacterial Names" (Skerman *et al.*, 1980), the subsequent list of nomenclatural changes between 1980 and 1985 (Moore *et al.*, 1985), and, later, "Validation Lists" published in the *International Journal of Systematic Bacteriology*. Only those species that were adequately described, and for which a type strain was available and deposited in a recognized and accessible collection, were retained as valid species. Any names not included in these lists have no standing in nomenclature. Species included at this time are listed in Tables I and II. Descriptions of new species must be published in the *International Journal of Systematic Bacteriology* or references to those descriptions must be published in a Validation List of that journal. The type strain of each species must be designated and made available as above.

3. PHENOTYPIC CHARACTERIZATION

3.1. Circumscription of the Genus *Clostridium*

Criteria essential for inclusion in the genus are principally morphological and may be dependent on culture age and media used. In some species (e.g., *C. perfringens, C. ramosum, C. malenominatum*), spores are seldom if ever seen. Although chopped meat agar slants or egg yolk agar plates incubated at 5–10°C below optimal temperature (Holdeman *et al.,* 1977) will support spore production of most species, other media better suited to the metabolism of the organisms in question may stimulate sporulation (e.g., addition of purine to agar medium for *C. acidurici, C. cylindrosporum,* and *C. purinilyticum;* cellulose for *C. cellobioparum;* cellobiose for *C. thermocellum;* cellobiose and rumen fluid for *C. polysaccharolyticum;* fructose for *C. aceticum*). When spores cannot be demonstrated visually, clostridia often survive alcohol treatment or heating at 80°C for 10 min although some may only survive 70°C for 10 min (e.g., *C. symbiosum, C. colinum, C. coccoides, C. clostridioforme*). Freshly isolated strains of some species may not form spores (e.g., *C. spiroforme, C. clostridioforme*), but repeated subculture, lyophilization, or other unfavorable treatment selects for spores, while vegetative cells die, and they can be demonstrated more easily in a higher proportion of the cells.

The early requirement that clostridia be gram-positive remains, although gram-variable species are accepted. In many species, gram-positive cells can only be seen in very young cultures (e.g., *C. aminovalericum, C. durum, C. formicaceticum, C. kluyveri, C. pasteurianum*). In some species, only gram-negative cells are ever seen, but in those species where thin sections have been observed by electron microscopy (e.g., *C. bryantii, C. magnum, C. papyrosolvens, C. polysaccharolyticum*), the cells have a single-layered cell wall typical of gram-positive organisms. When *C. lortetii,* an anaerobic spore-forming halophile, was shown to have a gram-negative cell wall structure, it was transferred to a new genus, *Sporohalobacter* (Oren *et al.,* 1987). This represents a rare case where a valid species has been transferred from, rather than to, the genus *Clostridium.*

There is flexibility in the degree of anaerobiosis required for inclusion in the genus. While *C. haemolyticum* and *C. novyi* A require strictly anaerobic

Table I. Species of *Clostridium* Described since Publication of *Bergey's Manual of Systematic Bacteriology*, Volume 2

Species	% G + C	Gelatin liquefied	Glucose fermented	Fermentation products[a]	Distinctive characteristics	Source
C. aerotolerans, van Gylswyk and van der Toorn, 1987	39–41	–	+	AFl, ethanol, hydrogen, carbon dioxide	Low redox potential not required; will not grow in well-aerated medium; ferments xylan, not cellulose	Corn stover, sheep rumen
C. bryantii, Stieb and Schink, 1985		–	–	A(P), hydrogen	Utilizes only fatty acids with 4–11 carbon atoms; requires hydrogen-utilizing anaerobe as coculture; ferments 2-methyl butyrate to AP, hydrogen	Marine and freshwater mud
C. cellulolyticum, Petitdemange, Caillet, Giallo, and Gaudin, 1984	41	–	+	ALF, ethanol, carbon dioxide, hydrogen	Digests cellulose; carbohydrates fermented; no growth below 25°C	Decayed grass compost
C. cellulosolvens, Sleat, Mah, and Robinson, 1984	26–27	–	+	ABFl, carbon dioxide, hydrogen, ethanol	Digests cellulose and xylan, fermentation products; central, subterminal, oblong spores	Fermenting poplar wood
C. cylindrosporum, Barker and Beck, 1942; Andreesen, Zindel, and Dürre, 1985	27–30	–	–	AF, carbon dioxide, ammonia	Ferments only purines. Only DNA–DNA hybridization, immunological methods, or 16S rRNA catalog will differentiate from C. acidurici	Soil, chicken intestines
C. disporicum, Horn, 1987	38–41	–	+	ALbs, ethanol	Cells may contain two spores; requires fermentable carbohydrate	Rat cecum

Species				Products	Characteristics	Habitat
C. ferridus, Patel, Monk, Littleworth, Morgan, and Daniel, 1987	39	NR	–	AiVvibbl, ethanol	Optimum temperature 68°C; ferments xylan, not cellulose	Hot spring
C. lentocellum, Murray, Hofmann, Campbell, and Madden, 1986	36	–	+	AL, ethanol, hydrogen, carbon dioxide	Cellulolytic; optimum temperature 40°C; very large colonies on cellulose agar	River bank mud
C. magnum, Schink, 1984	29	–	+	A	Degrades 2,3-butanediol; citrate utilized, not formate	Anoxic freshwater sediments
C. pfennigii, Krumholz and Bryant, 1985	38	–	–	B from benenoids, A from pyruvate or CO	Growth only with added methoxybenzenoids, pyruvate, or CO	Bovine rumen
C. populeti, Sleat and Mah, 1985	28	+ slow	+	BLA, hydrogen, carbon dioxide	Cellulose and xylan fermented; fermentation products; terminal oval spores	Fermenting poplar wood
C. scindens, Morris, Winter, Cato, Ritchie, and Bokkenheuser, 1985	45	–	+	A, hydrogen, ethanol	Degrades steroids; relatively aerotolerant; fimbriae present	Human feces
C. stercorarium, Madden, 1983[b]	39	–	+	AL, ethanol, hydrogen, carbon dioxide	Optimum temperature 65°C; degrades cellulose; requires fermentable carbohydrate; ferments pentoses	Compost heap

[a]A, acetic acid; B, butyric acid; F, formic acid; L, lactic acid; P, propionic acid; s, succinic acid; ib, isobutyric acid; iv, isovaleric acid; V, valeric acid. Products are listed in order of amounts produced; products in parentheses not always produced; major acid products, upper case; minor acid products, lower case; NR, not reported.
[b]A description of this species is given in *Bergey's Manual*, p. 1200, but it is not included in the keys or tables.

**Table II. Distribution of *Clostridium*
Species and Types to Phylogenetically
Defined Groups**[a]

Group I[b]
  Subgroup I-A
    C. beijerinckii
    C. botulinum types B, E, F (nonproteolytic)
    C. butyricum (*C. pseudotetanicum*)[c]
  Subgroup I-B
    C. aurantibutyricum
    C. baratii (*C. paraperfringens, C. perenne*)
    C. paraputrificum
  Subgroup I-C
    C. perfringens types A–D (*C. plagarum*)
  Subgroup I-D
    C. carnis
    C. chauvoei
    C. sartagoforme
    C. scatalogenes
    C. septicum
  Subgroup I-E
    C. fallax
  Subgroup I-F
    C. botulinum types A, B, F (proteolytic)
    C. putrificum
    C. sporogenes
  Subgroup I-G
    C. cadaveris
    C. oceanicum
  Subgroup I-H
    C. botulinum types C, D
    C. haemolyticum
    C. novyi types A–C
  Subgroup I-J
    C. acetobutylicum
    C. pasteurianum
    C. tyrobutyricum
  Subgroup I-K
    C. botulinum type G
    C. histolyticum
    C. limosum
    C. malenominatum
    C. subterminale
Group II
  Subgroup II-A
    C. bifermentans
    C. cellobioparum
    C. difficile
    C. ghoni
    C. glycolicum
    C. lituseburense
    C. mangenotii

Table II. *(Continued)*

Group II *(Continued)*
   Subgroup II-A *(Continued)*
      C. rectum
      C. sordellii
      C. tertium
      C. tetani
   Subgroup II-B
      C. aceticum
      C. acidurici
      C. formicaceticum
      C. purinilyticum
Group III
   C. aminovalericum
   C. indolis
   C. oroticum
   C. sphenoides
Group IV
   C. barkeri
Group V
   C. thermaceticum
   C. thermosaccharolyticum
Group VI
   C. innocuum
   C. ramosum

[a]From Johnson and Francis, 1975; Fox *et al.*, 1980; Stackebrandt and Woese, 1981; Tanner *et al.*, 1981, 1982; Woese, 1987.

[b]Within group I, organisms are arranged in homology groups as defined by Johnson and Francis (1975), who also indicate strain designation. Most of the species in subgroup II-A were in the Johnson and Francis homology group II.

[c]Organisms investigated by Johnson and Francis (1975) that are listed within parentheses have been omitted from the list of validly described species because they were subsequently shown to be later synonyms of the species with which they are listed. Species of uncertain phylogenetic affiliation are as follows: *C. absonum, C. arcticum, C. celatum, C. clostridioforme, C. cocleatum, C. coccoides, C. colinum, C. durum, C. felsineum, C. hastiforme, C. irregulare, C. kluyveri,* * *C. leptum,* * *C. nexile, C. papyrosolvens, C. polysaccharolyticum, C. propionicum, C. puniceum, C. putrefaciens, C. quercicolum,* * *C. roseum, C. saccharolyticum, C. sardiniense, C. spiroforme, C. sporosphaeroides, C. sticklandii,* * *C. symbiosum, C. thermautotrophicum,* ** *C. thermocellum, C. thermohydrosulfuricum,* ** *C. thermosulfurigenes,* ** *C. villosum* together with those listed in Table I. The phylogenetic positions of organisms marked with * and ** were recently elucidated by reverse transcriptase sequencing of 16S rRNA (*Zhao, H., Yang, D., Bryant, M., Woese, C. R. Poster 125, session 25; **Bateson, M., Ward, D. M., Poster R10, session 24, respectively. 88th Annual Meeting of the American Society for Microbiology, Miami Beach, May 8–13, 1988).

conditions and will not grow if even traces of oxygen are present, species such as *C. aerotolerans, C. histolyticum, C. carnis, C. tertium,* and *C. durum* will grow, but not produce spores, on freshly prepared blood agar plates incubated aerobically. In all cases, growth is enhanced and spores are formed only under anaerobic conditions.

There are no limitations to the morphological characteristics of the rods. Whereas most species have straight or curved rods, two have definite spirals (*C. spiroforme* and *C. cocleatum*), while cells of one (*C. coccoides*) are frankly round to oval. Cells may be thick or thin, long or short, with blunt, rounded, or pointed ends; motile or nonmotile spores may or may not swell the cells; they may be terminal, central, or subterminal (sometimes all three); two spores per cell are found in two described species (*C. oceanicum* and *C. disporicum*). The single requirement of spore formation has thus grouped together species of wide diversity, both phenotypic and genetic, in one genus.

3.2. Circumscription of Clostridial Species

It is in the classification to species, and identification thereof, that the most dramatic changes have occurred in the last 25 years. With the ability to isolate and study the genetic material of the cell (Marmur and Doty, 1961, 1962; Johnson and Ordal, 1968), a solid stable basis has now been found which is not dependent on the interpretation of specific phenotypic tests from which to determine the acceptable limits of a species. While knowledge of the overall DNA base composition (mol% G + C) is a valuable preliminary tool for determining the relatedness of a group of organisms, it has limited usefulness in bacterial classification since otherwise totally unrelated strains may have the same % G + C. This criterion can only be used to exclude a strain from a taxon, a difference between strains of up to 4% being considered acceptable at the species level. The most reliable method for differentiation of bacterial species is to show that the nucleotide sequences of the DNA of two strains are very similar, as determined by competition experiments. Methods for DNA and RNA isolation and determination of DNA base composition, as well as methods for comparing sequence similarities between organisms, were recently reviewed by Johnson (1985). Once a group of genetically similar strains has been established, one can select with confidence those phenotypic tests that will identify most easily and rapidly a member of the group, without recourse to lengthy and expensive homology experiments.

3.2.1. Metabolic Properties

The first criteria now being used in the genus *Clostridium*, both for classification to species and for identification, are related to metabolic activity: some species are saccharolytic, some proteolytic, some neither, some both. The divisions here are clear, reliable, and easily determined. Where organisms produce large amounts of ammonia from peptones (e.g., *C. sporogenes, C. sordellii*), early fermentation of carbohydrates may be quickly masked by ammonia production, but usually there will be indications of weak acid production, particularly from glucose.

One of the most stable properties that has been found is the formation of products of the fermentation of carbohydrates or utilization of peptones (Moore *et al.*, 1966; Hammann and Werner, 1980). The development of gas-liquid chromatographic (GLC) analyses has made possible the rapid detection and identification of short-chain volatile and nonvolatile fatty acids, alcohols, and some gases (hydrogen, carbon dioxide, and methane) produced in broth cultures. Because product formation involves many enzymes, it is predictive of other properties of the organism and is thus a more useful taxonomic characteristic than one that depends on one reaction such as indole formation.

3.2.2. Chemotaxonomic Properties

During their studies on the chemical composition of bacterial cell walls, Cummins and Harris (1956) found that the distribution of amino acids, and particularly diaminopimelic acid which formed the peptidoglycan in cell walls, was useful as a taxonomic marker. As determined by hydrolysis and paper chromatography, both the amino acid composition and the sugar and amino sugar components proved to be stable characteristics, not affected by variations in culture media or growth conditions. They proposed that the pattern of amino acid cell wall components was distinctive of genera, while species within genera could be distinguished by the sugar components of the cell walls. Cummins and Johnson (1971) studied a large group of strains phenotypically similar to *Clostridium butyricum*, the type species of the genus. Deoxyribonucleic acid (DNA) from each strain was isolated, and cell wall amino acids and sugars were determined. From DNA homology experiments two main groups emerged: the *Clostridium butyricum* group with glucose as its only wall sugar, and a group designated *C. beijerinckii* with glucose and galactose as its wall sugars.

For practical purposes cell wall composition has not been used in the classification and identification of clostridia up to this time, although the peptidoglycan types have been determined for many of the species (Weiss *et al.*, 1981). It seems probable that these data will have increasing significance in describing phylogenetic groupings. For excellent reviews on the subject, see Schleifer and Kandler (1972) and Schleifer and Stackebrandt (1983).

Polyacrylamide gel electrophoresis (PAGE) of soluble cellular bacterial proteins is a powerful tool in both bacterial classification (Fox and McClain,

1974; Kersters and De Ley, 1980) and bacterial identification (Moore *et al.*, 1980). It can be used not only to screen multiple isolates that are phenotypically similar to select strains for more expensive DNA–DNA homology studies, but also to verify the identification of strains, or for primary identification of isolates by comparison with patterns from type strains. In most of the studies that were done using PAGE, sodium dodecyl sulfate (SDS) was added to the system to increase the yield of protein from cells. It also causes dissociation of some proteins. The main result was the appearance of more protein bands when the gel was developed. These made visual comparison of the patterns more difficult and were found to be unnecessary for comparable and consistent results (Moore *et al.*, 1980).

Using the methods of Moore *et al.* (1980), protein patterns of up to 20 strains of each of 70 *Clostridium* species were determined (Cato *et al.*, 1982a). In agreement with the findings of Cummins and Johnson (1971), homologous strains of *C. butyricum* had patterns nearly identical to those of the other but quite distinct from those of homologous strains of *C. beijerinckii*. The variation within each group was minor and similar to that found in straight biochemical testing. The two species could thus be differentiated without waiting for three serial transfers in media containing biotin but no other vitamins.

The separation of *C. bifermentans* and *C. sordellii* has traditionally been a problem. This is not surprising since Nakamura *et al.* (1975) found that strains of the two species are 50–70% related by DNA homology studies. All of the criteria they selected to separate them—urease production, fermentation of mannose, deamination of arginine, and presence of mannose in the cell wall—were shown to be to some extent variable (Nakamura *et al.*, 1975, 1976). However, the cellular protein pattern of the two species are distinct.

In a related study, the type strains of *C. perenne* and *C. paraperfringens* (both listed in the 1980 Lists as valid species; Skerman *et al.*, 1980) were found to have protein patterns identical to that of the type strain of *C. baratii*. Confirmation that *C. perenne* and *C. paraperfringens* were subjective synonyms of *C. baratii* was obtained by DNA homology studies (Cato *et al.*, 1982b).

While these examples demonstrate the usefulness of the PAGE method, it does have definite limitations, particularly in primary identification of unknown isolates. Relative band position on the gels may be affected by culture age and conditions of culture. Any variation in room temperature and in density of gels will allow proteins to migrate at different rates to different spots on the gel. While the presence on each gel of an internal standard compensates for this discrepancy when gels are being scanned visually, only very reproducible gel patterns can be scanned by computer techniques (Kersters and De Ley, 1980).

The presence of cellular fatty acids has been proposed for many years as being of taxonomic significance in the classification and identification of clostridia. Moss and Lewis (1967) examined 41 strains representing 13 species and found that both the presence and relative amounts of fatty acids with a chain length of 10–20 carbon atoms (as determined by GLC) could distinguish readily between *C. perfringens, C. sporogenes,* and *C. bifermentans,* as well as separate these from the other 10 species tested.

In an expanded study, Elsden *et al.* (1980) mapped the fatty acid composition of the lipids of 23 species of proteolytic clostridia. Using capillary GLC and GLC/mass spectrometry, they identified 55 fatty acids in the C12 to C18 range. For the most part, the acids formed reflected the catabolic patterns found by Mead (1971) in his study of amino acids utilized and formed by 18 species of proteolytic clostridia.

The principal drawback to these studies was that too few strains of too few species could be profitably studied without becoming bogged down in the data generated. It would not have been possible to handle the wealth of data generated from the large number of strains studied (data necessary to provide a solid base for classification and identification), without the concurrent developments and improvements in computer capability. These have made it possible to sort, store, and retrieve information provided by modern techniques and instrumentation.

With the Microbial Identification System developed by Hewlett-Packard, it may eventually be possible to develop a comprehensive and precise library of reference patterns of cellular fatty acids for automated identification of anaerobic bacteria. This system includes a gas chromatograph, a flame ionization detector with integrator and computer, which can identify 102 compounds, including fatty acid methyl esters, dimethyl acetals, aldehydes, and unknown compounds that are distinctive for individual species. Results to date compare well with results from DNA–DNA homology studies (W. E. C. Moore, personal communication), but until a complete computerized library of reference patterns can be compiled, it must remain a research tool. A large mass of cells is required for each determination and, for the reference patterns to be reliable and accurate, at least 20 strains of each species (or 20 runs on different days with fresh media and reagents where fewer than 20 strains are available).

4. PHYLOGENETIC ASPECTS

4.1. Elucidation of Close Relationships by DNA–DNA Hybridization

Determination of DNA homologies is recommended for determination of high relationships at the intra- and interspecies level. In actual fact, the species is the only taxonomic unit that can be defined in phylogenetic terms (Johnson, 1973; Wayne *et al.,* 1987). A compilation of methods in use today was made by Schleifer and Stackebrandt (1983). Although representative strains of *Clostridium* were studied by these techniques as early as 1980, the number of species considered is still small. In this context, only two examples should serve to demonstrate the usefulness of this approach; others are found in the section "Taxonomic Comments" of the description of the genus *Clostridium* by Cato *et al.* (1986).

In his pioneering work, Johnson (1970) demonstrated the genetic heterogeneity of strains phenotypically similar to *C. butyricum*, which composed two homology groups. In a following extended study, Cummins and Johnson (1971) defined *C. butyricum* phylogenetically and united strains labeled *C. fallax* and *C. multifermentans* with *C. butyricum*. Strains of homology group II, together with other strains of *C. multifermentans* and those of *C. beijerinckii*, *C. rubrum*, *C. lacto-acetophilum*, and *C. aurantibutyricum*, were named *C. beijerinckii*. A second example for the genetic diversity of a phenotypically defined species is *C. botulinum* (Lee and Riemann, 1970; Wu *et al.*, 1972). In their studies, not only the unrelatedness of proteolytic and nonproteolytic strains was demonstrated, but for the first time the genetic closeness of toxic strains (*C. botulinum*) with nontoxic strains (*C. sporogenes*) was demonstrated as well. All other species used in these studies shared very little DNA similarity, which in the light of subsequent phylogenetic studies could be explained by their high genetic diversity (see Section 4.2.2.1).

4.2. Elucidation of Remote Relationships

4.2.1. rRNA Cistron Similarities

The most extensive phylogenetic study on the genus *Clostridium* is based on measurement of ribosomal (r) RNA homologies (Johnson and Francis, 1975). Advantages of this study were the inclusion of a large number of described *Clostridium* species (more than 50% of those included in section 13 of *Bergey's Manual of Systematic Bacteriology*, Volume 2, Cato *et al.*, 1986) and the extension of phylogenetic relationships from highly related organisms, detected by DNA–DNA pairing studies, to more remotely related species. Fifty-six species were examined by the competition method but some of them (Tables I and II) lost their rank in the following years as a result of phylogenetic studies. Organisms found to be highly related by DNA–DNA hybridization were also indistinguishable by DNA–rRNA pairing. The majority of strains defined by a low DNA G + C content of 22–33 mol% fell into two groups (I and II) with intergroup homologies of 20–40%. Within group I, members of subgroups I-A to I-E appeared slightly more closely related among each other than to members of subgroups I-F to I-K. Subgroup I-A contained the type species of the genus *C. butyricum*. Intragroup relationships were above 39 and 69% similarity for groups I and II, respectively. A third group (III) of six strains with a low DNA G + C content showed negligible similarity with *C. ramosum* ATCC 25582 as a reference (13–17%). These values were about as low as those found for these six strains and representatives of groups I and II (8–33%). 23S rRNA from *C. innocuum* ATCC 14501, exhibiting a high DNA G + C content of 44 mol%, hybridized insignificantly only to rRNA cistrons of four other clostridial species with a similar DNA G + C content (10–38%).

As Johnson and Francis (1975) pointed out, the distribution of strains into different homology groups (I-A to I-K) correlated only poorly with the distribution of those phenotypic properties used for their identification. This is true for spore location and gelatin liquefaction, glucose fermentation, cell morphology, cell wall structure, and fermentation acids. Unfortunately, no nonclostridial gram-positive reference organisms with low DNA G + C content were included in the DNA–rRNA pairing study. Although the authors observed the high degree of unrelatedness between most of the species, they consequently failed to detect the phylogenetic incoherency of the genus *Clostridium* (see Section 4.2.2). Thus, at that time, the situation was comparable to that of the genus *Pseudomonas* (Palleroni *et al.*, 1973) whose status as a genus harboring phylogenetically highly diverse taxa and misclassified strains could not be recognized for the same reason.

4.2.2. rRNA Sequences: The Phylogenetic Heterogeneity of the Genus *Clostridium*

Data on clostridia have also been accumulated by 16S rRNA cataloging and complete 5S rRNA sequences. These methods have been dealt with extensively by Fox *et al.* (1977) and Stackebrandt *et al.* (1985a) for 16S rRNA, and by Donis-Keller *et al.* (1980) and Peattie (1979) for 5S rRNA. Techniques developed recently include rapid sequencing of rRNA by reverse transcriptase (Lane *et al.*, 1985) and of cloned rDNA cistrons by standard techniques. Results of the latter approaches have not yet been published for clostridia; the phylogenetic tree based on complete 16S rRNA sequences (Woese, 1987; Fig. 1) depicts the phylogenetic position of the gram-positive eubacteria but does not show any detail of its intradivisional structure. [The terms "eubacteria" and "eubacterial" refer to members of the kingdom *Eubacteria*, as defined by Woese and Fox (1977), and not to members of the genus *Eubacterium*.]

Branching points of ancient lines of descent can be sorted out more reliably by rRNA sequences than by the determination of rRNA cistron similarities. The advantages of these approaches have been discussed in detail by Balch *et al.* (1979), Stackebrandt and Woese (1981), Fox and Stackebrandt (1987), and Woese (1987). Although the data accumulated by DNA-rRNA pairing (similarities) and 16S rRNA cataloging (S_{AB} values) show a linear correlation above S_{AB} values of 0.50 (see Fox and Stacke-

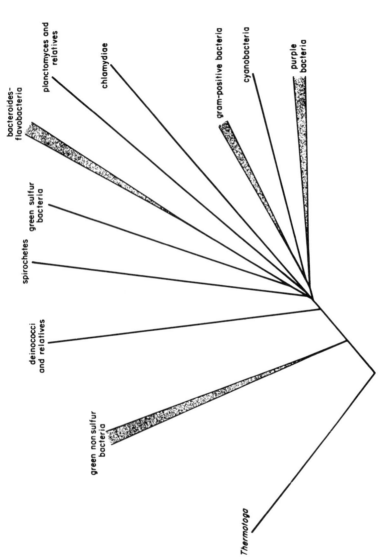

Figure 1. Eubacterial phylogenetic tree based on 16S rRNA comparisons (Woese, 1987, with permission of the author and the American Society for Microbiology). Depicted are the divisions (phyla). The genus *Clostridium* and allied taxa are enclosed in the gram-positive line of descent, which contains a deep bifurcation, separating the *Clostridium* and *Actinomyces* lines of descent. For tree construction method and source of sequences, see Woese (1987).

brandt, 1987 for references), the amount of scatter is such that even within regions of higher relationships it is difficult to merge the two types of data.

4.2.2a. 16S rRNA Cataloging.   The surprisingly high degree of phylogenetic heterogeneity of *Clostridium* became obvious when the 16S rRNAs of more than 120 gram-positive eubacterial species from 33 genera were analyzed by the 16S rRNA cataloging approach (Fox *et al.*, 1980; Stackebrandt and Woese, 1981; Tanner *et al.*, 1981, 1982; Woese *et al.*, 1985a,b; Stackebrandt and Fox, 1987; Ludwig *et al.*, 1988). Of the 18 species of *Clostridium* investigated, 7 were not included in the studies by Johnson and Francis (1975), which increases the number of validly described species that can today be placed phylogenetically to more than 60%. *Clostridium lortetii* (Oren *et al.*, 1983), on the other hand, was omitted from *Clostridium* (Oren *et al.*, 1987) because of the obvious lack of relationship to any member of this genus.

The genus *Clostridium* is a member of the *Clostridium/Bacillus* subdivision of gram-positive eubacteria (the other being the *Actinomyces* subdivision), whose main lines of descent are depicted in Figure 2. Subline I corresponds to homology group I (according to Johnson and Francis (see Table II), which harbors the majority of clostridial species, including the type strain. Subline II-A (homology group II) contains 12 species, while subline II-B is defined by 4 species, which were not included the study by Johnson and Francis (1975), i.e., *C. aceticum*, *C. acidurici*, *C. formicaceticum*, and *C. purinilyticum*. Subline III contains three species with a high G + C content of 41–44 mol % which could not be related by rRNA cistron homologies (*C. sphenoides*, *C. indolis*, *C. oroticum*; Woese and co-workers, unpublished) as well as *C. aminovalericum*, characterized by a lower G + C value (32 mol %). With average similarity coefficients (S_{AB} values) of 0.38, sublines I–III are equidistantly related, which confirms the low intergroup rRNA homologies. *Clostridium barkeri*, also defined by a high DNA G + C content of 44 mol %, constitutes subline IV, while *C. thermaceticum* and *C. thermosaccharolyticum* form subline V. *Clostridium innocuum* and *C. ramosum*, both being only remotely related, are defined as subgroup VI. Species which have not been included in phylogenetic studies as yet are listed in Table I and at the end of Table II.

The most exciting outcome of these studies was the finding that, in contrast to the present allocation of anaerobic, spore-forming, rod-shaped, and non-sulfur-reducing eubacteria into one genus, clostridial species did not form a phylogenetically homogeneous and exclusive group. Instead, these species were not only found in 6 independent and deep-branching sublines (abbreviated I–VI in Fig. 2), but most of these embraced non-clostridial species as well.

From this it follows that the genus *Clostridium* does not constitute a phylogenetically coherent taxon. In many examples, clostridial species are

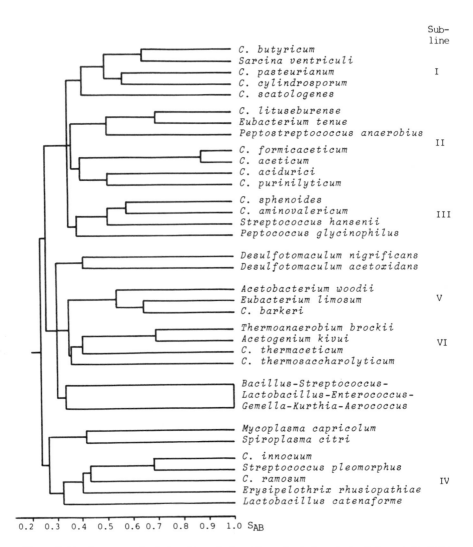

Figure 2. Dendrogram of relationships of clostridia and relatives derived from 16S rRNA catalogues (Fox *et al.*, 1980; Tanner *et al.*, 1981, 1982; Ludwig *et al.*, 1988). For the internal structure of the broad *Bacillus* cluster, see Stackebrandt *et al.* (1987) and Ludwig *et al.* (1988). S_{AB} is defined as the ratio twice the sum of bases of oligonucleotides (length greater than five) common to two catalogues A and B, to the sum of all bases (in oligonucleotides of length greater than five) in the two catalogues (Woese, 1987). As stated by Woese (1987), the cataloging approach is an imprecise measure on relationship below S_{AB} values of 0.4. Since no refined branching patterns have been published so far, this dendrogram presents the most comprehensive picture of the intrasubdivisional relationships.

more closely related to anaerobic non-spore-forming rods or anaerobic spore-forming spherical bacteria, or even wall-less gracilicutes, than they are related among themselves. Figure 2 depicts these intergeneric relationships of clostridia with members of *Sarcina, Peptococcus, Peptostreptococcus, Desulfotomaculum,* and *Eubacterium,* the facultatively anaerobic and aerobic taxa of the *Bacillus* subline of descent, mollicutes, and others. As published previously, certain eubacteria with gram-negative staining behavior, e.g., *Selenomonas, Sporomusa,* and *Megasphaera* (Stackebrandt *et al.,* 1985b), *Butyrivibrio* (Woese, 1987), and even the phototrophic *Heliobacterium chlorum* (Woese *et al.,* 1985a), are more closely related to clostridia then they are related to other gram-negative eubacteria.

4.2.2.b. 5S rRNA Sequences. Only five *Clostridium* species have so far been investigated with respect to 5S rRNA primary structures, namely, *C. bifermentans, C. butyricum, C. innocuum, C. pasteurianum,* and *C. tyrobutyricum* (Pribula *et al.,* 1976; Rogers *et al.,* 1985; Dams *et al.,* 1987). Their relationships with other gram-positives have been illustrated by Dams *et al.* (1987). Compared to 16S rRNA cataloging, data from 5S rRNA sequences demonstrate marked differences. These are (1) the intermixing of the gram-positive with a low DNA G + C content with actinomycetes, various subclasses of proteobacteria, and *Thermus* species; (2) the clustering of *C. tyrobutyricum* (rRNA homology subgroup I-J) with bacilli, and (3) the rather close relationship of *C. bifermentans* (subgroup II-A) with lactobacilli and staphylococci. Some of the discrepancies, caused by the small sample number and by differences in the evolutionary rate of 5S rRNAs, were discussed by Dams *et al.* (1987). A more plausible branching order of sublines is depicted by the analysis of 16S rRNA catalogues, but even this pattern will be improved when complete 16S rRNA sequences become available.

5. EVOLUTIONARY AND TAXONOMIC ASPECTS

Since clostridial species are present in all major sublines of the *Clostridium/Bacillus* subdivision, rod-shaped morphology, spore formation, and fermentative type of nutrition appear to be the dominating phenotypes of that subdivision. Moreover, since *Clostridium* species are the deepest branching organisms within the individual sublines, these characteristics have to be viewed as ancient features present in the ancestor of *Clostridium.* Assuming a rather constant evolutionary rate for RNAs, the strong radiation of clostridial species into various independent sublines must have occurred within a rather short period, after the ancestors of the *Clostridium/Bacillus* and the *Actinomyces* subdivisions separated from each other. Clostridia and related taxa belong to the so-called age group I, i.e., those genera whose lowest similarity coefficient separating species indicates an early evolutionary origin (Stackebrandt, 1988). The ancestral form evolved during the anaerobic phase of evolution and its descendants still live in niches, which may reflect their ancient conditions. Nonclostridial morpho- and physiotypes derived there independently from each other in different sublines but certain other features were retained, e.g., in *Desulfotomaculum* (morphology, anaerobic, spore formation), *Eubacterium* (morphology, anaerobic, metabolism), or *Acetobacterium*, and the various genera harboring spherical forms (anaerobic). The clostridia should therefore not be considered oxygen-sensitive mutants of *Bacillus* as postulated by Lee and Riemann (1970).

There is no question that the genus *Clostridium* as defined today is highly diverse and workers in the field would easily accept an improved taxonomic treatment. One of the major advantages of phylogenetic studies is that a hierarchical system that is compatible with natural relationships can be approached. This topic and the resulting problems with regard to feasibility and practicality have been dealt with by Balch *et al.* (1979), Stackebrandt and Woese (1984), and Stackebrandt (1988). The question of whether to use exclusively the phylogenetic data for the delineation of taxa or to have a combined approach which includes phylogeny and phenotype is under discussion.

At present it seems unwise to base ranks exclusively on geometric parameters, e.g., reassociation values or similarity coefficients, but molecular-genetic and phenotypic data should be interfaced. Using the phylogenetic tree as a skeleton, ranks may be defined by a broad range of phylogenetic parameters which are then adjusted to coincide with phenotypic features consistent with the genealogy of the clostridia. Since, from a practical point of view, the lower taxa like species and genera are the most important ones, these ranks should be phenotypically coherent. Families and higher taxa may reflect a higher degree of phenotypic diversity, such as presence or absence of spores and morphological differences. Examples of this strategy (Table III) have already been informally proposed by Fox and Stackebrandt (1987). Consequently, the genus *Clostridium* has to be retained for organisms forming group I, embracing the type species *C. butyricum*, but the parameters of the revised genus have yet to be defined. Species forming groups II–VI will lose their status as members of *Clostridium* and each of them will independently await new taxonomic treatment.

6. TOXINS, PHAGES, AND BACTERIOCINS OF CLOSTRIDIA

References for phages and bacteriocin may be found in the excellent reviews of Mahony (1979) and Reanney and Ackermann (1982). Most studies on clostridial phages have been performed on pathogenic species, es-

Table III. Possible Hierarchical Structure for the Gram-Positive Bacteria, as Defined by Fox et al., 1980[a]

Division: Gram-positive
  Class 1: *Clostridia*
    Order 1: Clostridiales
      Family 1: Clostridiaceae I
        Genera: *Clostridium butyricum* and relatives, (Group I according to Johnson and Francis, 1975) *Sarcina*
      Family 2: Clostridiaceae II
        Genera: *Clostridum lituseburense* and relatives, (Group II according to Johnson and Francis, 1975) *Peptostreptococcus anaerobius, Eubacterium tenue*
      Family 3: Clostridiaceae III
        Genera: *Clostridium sphenoides* and relatives, *Peptococcus glycinophilus, Streptococcus hansenii*
    Order 2: Desulfotomaculales
      Family 1: Desulfotomaculaceae
        Genera: *Desulfotomaculum, Ruminococcus*
          Uncertain affiliation: *Peptococcus aerogenes*
    Order 3: Mycoplasmatales
      Family 1: Mycoplasmataceae (true mycoplasma)
        Genera: *Mycoplasma, Spiroplasma, Acholeplasma, Anaeroplasma, Ureaplasma*
      Family 2: Clostridiaceae IV
        Genera: *Clostridium innocuum, Clostridium ramosum, Lactobacillus catenaforme, Erysipelothrix, Streptococcus pleomorphus*
    Order 4: Acetobacteriales
      Family 1: Acetobacteriaceae
        Genera: *Acetobacterium, Eubacterium limosum, Clostridium barkeri*
      Family 2: Thermoanaerobiaceae
        Genera: *Thermoanaerobium, Acetogenium, Clostridium thermaceticum, Clostridium thermosaccharolyticum*
    Order 5: Bacillales
      Family 1: Bacillaceae
        Genera: *Bacillus, Staphylococcus, Brochothrix, Listeria, Gemella, Thermoactinomyces, Caryophanon, Filibacter, Planococcus, Sporolactobacillus, Sporosarcina*
      Family 2: Streptococcaceae
        Genera: *Streptococcus, Enterococcus, Lactococcus*
      Family 3: Lactobacillaceae
        Genera: *Lactobacillus, Pediococcus, Leuconostoc*
      Of uncertain affiliation: *Aerococcus, Kurthia*
    Of uncertain affiliation: *Lactobacillus minutus, Heliobacterium chlorum, Heliobacillus mobilis, Butyrivibrio, Sporomusa, Selenomonas*
  Class 2: Actinomycetes

[a]Updated version of the informal proposal of Fox and Stackebrandt (1987), including some organisms whose phylogenetic position was recently determined.

pecially those associated with gas gangrene (*C. perfringens, C. histolyticum, C. sporogenes*), botulism (*C. butlinum*), and *C. tetani*. Historically, toxicity and pathogenicity have played a major role in the classification and, hence, taxonomy of the pathogenic or toxigenic clostridia. Such species or toxin types were defined by neutralization of the toxic or pathogenic effect by antisera which were usually prepared with only partially purified immunogens. This approach led to some unusual classification and nomenclature, from a phylogenetic point of view. The most obvious example is that of the relationships among the toxin types of C. *botulinum*, where the species has been defined by its production of a neurotoxin with certain pharmacological effects (Smith, 1977). In some cases (e.g., proteolytic strains of C. *botulinum*), the same genotype produces different toxins; in other cases, the same toxin is produced by different genotypes (e.g., C. *botulinum* type B toxin by proteolytic and nonproteolytic strains). Although the general trend has been to preserve the name C. *botulinum* as *nomen periculosum* (Minute 7, Judicial Commission of the International Committee on Systematic Bacteriology, 1987, *Int. J. Syst. Bacteriol.* **37**:85), Suen *et al.* (1988) proposed a new species for strains currently classified as C. *botulinum* type G. The new species, C. *argentinense,* also includes some nontoxigenic strains previously identified as C. *subterminale* or C. *hastiforme.* Since C. *botulinum* type G is relatively rare, this change in nomenclature will probably be well accepted. The major significance of this change is that it sets a precedent for more extensive changes in the nomenclature of the other genogroups presently recognized as members of C. *botulinum.*

The nomenclature of nontoxigenic variants of toxic species also has varied within the genus. Some nontoxigenic variants, e.g., C. *sporogenes* and C. *novyi* type C, were recognized as taxonomic entities. Other nontoxic or nonpathogenic "variants" received little attention, probably because they were unimportant in the medical situations in which they occurred.

The value of reliance on neutralization tests for "correct identification" of clostridia first came under serious question when it was determined that the toxins of C. *sordellii* and C. *difficile* were antigenically related. More recently, it was reported that one strain resembling C. *butyricum* produced C. *botulinum* type E toxin (Aureli *et al.,* 1986) and one strain resembling C. *baratii* produced C. *botulinum* type F toxin (Hall *et al.,* 1985). The "correct" nomenclature for such strains has not yet been determined.

The dependence of toxigenicity of C. *botulinum* types C and D on the presence of bacteriophages was discovered by Inoue and Iida (1968) and Ecklund *et al.* (1971). Nontoxigenic strains of C. *botulinum* types C and D are converted to toxigenic strains by infection with specific Tox+ bacteriophages and toxigenicity is concomitantly lost with the curing of phage (Ecklund *et al.,* 1971, 1972; Inoue and Iida, 1971). The nucleic acids of converting phages have been characterized and the existence of the structural genes for the toxins confirmed (Fujii *et al.,* 1988). In addition, Eklund *et al.* (1976) showed that production of a toxin of C. *novyi* types A and B was also influenced by bacteriophages. The cured nontoxic type A organism more closely resembled C. *botulinum* types C and D than the other types of C. *novyi.* Again, this finding is not surprising, considering the extremely high genetic and phenotypic relationship between C. *botulinum* types C and

D, and *C. novyi* types A–C (Johnson and Francis, 1975; see Table II; Holdeman and Brooks, 1970).

In a broad survey of 41 toxigenic strains of *C. botulinum*, including representatives of types A–F, Dolman and Chang (1972) were able to characterize four phage types, today classified (Reanney and Ackermann, 1982) as morphotypes A1 (Myoviridae), B1, B2 (Sytloviridae), and C1 (Podoviridae). While most strains harbored a single defined morphotype, others were found to contain two different phages. Phages of proteolytic (types A, B, F) and nonproteolytic (types B, E, F) *C. botulinum* strains showed a high degree of variation both within and between the types. A similar distribution of phage morphotypes also has been found in other genera of gram-positive bacteria (e.g., *Bacillus*, *Streptococcus*, *Staphylococcus*) and in gram-negative taxa. No phage-typing system for *C. botulinum* is available, but, as pointed out by Mahony (1979), phage host ranges seem to be restricted to nonproteolytic types B, E, and F and nontoxigenic "type E" strains.

While phage morphotypes in general are poor indicators of relatedness above the species level, the exclusive finding of the rare 3A type (prolate heads and 33-μm-long tails) in members of *Caryophanon*, *Staphylococcus*, *Streptococcus*, and *C. botulinum* (cf. Reanney and Ackermann, 1982) may reflect a common origin of both hosts and phages.

A number of reports have been published on temperate and virulent phages isolated from *C. perfringens* types A–F, but the results were somewhat contradictory. No phage-typing scheme exists for this species because, as indicated by Mahony (1979), the host range of phages is too narrow. This view is in accord with the genetic relationship of those types that are indistinguishable by DNA–rRNA pairing (Johnson and Francis, 1975; see Table II).

Most work on bacteriocins has also been performed on *C. perfringens* and *C. botulinum*, but bacteriocins have also been detected and analyzed in *C. septicum*, *C. butyricum*, and *C. sporogenes* (cf. Reanney and Ackermann, 1982). Bacteriocins from *C. perfringens* ("perfringocins" or "welchicins") seem to represent a heterogeneous group of molecules, chemically definable as proteins or protein–polysaccharide complexes. They are either sensitive against temperature, proteolytic enzymes, and extreme pH values or they are resistant. Some are spontaneously produced while others can be induced. No bacteriophage-like particles have been found with bacteriocin activity. The host range may either be restricted to strains of *C. perfringens*, or they are active ("welchicin" A and B) not only against a wider range of clostridial species, but against bacilli, staphylococci, streptococci, and even *Corynebacterium diphtheriae*.

Not unexpected is the broad host range of bacteriocins of *C. botulinum* strains, for which typing might be possible: those active only on nonproteolytic strains (isolated from nontoxic type E strains); those active only on proteolytic strains (isolated from proteolytic type B strains); and those which are not restricted to one class or the other (isolated from types A, B, and C strains) (cf. Mahony, 1979). Boticin P, produced by a nontoxigenic type E strain, is probably a mutant of a normal phage consisting of a heat- and pH-sensitive tail-like structure with a contracted sheath (like colicins O and M) (Lau *et al.*, 1974). Boticin C, on the other hand, which was also isolated from a nontoxigenic type E strain (Ellison and Kautter, 1970), was heat-stable and nonsedimentable by ultracentrifugation. Dolman and Chang (1972) reported on a killer particle (PBSX type) in *C. botulinum*, which is a phage-like entity defined by a small head, a long, thick contrac- tile tail with suspicous striations, and a base plate. Such killer particles and related structures have been found in other gram-positive eubacteria, e.g., bacilli, streptococci, and actinomycetes (Reanney and Ackermann, 1982).

ACKNOWLEDGMENTS. We are indebted to Lilian V. H. Moore and W. E. C. Moore for valuable discussions.

REFERENCES

Aureli, P., Fenicia, L., Pasolini, B., Gianfranceschi, M., McCroskey, L. M., and Hatheway, C. L., 1986, Two cases of type E infant botulism in Rome caused by neurotoxigenic *Clostridium butyricum*, *J. Infect. Dis.* **154:**207–211.

Andreesen, J. R., Zindel, V., and Dürre, P., 1985, *Clostridium cylindrosporum* (ex Barker and Beck 1942) nom. rev., *Int. J. Syst. Bacteriol.* **35:**206–208.

Balch, W. E., Fox, G. E., Magrum, L. J., Woese, C. R., and Wolfe, R. S., 1979, Methanogens: Reevaluation of a unique biological group, *Microbiol. Rev.* **43:**260–296.

Barker, H. A., and Beck, J. V., 1942, *Clostridium acidi-urici* and *Clostridium cylindrosporum*, organisms fermenting uric acid and some other purines, *J. Bacteriol.* **43:**291–304.

Bergey, D. H., Harrison, F. C., Breed, R. S., Hammer, B. W., and Huntoon, F. M., 1923, *Bergey's Manual of Determinative Bacteriology*, Williams and Wilkins, Baltimore, pp. 316– 337.

Cato, E. P., Hash, D. E., Holdeman, L. V., and Moore, W. E. C., 1982a, Electrophoretic study of *Clostridium* species, *J. Clin. Microbiol.* **15:**688–702.

Cato, E. P., Holdeman, L. V., and Moore, W. E. C., 1982b, *Clostridium perenne* and *Clostridium paraperfringens:* Later subjective synonyms of *Clostridium barati*, *Int. J. Syst. Bacteriol.* **32:**77– 81.

Cato, E. P., George, W. L., and Finegold, S. M., 1986, Genus *Clostridium*, in: *Bergey's Manual of Systematic Bacteriology* (P. H. A. Sneath, N. S. Mair, M. E. Sharpe, and J. G. Holt, eds.), Williams and Wilkins, Baltimore, pp. 1141–1200.

Chester, F. D., 1901, *A Manual of Determinative Bacteriology*, Macmillan, New York, pp. 295– 394.

Cummins, D. S., and Harris, H., 1956, The chemical composition of the cell wall in some gram-positive bacteria and its possible value as a taxonomic character, *J. Gen. Microbiol.* **14:**583–600.

Cummins, C. S., and Johnson, J. L., 1971, Taxonomy of the clostridia: Wall composition and DNA homologies in *Clostridium butyricum* and other butyric acid clostridia, *J. Gen. Micro- biol.* **67:**33–46.

Dams, E., Huysmans, E., Vandenberghe, A., and De Wachter, R., 1987, Structure of clostridial 5R ribosomal RNAs and bacterial evolution, *System. Appl. Microbiol.* **9:**54–61.

Dolman, C. E., and Chang, E., 1972, Bacteriophages of *Clostridium botulinum, Can. J. Microbiol.* **18:**67–76.

Donis-Keller, H., Maxam, A. M., and Gilbert, W., 1977, Mapping adenines, guanines and pyrimidines in RNA. *Nucl. Acids Res.,* **8:**2527–2537.

Ecklund, M. W., Poysky, F. T., Reed, and Smith, C. A., 1971, Bacteriophage and the toxigenicity of *Clostridium botulinum* type C. *Science* **172:**480–482.

Ecklund, M. W., Poysky, F. T., and Reed, S. M., 1972, Bacteriophage and the toxigenicity of *Clostridium botulinum* type D. *Nature New. Biol.* **235:**16–17.

Ecklund, M. W., Poysky, F. T., Peterson, M. E., and Myers, J. A., 1976, *Infect. Immun.* **14:**793–803.

Ellison, J. S., and Kautter, J. A., 1970. Purification and some properties of two boticins, *J. Bacteriol.* **104:**19–26.

Elsden, S. R., Hilton, M. G., Parsley, K. R., and Self, R., 1980, The lipid fatty acids of proteolytic clostridia, *J. Gen. Microbiol.* **118:**115–123.

Fox, R. H., and McClain, D. E., 1974, Evaluation of the taxonomic relationship of *Micrococcus cryophilus, Branhamella catarrhalis* and *Neisseriae* by comparative polyacrylamide gel electrophoresis of soluble proteins, *Int. J. Syst. Bacteriol.* **24:**172–176.

Fox, G. E., and Stackebrandt, E., 1987, The application of 16S rRNA cataloguing and 5S rRNA sequencing in bacterial systematics, in: *Methods in Microbiology,* Vol. 19 (R. R. Colwell and R. Grigorova, eds.), Academic Press, Orlando, pp. 405–458.

Fox, G. E., Pechman, K. J., and Woese, C. R., 1977, Comparative cataloguing of 16S ribosomal ribonucleic acid: Molecular approach to prokaryotic systematics. *Int. J. Syst. Bacteriol.,* **27:** 44–57.

Fox, G. E., Stackebrandt, E., Hespell, R. B., Gibson, J., Maniloff, J., Dyer, T. A., Wolfe, R. S., Balch, W. E., Tanner, R. S., Magrum, L. J., Zablen, L. B., Blakemore, R., Gupta, R., Bonen, L., Lewis, B. J., Stahl, D. A., Luehrsen, K. R., Chen, K., and Woese, C. R., 1980, The phylogeny of prokaryotes, *Science* **209:**457–463.

Fujii, N., Oguma, K., Yokosawa, N., Kimura, K., and Tsuzuki, K., 1988, Characterization of bacteriophage nucleic acids obtained from *Clostridium botulinum* types C and D, *Appl. Environ. Microbiol.* **54:**69–73.

Hall, J. D., McCroskey, L. M., Pinkomb, B. J., and Hatheway, C. L., 1985, Isolation of an organism resembling *Clostridium baratii* which produces type F botulinal toxin from an infant with botulism, *J. Clin. Microbiol.* **21:**654–655.

Hammann, R., and Werner, H., 1980, Fermentation products (using g.l.c.) in the differentiation of non-sporing anaerobic bacteria, in: *Microbial Classification and Identification* (M. Goodfellow and R. G. Board, eds.), Academic Press, London, pp. 257–271.

Handuroy, P., Ehringer, G., Urbain, A., Buillot, G., and Magron, J., 1937, *Dictionnaire des Bacteries Pathogenes,* Masson et Cie, Paris, pp. 89–144.

Holdeman, L. V., and Brooks, J. B., 1970, Variation among strains of *Clostridium botulinum* and related clostridia, in: *Proceedings of the Botulism Conference,* (M. Herzberg, ed.), UJNR Conference on Toxic Microorganisms, U.S. Govt. Printing Office, Washington, D.C., pp. 278–286.

Holdeman, L. V., Cato, E. P., and Moore, W. E. C. (eds.), 1977, *Anaerobic Laboratory Manual,* 4th ed., Virginia Polytechnic Institute and State University, Blacksburg, VA.

Horn, N., 1987, *Clostridium disporicum* sp. nov., a saccharolytic species able to form two spores per cell, isolated from a rat cecum, *Int. J. Syst. Bacteriol.* **37:**398–401.

Inoue, K., and Iida, 1968, Bacteriophages of Clostridium *botulinum, J. Virol.* **2:**537–540.

Inoue, K., and Iida, H., 1971, Phage-conversion of toxigenicity in *Clostridium botulinum* types C and D. *Jap. J. Med. Sci. Biol.* **24:**53–56.

Johnson, J. L., 1970, Relationship of deoxyribonucleic acid homologies to cell wall structure, *Int. J. Syst. Bacteriol.* **20:**421–424.

Johnson, J. L., 1973, Use of nucleic-acid homologies in the taxonomy of anaerobic bacteria, *Int. J. Syst. Bacteriol.* **23:**308–315.

Johnson, J. L., 1985, DNA reassociation and RNA hybridization of bacterial nucleic acids, in: *Methods in Microbiology*, Vol. 18 (G. Gottschalk, ed.), Academic Press, London, pp. 34–74.

Johnson, J. L., and Ordal, E. J., 1968, Deoxyribonucleic acid homology in bacterial taxonomy: Effect of incubation temperature on reaction specificity, *J. Bacteriol.* **95:**893–900.

Johnson, J. L., and Francis, B. S., 1975, Taxonomy of the clostridia: Ribosomal ribonucleic acid homologie among the species, *J. Gen. Microbiol.* **88:**229–244.

Kersters, K., and De Ley, J., 1980, Classification and identification of bacteria by electrophoresis of their proteins, in: *Microbial Classification and Identification* (Goodfellow, M., and Board, R. G., eds.), Academic Press, London, pp. 273–297.

Krumholz, L. R., and Bryant, M. P., 1985, *Clostridium pfennigii* sp. nov. uses methoxyl groups of monobenzenoids and produces butyrate, *Int. J. Syst. Bacteriol.* **35:**454–456.

Lane, D. J., Pace, B., Olsen, G. J., Stahl, D. A., Sogin, M. L., and Pace, N. R., 1985, Rapid determination of 16S ribosomal RNA sequences for phylogenetic analysis. *Proc. Natl. Acad. Sci. USA* **82:**6955–6959.

Lapage, S. P., Sneath, P. H. A., Lessel, E. F., Skerman, Y. B. D., Seeliger, H. P. R., and Clark, W. A. (eds.), 1975, *International Code of Nomenclature of Bacteria*, American Society for Microbiology, Washington, D.C.

Lau, A. H. S., Hawirko, R. Z., and Chow, C. T., 1974, Purification and properties of boticin P produced by *Clostridium botulinum*, *Can. J. Microbiol.* **20:**385–390.

Lee, W. H., and Riemann, H., 1970, Correlation of toxic and nontoxic strains of Clostridium botulinum by DNA composition and homology, *J. Gen. Microbiol.* **60:**117–123.

Ludwig, W., Weizenegger, M., Kilpper-Bälz, R., and Schleifer, K. H., 1988, Phylogenetic relationships of anaerobic streptococci, *Int. J. Syst. Bacteriol.* **38:**15–18.

Madden, R. H., 1983, Isolation and characterization of *Clostridium stercorarium* sp. nov., cellulolytic thermophile, *Int. J. Syst. Bacteriol.* **33:**837–840.

Mahony, D. E., 1979, Bactericions, bacteriophage and other epidemiological typing methods for the genus *Clostridium*, in: *Methods in Microbiology*, Vol. 13 (T. Bergan and J. R. Norris, eds.), Academic Press, New York, pp. 1–30.

Marmur, J., and Doty, P., 1961, Thermal renaturation of deoxyribonucleic acids, *J. Mol. Biol.* **3:**585–594.

Marmur, J., and Doty, P., 1962, Determination of the base composition of deoxyribonucleic acid from its thermal denaturation temperature, *J. Mol. Biol.* **5:**109–118.

Mead, G. C., 1971, The amino acid-fermenting clostridia, *J. Gen. Microbiol.* **67:**47–56.

Moore, W. E. C., Cato, E. P., and Holdeman, L. V., 1966, Fermentation patterns of some *Clostridium* species, *Int. J. Syst. Bacteriol.* **16:**383–415.

Moore, W. E. C., Hash, D. E., Holdeman, L. V., and Cato, E. P., 1980, Polyacrylamide slab gel electrophoresis of soluble proteins for studies of bacterial floras, *Appl. Environ-Microbiol.* **39:**900–907.

Moore, W. E. C., Cato, E. P., and Moore, L. V. H., 1985, Index of the bacterial and yeast nomenclatural changes published in the *Int. J. Syst. Bacteriol.* since the 1980 approved lists of bacterial names (January 1, 1980 to January 1, 1985), *Int. J. Syst. Bacteriol.* **35:**382–407.

Morris, G. N., Winter, J., Cato, E. P., Ritchie, A. E., and Bokkenheuser, V. D., 1985, *Clostridium scindens* sp. nov., a human intestinal bacterium with desmolytic activity on certicoids, *Int. J. Syst. Bacteriol.* **35:**478–481.

Moss, C. W., and Lewis, V. J., 1967, Characterization of clostridia by gas chromatography. 1. Differentiation of species by cellular fatty acids, *Appl. Microbiol.* **15:**390–397.

Murray, W. D., Hofmann, L., Campbell, N. L., and Madden, R. H., 1986, *Clostridium lentocellum* sp. nov., a cellulolytic species from river sediment containing paper-mill waste, *System. Appl. Microbiol.* **8:**181–184.

Nakamura, S., Shimamura, T., Hayashi, H., and Nishida, S., 1975, Reinvestigation of the taxonomy of *Clostridium bifermentans* and *Clostridium sordelli*, *J. Med. Microbiol.* **8:**299–309.

Nakamura, S., Shimamura, T., and Nishida, S., 1976, Urease-negative strains of *Clostridium sordellii*, *Can. J. Microbiol.* **22**:673–676.

Oren, A., Weisburg, W. G., Kessel, M., and Woese, C. R., 1984, *Halobacteroides halobius* gen. nov, spec. nov., a moderately halophilic, obliatory anaerobic bacterium from the bottom sediments of the Dead Sea. *Syst. Appl. Microbiol.* **5**:58–70.

Oren, A., Pohla, H., and Stackebrandt, E., 1987, Transfer of *Clostridium lortetii* to a new genus *Sporohalobacter* gen. nov as *Sporohalobacter lortetii* comb. nov., and description of *Sporohalobacter marismortui* sp. nov., *System. Appl. Microbiol.*, **9**:239–246.

Palleroni, N. J., Kunisawa, R., Contopoulou, R., and Doudoroff, M., 1973, nucleic and homologies in the genus *Pseudomonas. Int. J. Syst. Bacteriol.* **23**:333–339.

Patel, K. C., Monk, C., Littleworth, H., Morgan, H. W., and Daniel, R. M., 1987, *Clostridium fervidus* sp. nov., a new chemoorganotrophic acetogenic thermophile, *Int. J. Syst. Bacteriol.* **37**:123–126.

Peattie, D. A., 1979, Direct chemical method for sequencing RNA. *Proc. Natl. Acad. Sci USA* **76**:1760–1764.

Petitdemange, E., Caillet, F., Giallo, J., and Gaudin, C., 1984, *Clostridium cellulolyticum* sp. nov., a cellulolytic, mesophilic species from decayed grass, *Int. J. Syst. Bacteriol.* **34**:155–159.

Prazmowski, A., 1880, Untersuchung über die Entwicklungsgeschichte und Fermentwirkung einiger Bakterien, Arten. Inaug. Diss. Hugo Voigt, Leipzig, pp. 1–58.

Prévot, A. R., 1938, Études de systématique bactérienne IV, Critique de la conception actuelle du genre *Clostridium, Ann. Inst. Pasteur* (Paris) **61**:72–91.

Prévot, A. R., 1940, *Manual de Classification et de Détermination des Bactéries Anaérobies*. Masson et Cie, Paris, pp. 87–169.

Pribula, C. D., Fox, G. E., and Woese, C. R., 1976, Nucleotide sequence of *Clostridium pasteurianum* 5S rRNA, *FEBS Lett.*, **64**:350–352.

Reanney, D. C., and Ackermann, H.-W., 1982, Comparative biology and evolution of bacteriophages, in: *Advances in Virus Research*, Vol. 27 (M. A. Lauffes, F. B. Bang, K. Gavamorosch, and K. G. Smith, eds.), Academic Press, New York, pp. 205–280.

Riemann, H., and W. H. Lee, 1970, The genetic relatedness of proteolytic *Clostridium botulinum* strains, *J. Gen. Microbiol.* **64**:85–90.

Rogers, M. J., Simmons, J., Walker, R. T., Weisburg, W. G., Woese, C. R., Tanner, R. S., Robinson, I. M., Stahl, D. A., Olsen, G. J., Leach, R. H., Maniloff, J., 1985, Construction of the mycoplasma evolutionary tree from 5S rRNA sequence data, *Proc. Natl. Acad. Sci. USA* **82**:1160–1164.

Schink, B., 1984, *Clostridium magnum* sp. nov., a non-autotrophic homoacetogenic bacterium. *Arch. Microbiol.* **137**:250–255.

Schleifer, K. H., and Kandler, O., 1972, Peptidoglycan types of bacterial cell walls and their taxonomic implications. *Bacteriol. Rev.* **36**:407–477.

Schleifer, K. H., and E. Stackebrandt, 1983, Molecular systematics of prokaryotes, *Ann. Rev. Microbiol.* **37**:143–187.

Skerman, V. B. D., McGowan, V., and Sneath, P. H. A. (ed.), 1980, Approved lists of bacterial names, *Int. J. Syst. Bacteriol.* **30**:225–420.

Sleat, R., and Mah, R. A., 1985, *Clostridium populeti* sp. nov., a cellulolytic species from a woody-biomass digestor, *Int. J. Syst. Bacteriol.* **35**:160–163.

Sleat, R., Mah, R. A., and Robinson, R., 1984, Isolation and characterization of an anaerobic, cellulolytic bacterium, *Clostridium cellulovorans* sp. nov., *Appl. Environ. Microbiol.* **48**:88–93.

Smith, L. DS., 1977, *Botulism: the Organism, Its Toxins, The Disease*, Charles C. Thomas, Springfield, Ill.

Stackebrandt, E., 1988, Phylogenetic relationships vs. phenotypic diversity: how to achieve a phylogenetic classification system of the eubacteria, *Can. J. Microbiol.* **34**:552–556.

Stackebrandt, E., and Woese, C. R., 1981, The evolution of prokaryotes, in: *Molecular and Cellular Aspects of Microbial Evolution* (M. I. Carlile, I. F. Collins, and B. E. B. Moseley, eds.), Cambridge University Press, Cambridge, pp. 1–31.

Stackebrandt, E., and Woese, C. R., 1984, The phylogeny of prokaryotes. *Microbiol. Sci.* **1:** 117–122.

Stackebrandt, E., Ludwig, W., and Fox, G. E., 1985a, 16S ribosomal RNA oligonucleotide cataloguing, in: *Methods in Microbiology,* Vol. 18 (G. Gottschalk, ed.), Academic Press, Orlando, pp. 75–107.

Stackebrandt, E., Pohla, H., Kroppenstedt, R., Hippe, H., and Woese, C. R., 1985b, 16S rRNA analysis of *Sporomusa, Selenomonas,* and *Megasphaera:* On the phylogenetic origin of gram-positive eubacteria. *Arch. Microbiol.* **143:**270–276.

Stieb, M., and Schink, B., 1985, Anaerobic oxidation of fatty acids by *Clostridium bryantii* sp. nov., a sporeforming, obligately syntrophic bacterium, *Arch. Microbiol.* **140:**387–390.

Suen, J. C., Hatheway, C. L., Steigerwalt, A. G., and Brenner, D. J., 1988, *Clostridium argentinense* sp. nov.: A genetically homogenous group composed of all strains of *Clostridium botulinum* toxin type G, and some nontoxigenic strains previously identified as *Clostridium subterminale* or *Clostridium hastiforme, Int. J. Syst., Bacteriol.* **38:**375–381.

Tanner, R. S., Stackebrandt, E., Fox, G. E., and Woese, C. R., 1981, A phylogenetic analysis of *Acetobacterium woodii, Clostridium barkeri, Clostridium butyricum, Clostridium lituseburense, Eubacterium limosum,* and *Eubacterium tenue, Curr. Microbiol.,* **5:**35–38.

Tanner, R. S., Stackebrandt, E., Fox, G. E., Gupta, L. J., Magrum, L. J., and Woese, C. R., 1982, A phylogenetic analysis of anaerobic eubacteria capable of synthesizing acetate from carbon dioxide., *Curr. Microbiol.* **7:**127–132.

van Gylswyk, N. O., and van der Toorn, J. J. T. K., 1987, *Clostridium aerotolerans* sp. nov., a xylanolytic bacterium from corn stover and from the rumina of sheep fed corn stover, *Int. J. Syst. Bacteriol.* **37:**102–105.

Wayne, L. G., Brenner, D. J., Colwell, R. R., Grimont, P. A. D., Kandler, O., Krichevsky, M. I., Moore, L. H., Moore, W. E. C., Murray, R. G. E., Stackebrandt, E., Starr, M. P., and Trüper, H. G., 1987, Report of the ad hoc committee on reconciliation of approaches to bacterial systematics., *Int. J. Syst. Bacteriol.* **37:**463–464.

Weinberg, M., Nativelle, R., and Prévot, A. R., 1937, Les *Microbes Anaérobies,* Masson et Cie, Paris, pp. 120–515.

Weiss, N., Schleifer, K. H., and Kandler, O., 1981, The peptidoglycan types of grampositive anaerobic bacteria and their taxonomic implications, *Rev. Inst. Pasteur de Lyon* **14:**3–12.

Woese, C. R., 1987, Bacterial evolution, *Microbiol. Rev.* **51:**221–271.

Woese, C. R., and Fox, G. E., 1977, Phylogenetic structure of the prokaryotic domain: The primary kingdoms. *Proc. Natl. Acad. Sci. USA* **74:**5088–5090.

Woese, C. R., Debrunner-Vossbrinck, B., Oyaizu, H., Stackebrandt, E., and Ludwig, W., 1985a, Gram-positive bacteria: Possible photosynthetic ancestry, *Science* **229:**762–765.

Woese, C. R., Stackebrandt, E., and Ludwig, W., 1985b, What are mycoplasmas: The relationship of tempo and mode in bacterial evolution, *J. Molec. Evol.* **21:**305–316.

Wu, J. I. J., Riemann, H., and Lee, W. H., 1972, Thermal stability of the deoxyribonucleic acid hybrids between proteolytic strains of *Clostridium botulinum* and *Clostridium sporogenes. Can. J. Microbiol.* **18:**97–99.

Introduction to the Physiology and Biochemistry of the Genus *Clostridium*

2

JAN R. ANDREESEN, HUBERT BAHL, and
GERHARD GOTTSCHALK

1. INTRODUCTION

The genus *Clostridium* was created by Prazmowski in 1880. Four criteria presently classify an organism as a clostridium: (1) the ability to form endospores; (2) restriction to an anaerobic energy metabolism; (3) the inability to carry out a dissimilatory reduction of sulfate; and (4) the possession of a gram-positive cell wall, which may react gram-negative. These criteria are met by an otherwise diverse assembly of microorganisms, and the genus *Clostridium* has grown to be one of the largest genera among prokaryotes. A total of 83 species are listed in *Bergey's Manual of Systematic Bacteriology* (Cato *et al.*, 1986). Since this list was compiled, a number of new species have been described, while others, such as *C. tetanomorphum* and *C. cylindrosporum*, have been omitted (see Chapter 1). In this chapter the span of properties found among the clostridia will be outlined. Additional information on the general taxonomy, the general properties of clostridia, and clostridial fermentations may be found in a number of recent reviews (Barker, 1961, 1978, 1981; Wood, 1961; Thauer *et al.*, 1977; Gottschalk and Andreesen, 1979; Gottschalk *et al.*, 1981; Booth and Mitchell, 1987).

2. GENERAL PROPERTIES OF THE CLOSTRIDIA

2.1. Cell Shape, Spores, and Motility

Cells of most species are rod-shaped with round or, in some cases, pointed ends (e.g., *C. clostridiiforme*). Rods may be cylindrical or "cigar-

---

JAN R. ANDREESEN, HUBERT BAHL, and GERHARD GOTTSCHALK ● Institut für Microbiologie, Georg-August-Universität, D-3400 Göttingen, Federal Republic of Germany.

shaped" (e.g., *C. sphenoides*); they are straight or slightly curved, and measure 0.5–2 μm in diameter and up to 30 μm in length. Species which differ extensively from the general shape described above are *C. coccoides*, *C. cocleatum* (semicircular), and *C. spiroforme*.

Clostridia form oxygen-, heat-, and alcohol-resistant endospores. Depending on the species they occur in either a central, subterminal, or terminal position, and cells containing spores are often swollen. The heat resistance of the spores from a number of clostridial species is remarkable. The record is held by *C. thermohydrosulfuricum*, the spores of which exhibit 10% survival when heated for 20 hr at 100°C or for 1 min at 120°C (Wiegel and Ljungdahl, 1979).

The ability to form spores is difficult to demonstrate for a number of clostridial species. Some apparent non-spore formers have had to be reclassified following the detection of spores in them (e.g., *C. clostridiiforme*). A general procedure for the initiation of spore formation cannot be given. The most reliable sporulation medium for many species is chopped meat agar (Holdeman *et al.*, 1977). Some species sporulate in media rich in carbohydrates, others when grown on the surface of agar plates, at suboptimal temperature, or in the presence of methyl xanthines (Perkins, 1965; Pheil and Ordal, 1967; Roberts, 1967; Bergère and Hermier, 1970; Marks and Freese, 1987).

Most clostridial species are motile and are peritrichously flagellated. For a number of species motility has never been reported. These species include *C. barkeri*, *C. cocleatum*, *C. innocuum*, *C. perfringens*, *C. ramosum*, and *C. thermosaccharolyticum* (Cato *et al.*, 1986).

2.2. Mol % G + C, DNA–DNA, and rRNA Homologies

An indication of the heterogeneity exhibited by the genus *Clostridium* is apparent from the mol % G + C values found. These range from 22 mol % to 55 mol %. Many species cluster around 28%. Other smaller clusters center around 35, 45, and 52%.

DNA–DNA homology studies have been undertaken with certain butyric acid-producing species, with glutamate-utilizing strains, and with thermophilic species (Johnson, 1973; Kaneuchi *et al.*, 1979; Matteuzi *et al.*, 1978). Oligonucleotide cataloguing of 16S rRN17 revealed that some clostridial species are more closely related to certain non-spore formers than to other clostridia (Chapter 1). However, the incomplete nature of the available data does not allow conclusions to be drawn for the whole genus. Similarly, data obtained from gel electrophoretic analyses of soluble cellular proteins—useful to detect differences or correspondences of closely related species or strains—are not adequate for a reorganization of the taxonomy of the gram-positive anaerobic spore formers.

2.3. Walls and Membranes

The wall of most clostridia consists of peptidoglycan of the meso-diaminopimelic acid direct-linked type (Cummins, 1970; Cummins and Johnson, 1971; Schleifer and Kandler, 1972; Weiss *et al.*, 1981). The presence of teichoic acids in a number of species has been demonstrated. Thermophilic species contain a proteinaceous S layer as the outermost layer (Sleytr and Glauert, 1976); it is very regularly structured and hexameric building blocks can be seen under the electron microscope. Cytoplasmic membranes of *Clostridia* have scant attention. In most species they are probably devoid of redox carriers, such as quinones or cytochromes. However, the latter have been found in species like *C. thermoaceticum, C. aceticum,* and *C. formicoaceticum*. Some information on the phospholipids of clostridial membranes is also available (Baumann *et al.*, 1965; Vollherbst-Schneck *et al.*, 1984; Oulevey *et al.*, 1986; Johnston and Goldfine, 1988).

The possession of a Gram-positive cell wall (the lack of an outer membrane) is now a property of taxonomic value because the members of the recently described genus *Sporomusa* are also strict anaerobes but contain a gram-negative cell wall (Möller *et al.*, 1984).

2.4. Plasmids and Gene Transfer

Although numerous plasmids have been detected in clostridial species (Lee *et al.*, 1987; Rogers, 1986), the majority are functionally cryptic. Those plasmids to which a function has been assigned are principally confined to pathogenic *Clostridium* spp. Thus plasmids have been identified in *C. perfringens* which mediate resistance to antibiotics (Brefort *et al.*, 1977; Abraham *et al.*, 1985) and bacteriocin production (Garnier and Cole, 1988), a *C. tetani* plasmid has been shown to encode toxin production (Finn *et al.*, 1984), and a plasmid found in *C. cochlearium* has been strongly implicated in the decomposition of methyl mercury (Pan-Hou *et al.*, 1980). Therefore it seems to be just a question of time until further plasmid-born clostridial activities are elucidated.

Conjugal transfer of genetic material has been reported between strains of *C. perfringens* and of *C. difficile* and from *Bacillus subtilis* to *C. acetobutylicum* (Brefort *et al.*, 1977; Oultram and Young, 1985). Transformation was successfully carried out with autoplasts of *C. perfringens* (Heefner *et al.*, 1984) and with protoplasts of *C. acetobutylicum* (Reid *et al.*, 1983; Lin and Blaschek, 1984). Thus, tools for changing certain properties of *Clostridia* by genetic techniques are becoming increasingly available. Chapter 3 of this book gives detailed information on the genetics of clostridia.

2.5. Pathogeneity

More than 30 clostridial species have been isolated from human clinical specimens. Most of these species are considered to be nonpathogens; they enter the organism via the gastrointestinal, respiratory, or genitourinary tracts, or via wounds. Several of them are "opportunistic" and may develop in debilitated, immunosuppressed, or otherwise compromised patients. The important clostridial pathogens of man and animals are listed in Table I and are the subject of Chapter 8. Their pathogenic determinants are proteinaceous toxins. *Clostridium botulinum* and *C. tetani* produce neurotoxins that are lethal in extremely low concentrations. 10–100 μg of tetanospasmin and approximately 1 μg of the *C. botulinum* neurotoxin represent the lethal human dose (Schantz and Sugiyama, 1974; Habermann and Dreyer, 1986).

Clostridium perfringens and, to a lesser extent, *C. novyi* and *C. septicum* are responsible for gas gangrene which occurs in wounds or after compound fractures of bones. The disease develops by growth of the *Clostridia* within the infected area. Toxins are produced which degrade phospholipids, collagen, and other proteins. This results in further necrosis and bacterial growth. Finally, the bacteria gain access to the blood and give rise to bacteremia (Willis, 1977).

2.6. Growth

2.6.1. Oxygen Sensitivity

Clostridia carry out a fermentative energy metabolism; they do not require O_2 for ATP production. Moreover, they exhibit a more or less pronounced sensitivity toward O_2. *Clostridium haemolyticum* requires a pO_2 smaller than 0.5% for growth; *C. novyi* type A tolerates a pO_2 up to 3%

Table I. *Clostridium* Species Causing Diseases

Species	Disease
C. botulinum	Botulism
C. chauvoei	Blackleg (cattle, swine)
C. difficile	Pseudomembranous entero-colitis
C. histolyticum	Gas gangrene
C. novyi	Gas gangrene
C. perfringens	Gas gangrene
C. septicum	Gas gangrene
C. tetani	Tetanus

during growth (Loesche, 1969). The time of survival during exposure to air is also different for the various species; *C. haemolyticum* and *C. novyi* type B survive for minutes and *C. acetobutylicum* and *C. butyricum* for hours when exposed to air. Known as oxygen-tolerant species are *C. carnis, C. durum, C. tertium, C. histolyticum,* and *C. aerotolerans* (van Gylswyk and van der Toorn, 1987).

Two reasons can be considered for the oxygen sensitivity of the clostridia: lack or shortage of defense mechanisms against toxic byproducts of oxygen metabolism (hydroxyl radical, superoxide anion, and hydrogen peroxide), and interference with the intermediary and biosynthetic metabolism (Morris, 1980). Superoxide dismutase is not generally absent in clostridia; it has been detected in *C. perfringens, C. sporogenes,* and *C. ramosum* (Gregory *et al.*, 1978). It is not known whether catalase is present in some clostridia. At least it can be concluded that the defense mechanisms, if partially present, are not sufficient for a removal of toxic products derived from O_2.

Interference of O_2 with the intermediary metabolism may be caused by reaction with NADH as catalyzed by NADH oxidases present in many clostridial species in high activities. Such a reaction would lead to NADH depletion and a collapse of metabolism.

2.6.2. pH and Temperature Range

For most clostridia, growth is optimal at pH 6.5–7.0 and temperatures between 30 and 37°C. The span of optimal growth temperatures reaches from 15 to 69°C (Cato *et al.*, 1986). In Table II thermophilic species are listed, some of which received special attention in recent years in connec-

Table II. Thermophilic *Clostridium* Species

Organism	Growth temperature (°C)		
	Minimum	Maximum	Optimum
C. fervidus	37	80	68
C. stercorarium	ND	ND[a]	65
C. thermoaceticum	47	65	56–60
C. thermoautotrophicum	42	66	60–62
C. thermocellum	>37	ND	60–64
C. thermocopriae	47	74	60
C. thermohydrosulfuricum	42	78	67–69
C. thermolacticum	50	70	60–65
C. thermosaccharolyticum	37	65	55–62
C. thermosulfurogenes	35	75	60

[a]ND, not determined.

tion with the production of fuels and chemicals from biomass and of thermostable enzymes.

For a given organism the optimal pH for growth may vary from substrate to substrate, e.g., *C. aceticum* will only grow autotrophically between pH 8.1 and 8.5 (Wieringa, 1941; Braun *et al.*, 1981), whereas growth with carbohydrates requires a pH around 7. Even sugar utilization is influenced by the initial pH; both *C. aminobutyricum* and *C. aminovalericum* are described as saccharolytic species with an optimal growth between pH 7.4 and 7.7; on carbohydrates they will not grow below pH 7 (Hardman and Stadtman, 1960a,b). Using standard techniques for identification of anaerobes (Holdeman *et al.*, 1977), both organisms appear as nonsaccharolytic species because the routine media have initial pH values of 6.9–7.0 and utilization of sugars is evaluated on the basis of a decrease in pH due to acid production (below pH 6.0 = weak, below pH 5.5 = strong). In the case of the purinolytic clostridia, which produce large amounts of ammonia and raise the pH, slightly alkaline pH values of 8–9 are preferred (Schiefer-Ullrich *et al.*, 1984). Similarly, acid formation might be neutralized by ammonia production when amino acids are degraded by peptolytic organisms.

An interesting correlation was found between pH ranges tolerated and the temperature applied (Karlsson *et al.*, 1988). *Clostridium bifermentans* and *C. sporogenes* tolerated quite alkaline pH values of 12.2 and 11.7 at low temperatures, whereas at higher temperatures, lower pH values were much better endured. This dependence on the temperature might reflect a lower fluidity of the membrane and thus an altered permeability for ions at lower temperatures with consequences for pH homeostasis (Booth, 1985) and ion transport (Rosen and Silver, 1987).

2.6.3. Nutritional Requirements

There is a general tendency to use media which are easy to prepare, meet the nutritional requirements of many different species, and are recommended for diagnostic purposes. Therefore, PYG (peptone–yeast extract–glucose) or CMC (chopped meat carbohydrate) are recommended to cultivate clostridial species (Holdeman *et al.*, 1977). One reason for the avoidance of synthetic media is the tendency of many clostridia to exhibit rapid initiation of growth, shorter generation times, and much better growth yields in complex media. However, the media compilation of Koser (1968) clearly demonstrates that synthetic media have long been available for most of the known saccharolytic and peptolytic clostridia. Species highly specialized with respect to their carbon and energy source (e.g., *C. kluyveri* or *C. acidiurici*) are routinely grown in synthetic media (Bornstein and Barker, 1948; Rabinowitz, 1963), being unable to thrive in standard complex media. Common salt, trace elements, and vitamin solutions may be used to prepare a synthetic medium for many species; but the requirement for purine and pyrimidine bases and for amino acids can vary significantly. It is quite important to balance the concentration of the latter for a certain species in order to meet individual needs for these compounds as precursors for proteins or as energy-yielding substrates. Because of different regulatory patterns in different organisms, the surplus of one amino acid might affect the biosynthetic capacity for other amino acids, resulting in a disregulation and, finally, growth cessation. Thus, different species require amino acid mixtures of different compositions (Whitmer and Johnson, 1988). There are a few species which also require unknown factors or a certain peptide for growth. Trypticase-peptone seems to be a rather rich source of such compounds.

To avoid all the problems mentioned, complex ingredients are routinely added to the media used for cell propagation. This might mask certain capabilities of the organisms due to repression or lack of induction of enzyme synthesis. For example, C. thermoaceticum has been isolated and grown for decades as a saccharolytic organism in the presence of yeast extract, peptone, and tomato juice (Andreesen et al., 1973). The eventual demonstration of its close relationship with C. thermoautotrophicum (as judged by DNA–DNA homology, murein composition, and substrate utilization) led to the rediscovery of its requirement for just two vitamins, besides trace elements (Lundie and Drake, 1984), and of hydrogenase activity, with the concomitant capacity of autotrophic growth (Kerby and Zeikus, 1983).

The efficiency of utilization of a substance may be greatly affected by its concentration within the growth medium. The realization that substrate concentrations can be inhibitory to growth has led to the use of low concentrations, particularly with more exotic substrates such as ferulic acid, which yields methanol as the actual substrate (Bache and Pfennig, 1981). To achieve the degradation of more reduced compounds such as fatty acids by C. bryantii, other hydrogen-scavenging organisms have to be present and the phosphate concentration has to be low (Stieb and Schink, 1985). On the other hand, organisms carrying out the Stickland reaction have a growth requirement for substrate combinations such as two amino acids. In other instances, media may need to contain bicarbonate, and even acetate, as building blocks for biosynthetic purposes or as an electron acceptor (e.g., for C. kluyveri and the acetogens).

The importance of the addition of certain trace elements for growth and substrate utilization is generally more pronounced in synthetic media (Schiefer-Ullrich et al., 1984); however, for some elements it can also be observed in the presence of yeast extract (Imhoff and Andreesen, 1979). A requirement for much higher concentrations of molybdate, tungstate, and selenite was observed in complex than in synthetic media (Andreesen et al., 1974; Leonhardt and Andreesen, 1977). Hence, a specific transport system with a high affinity for these trace elements might become repressed by components present in complex media, leaving the transport to a more general system with much lower affinity (Wagner and Andreesen, 1987).

To summarize, complex media are convenient to use for the cultivation of clostridia. However, their components might mask abilities for biosynthesis and degradation. Synthetic media are tedious to prepare when they consist of many components.

2.6.4. Growth Rate

Following the general introduction of the Hungate technique to obtain truly anaerobic conditions (Bryant, 1972; Balch *et al.*, 1979), by the availability of a variety of reducing agents [cysteine, sulfide, dithionite, titanium(III), thioglycolate], by the formulation of new trace element solutions, and by the establishment of defined growth requirements, most clostridial species will go through an exponential growth phase when cultivated. Thus, doubling times can be calculated. Saccharolytic species like *C. butyricum* may double their mass within 30–40 min. This indicates a very active general metabolic flux because clostridial glucose fermentation might yield only about 3.4 mol ATP per mol glucose (Jungermann *et al.*, 1973) as compared to 25–38 mol ATP conserved by aerobic organisms (minimal doubling time 15–20 min).

The purine-degrading clostridia are a good example of organisms with a high metabolic flux. In mineral media, they exhibit generation times of about 60–80 min (Dürre *et al.*, 1981; Schiefer-Ullrich *et al.*, 1984) despite the low ATP yield of less than two per purine and the need to synthesize all cell material starting from C_2 and C_1 carbon units (Stouthammer, 1979). Xanthine dehydrogenase as the enzyme responsible for initial substrate attack is extremely active in these organisms due to the presence of the trace element selenium as part of the active center (Wagner *et al.*, 1984). A similar situation exists for formate dehydrogenase, which is generally more active (but also more unstable) when the enzyme composition is more complex (Adams and Mortenson, 1985). The examples given demonstrate that growth and especially growth rates depend on several parameters including the nature of the substrate and the availability of certain growth factors in proper concentrations.

2.7. Handling and Storage

The oxygen intolerance of the clostridia necessitates the employment of methods to exclude oxygen from the culture media. The so-called Hungate technique is widely used for isolation of clostridial species and for their cultivation in small volumes. Glove boxes and anaerobic jars are used for growing clostridia on agar plates. These facilities now allow plating with high efficiencies, a prerequisite for genetic work with these organisms.

Large-scale cultivation is done in conventional bioreactors, in most cases under an atmosphere of N_2. Clostridia will not start growing when the E_h of their environment is above approximately -150 mV (O'Brien and Morris, 1971). Therefore, it is not only necessary to free the media from oxygen; they have to be reduced. This is usually done by the addition of small amounts of sodium thioglycolate, L-cysteine, sodium sulfide, or sodium dithionate.

Strains are preferentially stored as spores. Clostridia, however, differ widely in their readiness to produce spores and special sporulation media had to developed for a number of clostridial species (see Section 2.1). Spores are generally aerotolerant and will not germinate during exposure to air. Germination requires anaerobic conditions and a nutrient supply (Marks and Freese, 1987).

3. GENERAL FERMENTATION STRATEGY

3.1. Degradation of Polysaccharides

The most important substrates for the clostridia in nature are polymers such as cellulose, starch, pectins, other polysaccharides, and proteins. Typical degraders of these macromolecules are given in Table III. With the exception of *C. thermocellum, C. thermolacticum, C. thermocopriae,* and *C. stercorarium,* all cellulolytic species are mesophiles. Most of them produce ethanol/acetate mixtures as products. Only *C. populeti* and *C. cellulovorans* form predominantly butyrate (Sleat *et al.,* 1984; Sleat and Mah, 1985).

Cellulose breakdown is accomplished by a multienzyme complex which has been extensively studied in *C. thermocellum.* This complex has

Table III. Clostridial Species Degrading Polysaccharides

Cellulose	Starch	Pectin
C. cellobioparum	*C. acetobutylicum*	*C. beijerinckii*
C. cellulovorans	*C. aurantibutyricum*	*C. butyricum*
C. cellulolyticum	*C. beijerinckii*	*C. felsineum*
C. josui	*C. butyricum*	*C. puniceum*
C. lentocellum	*C. paraputrificum*	*C. thermocellum*
C. papyrosolvens	*C. perfringens*	*C. thermosulfurogenes*
C. populeti	*C. puniceum*	
C. stercorarium	*C. roseum*	
C. thermocellum	*C. thermohydrosulfuricum*	
C. thermocopriae	*C. thermosaccharoticum*	
C. thermolacticum	*C. thermosulfurogenes*	

been designated the cellulosome (Lamed *et al.*, 1983; Mayer, 1988). It remains associated with the cell surface and exhibits a molecular weight of 2 million. Various types of β-(1–4)-endoglucanases are present in these particles; they cleave cellulose into C_2–C_4 cellooligosaccharides, which before uptake by the cells may be degraded further by other enzyme components of the cellulosome. Genes coding for components of the cellulosome have been cloned and expressed in *Escherichia coli* (Cornet *et al.*, 1983; Schwartz *et al.*, 1986).

Starch is degraded by so many clostridial species that only a representative collection can be given in Table III. Recently, the thermophilic amylolytic species received special attention. They produce remarkably heat-stable α-amylase, pullulanase, and glucoamylase (Hyun and Zeikus, 1985; Madi *et al.*, 1987; Koch *et al.*, 1987). Usually, these enzymes remain cell-associated; however, conditions have been found (continuous culture under starch limitation) under which these enzymes are overproduced and secreted (Antranikian *et al.*, 1987).

Pectin degradation by *C. felsineum* and other clostridia has been important in retting of flax, hemp, or jute (Avrova, 1975). Some species expressing pectinolytic activities are given in Table III. In *C. thermosulfurogenes, C. roseum*, and *C. felsineum*, degradation of pectin involves a pectin esterase which cleaves the methyl ester bonds and a polygalacturonate hydrolase (Lund and Brocklehurst, 1978; Schink and Zeikus, 1983). Species like the former *C. multifermentans* (now: *C. beijerinckii* and *C. butyricum*) contain in addition a pectin lyase which removes monomers from pectin by trans-elimination (MacMillan and Vaughn, 1964).

Xylan and other hemicelluloses as well as other polysaccharides serve as growth substrates for a number of clostridial species (van Gylswyk and van der Toorn, 1987). Thus, polysaccharides are an important nutritive basis for this group of microorganisms.

Some of the proteolytic species are pathogenic (Table I). The digestion of meat and milk, liquefaction of gelatin, production of lecithinase and lipase, and hemolysis are taxonomically important characters for clostridia (Holdeman *et al.*, 1977; Cato *et al.*, 1986). Thus, specialists for protein and collagen utilization have been isolated—*C. proteolyticum* and *C. collagenovarans*—both unable to ferment single or mixtures of amino acids (Jain and Zeikus, 1988). The fermentation of the latter compounds is outlined in Section 3.6.

3.2. Ethanol, Acetate, and/or Butyrate as Major Fermentation Products

Clostridia employ the Embden–Meyerhof–Parnas pathway for breakdown of monosaccharides (Thauer *et al.*, 1977; Gottschalk, 1986). Sugar acids such as gluconate are degraded via a modified Entner–Doudoroff pathway (Andreesen and Gottschalk, 1969) in which gluconate is first dehydrated to 2-keto-3-deoxygluconate and then phosphorylated in position 6. The resulting 2-keto-3-deoxy-6-phosphogluconate is then cleaved by an aldolase to yield pyruvate and glyceraldehyde-3-phosphate. The central intermediate in the formation of fermentation products is pyruvate. In most clostridial fermentations it is converted to acetyl coenzyme A by pyruvate ferredoxin oxidoreductase. In addition to acetyl coenzyme A the products CO_2 and reduced ferredoxin are formed in this reaction. Because of the low redox potential of ferredoxin, electrons from the reduced form of this iron sulfur protein can be transferred via hydrogenase to protons, so that molecular hydrogen is evolved.

The fate of acetyl-CoA can be different depending on the clostridial species. It can be converted into a mixture of ethanol, acetate, and/or butyrate (Fig. 1). The ratio in which these products are formed depends on the amount of H_2 evolved. If it is produced only in connection with the conversion of pyruvate to acetyl coenzyme A, $2H_2$ can be evolved per glucose and the following fermentation equations can be formulated:

(a) Glucose + 3ADP + 3P$_i$ → butyrate + $2CO_2$ + $2H_2$ + 3ATP

(b) Glucose + 3ADP + 3P$_i$ → acetate + ethanol + $2H_2$ + 3ATP

The resulting product ratios (ethanol/acetate, 1 : 1; only butyrate, no acetate) are seldom found. In most instances clostridia producing butyrate also produce acetate (see Table IV). This becomes feasible because they produce not only $2H_2$/glucose but approximately 2.3 H_2/glucose. The additional $0.3H_2$ originates from NADH, which can be oxidized to a limited extent by NADH ferredoxin oxidoreductase (Fig. 1) (Jungermann et al., 1973). This extra H_2 evolution relieves the redox balance of the fermentation and allows the direct conversion of part of the acetyl coenzyme A via acetyl phosphate into acetate.

Clostridia producing ethanol/acetate mixtures very often evolve less than $2H_2$/glucose. Apparently, some of the reduced ferredoxin is used by these organisms to reduce NAD^+ to NADH, which is then employed for the reduction of acetyl coenzyme A to enthanol via acetaldehyde. As a consequence, more ethanol and less acetate is produced.

If organisms produce $2H_2$/glucose, the fermentation of glucose to ethanol/acetate or butyrate is coupled with the phosphorylation of 3ADP to 3ATP. Evolution of more H_2 leads to an increase of the ATP yield and evolution of less H_2 to a corresponding decrease. The fact that a number of clostridia produce more ethanol than acetate indicates that fermentations are not always optimized with respect to the ATP yield. Other parameters may become more important to the organisms, parameters such as fermentation rates, inhibitory effects of the products formed, or pH.

A special group of clostridia, the homoacetogenic species, will be dis-

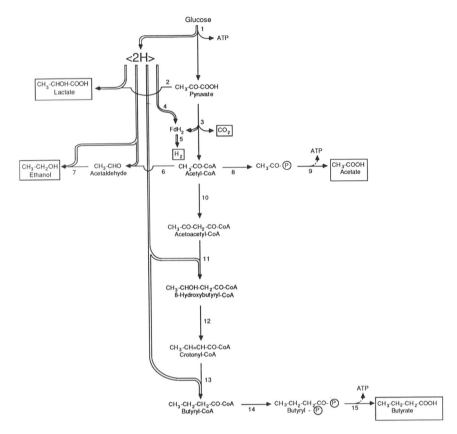

Figure 1. Schematic diagram of the reactions in *Clostridium* species leading to the formation of ethanol, organic acids, CO_2 and H_2. →, carbon flow; ◊, electron flow; ---▶, energy-yielding reactions. 1, reactions of the Embden–Meyerhof–Parnas pathway; 2, lactate dehydrogenase; 3, pyruvate:ferredoxin oxidoreductase; 4, ferredoxin:NAD^+ oxidoreductase; 5, hydrogenase; 6, acetaldehyde dehydrogenase; 7, alcohol dehydrogenase; 8, phosphotransacetylase; 9, acetate kinase; 10, β-ketothiolase; 11, β-hydroxybutyryl-CoA dehydrogenase; 12, crotonase; 13, butyryl-CoA dehydrogenase; 14, phosphotransbutyrylase; 15, butyrate kinase.

cussed in Chapter 5. These bacteria produce almost 3 mol acetate from 1 mol glucose or fructose. In addition, *C. aceticum* and *C. thermoautotrophicum* produce acetate from H_2 and CO_2. The principal flux of carbon from the hexoses degraded to acetate is similar to that in other clostridia. What is special is the disposal of reducing equivalents. They are transferred not to acetyl-CoA or to pyruvate but to CO_2.

3.3. Fermentation Products Formed under Stress Conditions

Under certain conditions clostridial species shift their fermentation pattern, and other products are produced. Some of the products formed

Table IV. Fermentation Products of a Few *Clostridium* Species[a]

Products	Amounts formed in mol/100 mol glucose fermented		
	C. butyricum	*C. perfringens*	*C. acetobutylicum*
Acetate	42	60	14
Butyrate	76	34	4
Lactate	—	33	—
CO_2	188	176	221
H_2	235	214	135
Acetone	—	—	22
Butanol	—	—	56
Ethanol	—	26	6

[a]Taken from Wood, 1961.

and the organisms are given in Table V. Although many of these organisms contain lactate dehydrogenase (Fig. 1), lactate is not a major product of clostridial fermentations. This enzyme requires elevated concentrations of fructose-1,6-bisphosphate for activity (Freier and Gottschalk, 1987). Thus, lactate is produced only if there is some blockage of the usual fermentation pathways. Such a situation arises if further breakdown of pyruvate is inhibited by iron limitation (Bahl *et al.*, 1986) or by carbon monoxide (Kubowitz, 1934). Likewise, the availability of substrate in excess may lead to an increase in the intracellular concentration of fructose-1,6-bisphosphate and hence to lactate production.

Other products that are formed under special conditions are the solvents acetone, butanol, and isopropanol (see also Chapter 4). Species producing these products are listed in Table V. Their formation is also dependent on certain conditions. In *C. acetobutylicum* a certain butyrate concentra-

Table V. Solvent Production by Various *Clostridum* Species from Glucose

Organism	Products formed (g/100 g glucose fermented)[a]				
	Acetone	Butanol	Ethanol	Isopropanol	1,2-Propandiol
C. acetobutylicum	10	21	3	—	—
C. aurantibutyricum	6	17	ND[b]	1	—
C. beijerinkii	2	25	ND	—	—
	—	17	ND	3	—
C. puniceum	3	32	2	—	—
C. tetanomorphum	—	17	10	—	—
C. sphenoides	—	—	37	—	21
C. thermosaccharolyticum	—	—	4	—	27

[a]Data calculated from Beesch (1952), George *et al.* (1983), Holt *et al.* (1988), Gottwald *et al.* (1984), Tran-Dinh and Gottschalk (1985), Cameron and Cooney (1986).
[b]ND, not determined.

tion is a prerequisite. Furthermore, it also requires a pH lower than 5 and a physiological intracellular pH that allows the synthesis of the enzymes required for the formation of butanol and acetone (Bahl and Gottschalk, 1984; Gottwald and Gottschalk, 1985).

Another product belonging to this category is 1,2-propanediol, which is produced by *C. sphenoides* under conditions of phosphate limitation (Tran-Dinh and Gottschalk, 1985). The latter is not a prerequisite for *C. thermosaccharolyticum* (Cameron and Cooney, 1986). The so-called methyl glyoxal pathway is responsible for the formation of this compound. It allows the further conversion of dihydroxyacetone phosphate without any involvement of a phosphorylated compound.

3.4. Fermentation of Organic Acids

Both malate and fumarate are fermented by the acetogenic *C. formicoaceticum, C. aceticum,* and *C. magnum* (Andreesen *et al.,* 1970; Schink, 1984a). Despite the fact that they are interconvertible by the enzyme fumarase, the products derived from malate and fumarate by *C. formicoaceticum* are different: fumarate is partially reduced to succinate, whereas malate is converted to acetate and bicarbonate (Dorn *et al.,* 1978a). The enzyme fumarate reductase seems to face the periplasm of this organism. As a consequence, it is inaccessible to the fumarate generated intracellularly (Dorn *et al.,* 1978b).

Although generally thought to be inert under anaerobic dark conditions, succinate can be metabolized by *C. kluyveri,* together with ethanol, to yield acetate (Keanealy and Waselefsky, 1985). Other anaerobes can ferment succinate as the sole substrate, e.g., *Propiogenium modestum,* which derives energy from a sodium gradient formed by decarboxylation (Dimroth, 1987). In contrast, citrate is known to serve as a single substrate for a variety of species such as *C. sphenoides, C. sporosphaeroides, C. sporogenes, C. subterminale, C. symbiosum,* and *C. magnum* (Antranikian *et al.,* 1984; Schink, 1984b), which are mostly known as amino acid fermenters. Degradation of citrate is initiated by the enzyme citrate lyase, the activity of which is thoroughly regulated by chemical modification (Antranikian and Giffhorn, 1987). Of the products formed, acetate and oxaloacetate, only the latter can be utilized as an energy source by established pathways.

Tartrate has been the selective substrate of "*C. tartarivorum,*" now regarded as a biotype of *C. thermosaccharolyticum* (Matteuzzi *et al.,* 1978). Mesophilic enrichment cultures led to the isolation of a variety of non-spore-forming anaerobes assigned to the genera *Bacteroides, Acetivibrio, Ruminococcus,* and *Ilyobacter.* By tartrate dehydratase activity, specific for the respective enantiomers, oxaloacetate is formed (Schink, 1984b).

Lactate fermentation can be accomplished by certain clostridia if an additional electron acceptor is provided. This can be acetate, as in the case of *C. beijerinckii* ("*C. lactoacetophilum*") (Bhat and Barker, 1948) and *C. acetobutylicum* (Freier and Gottschalk, 1987), or carbonate, as in the case of *C. formicoaceticum*, which converts 1 lactate to 1.25 acetate (Andreesen *et al.*, 1970; Giffhorn, 1980). *Clostridium propionicum* forms acrylate from lactate, which can serve as electron acceptor (Akedo *et al.*, 1983).

3.5. Fermentation of Alcohols

The fermentation of primary alcohols, diols, and acetoin is primarily accomplished by the anaerobic genera *Acetobacterium* and *Pelobacter* (Eichler and Schink, 1985). Diols such as ethylene glycol and glycerol are rearranged by, for example, *C. glycolicum* or *C. acetobutylicum*. The former organism employs a radical mechanism, not a cobamide coenzyme-dependent diol dehydratase, to give the corresponding aldehyde (Hartmanis and Stadtman, 1986). In the case of glycerol, the 3-hydroxypropionaldehyde formed is subsequently reduced to 1,3-propanediol. *Clostridium kluyveri* represents the first well-studied ethanol utilizer requiring acetate as electron acceptor. It forms, depending on the ratio of supplied ethanol to acetate, varying amounts of butyrate, caproate, and molecular hydrogen. Bicarbonate is required for the synthesis of cellular constituents. The organism served as model for fatty acid and amino acid biosynthesis, stressing the role of CoA esters, of carboxylation reactions, and of the stereospecificity of enzymatic reactions (Barker, 1978). Its enzymes and those from certain other clostridial species are exploited for producing chiral compounds (Simon *et al.*, 1985). *Clostridium kluyveri* has now been shown to utilize 1-propanol besides ethanol (as alcohol) and succinate, in addition to acetate, propionate, butyrate, valerate, 2-butenoate, 3-butenoate (crotonate), and 4-hydroxybutyrate as electron acceptor (Keanealy and Waselefsky, 1985).

Some clostridial species, such as *C. formicoaceticum* and *C. aceticum*, are able to utilize alcohols like methanol and ethanol by using bicarbonate as electron acceptor for acetate formation (Giffhorn, 1980; Schink, 1984a). Instead of CO_2 reduction to acetate, propionate formation from ethanol and CO_2 can be observed in certain cases to obtain a redox balance (Schink *et al.*, 1985). *Clostridium pfennigii* is specialized on the methyl carbon of methoxylated compounds such as ferulate found as precursor of lignin biosynthesis; it is even unable to utilize methanol (Krumholz and Bryant, 1985). The aromatic basal structure resorcinol (1,3-dihydroxybenzene) is fermentatively degraded by a *Clostridium* sp. (Tschech and Schink, 1985).

Clostridium bryantii stands at the end of the degradative food chain exerted by clostridial species, since it can even oxidize fatty acids anaerobically in an obligately syntrophic association with H_2-utilizing bacteria: even-numbered fatty acids (with up to 10 carbon atoms) to acetate and hydrogen, odd-numbered fatty acids (with up to 11 carbon atoms) and 2-methylbutyrate (formed from isoleucine) to acetate, propionate, and H_2 (Stieb and Schink, 1985). 3-Methylbutyrate (formed from leucine) was the only substrate attacked by an unclassified rod in coculture with H_2 scavengers using the pathway as shown for aerobic leucine metabolism (Stieb and Schink, 1986).

3.6. Fermentation of N-Containing Substrates

Amino acids, purines, and pyrimidines constitute an important part of living matter as expressed by the general formula for cell mass of $C_4H_8O_2N$. The degradation (and biosynthesis) of purines and pyrimidines interacts with that of certain amino acids such as glycine, β-alanine, and aspartate (Fig. 2). On the other hand, the pyridine nicotinic acid, which is part of the coenzymes NAD^+ or $NADP^+$, is degraded via a completely independent pathway. In many cases the route for degradation is quite similar to that for biosynthesis. For some central amino acids such as glutamate, aspartate, ornithine, and threonine, different pathways have evolved; their occurrence is restricted to anaerobic bacteria. To avoid confusion within the cells between incorporation into proteins and degradation, the racemization into the corresponding D-amino acids as found for lysine, ornithine, and proline is employed as the first step in catabolism. Due also to the importance of N-containing compounds in biotechnological terms, the strategy of their fermentation is dealt with in more detail.

Most of the amino acids can serve as sole carbon, nitrogen, and energy sources. However, the Stickland reaction is quite typical for the amino acid decomposition by clostridia. This normally means that one amino acid only serves as electron donor, another as electron acceptor. Certain amino acids (e.g., phenylalanine) can serve both purposes; one molecule gets oxidized, another reduced. This is the simplest way to ferment amino acids and to provide the cell with about 0.5 ATP per amino acid transformed (Fig. 3A, B, D, E).

3.6.1. Branched Chain Amino Acids

The branched chain and aromatic (see Section 3.6.2) amino acids are typically attacked in a Stickland reaction. Usually they serve only as electron donors (Elsden *et al.*, 1976; Elsden and Hilton, 1978), as indicated by the formation of branched chain or aromatic fatty acids shorter by one carbon atom as compared to the original amino acid. Breakdown is initiated by transamination (Fig. 3A) or direct oxidation (Fig. 3B) to yield the corresponding 2-oxo acid which is oxidatively decarboxylated to CO_2, the

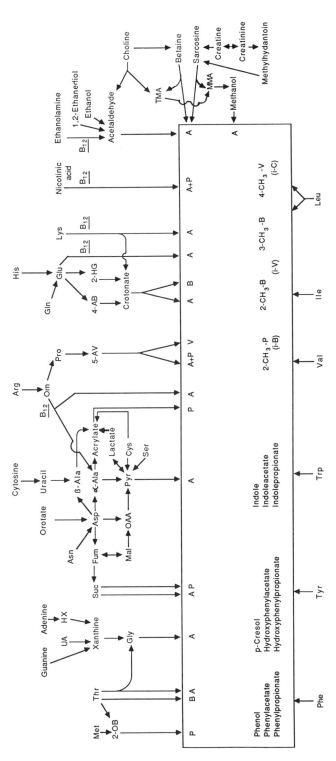

Figure 2. Generalized carbon flow of N-containing compounds to fatty acids in clostridia. Not shown are CO_2, formate, H_2, NH_3, nor simple redox intercorversions such as ethanol or butyrate formation from acetate. Abbreviations used: A = acetate, P = propionate, B = butyrate, V = valerate, 4-CH_3-V = 4-methylvalerate, i-C = isocaproate, 2-OB = 2-oxobutyrate, 5-AV = 5-aminovalerate, 2-HG = 2-hydroxyglutarate, TMA = trimethylamine, UA = uric acid, B_{12} = involvement of cobamide coenzyme.

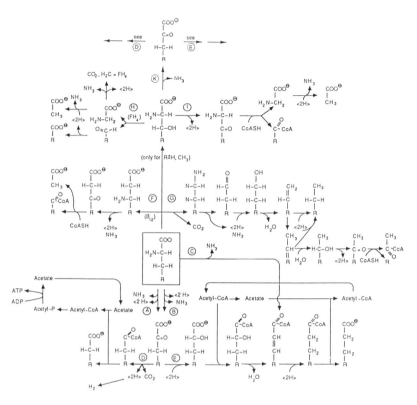

Figure 3. Principal routes of amino acid fermentation. A, formation of 2-oxo acid by trans-amination and reoxidation of acceptor by, for example, glutamate dehydrogenase; B, direct oxidation of amino acid by a dehydrogenase; C, desaturative deamination to enoate, and the product shown is formed after CoA transferase reaction; D, oxidative decarboxylation, hydrogen production, and energy conservation; E, reduction to 2-hydroxy acid, activation by CoA-transferase, dehydration to enoyl compound, reduction, CoA ester transfer; F, formation of β-amino acid by (B_{12} = cobamide coenzyme-dependent) migration of amino group, oxidation to 3-oxo amino acid, and thiolytic cleavage; G, decarboxylation to the amine, transformation to the hydroxy compound, dehydration (migration of double bond), and reduction (or hydration of double bond, oxidation, and thiolytic cleavage); H, tetrahydrofolate (FH_4)-dependent cleavage of 3-hydroxyamino acids (serine, threonine) to an aldehyde and glycine, oxidation or reduction of the latter; I, oxidation to 3-oxoamino acid, and thiolytic cleavage to acyl-CoA and glycine; K, deaminase reaction forming a 2-oxo acid.

acyl-CoA, and reduced ferredoxin (Fig. 3D). The latter can be reoxidized by hydrogenase to yield hydrogen gas. On the other hand, some species are able to form a 2-oxocarboxylate reductase (Giesel and Simon, 1983), which produces the corresponding 2-hydroxy acid (Fig. 3E). A dehydratase yields the corresponding enoate (Machacek-Pitsch *et al.*, 1985), which can be reduced directly (or after activation by a CoA transferase) by a relatively unspecific enoate reductase (Simon *et al.,* 1985). However, the enzyme requires a hydrogen atom in position 3. Thus, a steric hindrance is indicated for the unsaturated branched chain fatty acids derived from valine and isoleucine. Therefore, these latter two amino acids can only be oxidized to 2-methylpropionate and 2-methylbutyrate, respectively. Leucine, however, can be oxidized to 3-methylbutyrate as well as reduced to 4-methylvalerate (isocaproate), as typical for *C. sporogenes* and related organisms (Elsden and Hilton, 1978; Britz and Wilkinson, 1982; Giesel and Simon, 1983). The degradation of the individual branched chain amino acids is reflected in the fatty acid composition of the lipids of proteolytic clostridia (Elsden *et al.,* 1980). Anteiso acids, with odd numbers of carbon atoms, can only be formed from isoleucine, whereas iso acids with even and odd numbers of carbons reflect the degradation of valine and leucine, respectively.

3.6.2. Aromatic Amino Acids

The three aromatic amino acids phenylalanine, tyrosine, and tryptophan create no problems for enoate reductase due to the general $-CH_2-CHNH_2-COOH$ moiety. It follows that these amino acids can be oxidized as well as reduced to the respective arylacetic or arylpropionic acid. However, there seems to be a species-specific preference since an oxidation is routinely observed, e.g., *C. sticklandii, C. subterminale,* and similar organisms (Mayraud and Bourgeau, 1982), whereas a reduction occurs preferentially with *C. sporogenes, C. botlinum* type A, B, and F, *C. bifermentans,* and others (Moss *et al.,* 1970; Elsden *et al.,* 1976, 1979; Jellet *et al.,* 1980; Giesel and Simon, 1983). *Clostridium bifermentans, C. sordellii,* and *C. difficile* can even employ phenylalanine as both electron donor and acceptor at the same time (Elsden *et al.,* 1976). A more detailed study including *C. sporogenes* and related organisms indicated (Fryer and Mead, 1979) that only *C. sporogenes* can actually grow at the expense of the disproportionation of phenylalanine. In addition, this amino acid can act as electron donor for glycine and proline reduction or as electron acceptor of alanine or branched chain amino acid oxidation.

As a peculiarity, the complete removal of the side chain from the aromatic ring system was reported for *C. tetani* and related organisms which form phenol and indole from phenylalanine and tryptophan, respectively. These organisms also ferment glutamate via β-methylaspartate and do not incorporate the carbons of branched chain amino acids into cellular fatty acids (Elsden *et al.,* 1976, 1980). Another specialty is the formation of *p*-cresol from tyrosine via hydroxyphenylacetate by *C. difficile* (D'Ari and Barker, 1985).

Energy is generally conserved by transforming the acyl-CoA esters (generated by oxidative decarboxylation of the 2-oxo acid) into, for example, acetyl-CoA by CoA transferase (Bader *et al.*, 1982), which finally yields acetate and ATP by phosphate acetyltransferase and acetate kinase reactions (Fig. 3D). The extent of energy conservation during the reduction of the respective enoates (Fig. 3E, G) might differ to a high degree (Bader and Simon, 1983; Tschech and Pfennig, 1984).

3.6.3. Glycine and Alanine

The simplest amino acid, glycine, most often functions only as an electron acceptor yielding acetate plus NH_3 as products of *C. sporogenes* and *C. sticklandii* (Seto, 1980). *Clostridium histolyticum* and the purine degrading clostridia can grow solely on glycine if complex ingredients or some purines are also provided (Dürre and Andreesen, 1982a; Lebertz, 1984; Lebertz and Andreesen, 1988). A quarter mole of glycine will be oxidized to CO_2 by glycine decarboxylase, whereas three-quarters will be directly reduced to acetate by the glycine reductase system (Fig. 3H). The enzyme complex includes a genuine selenocysteine-containing protein which is responsible for the generally observed selenium dependence of growth on glycine (Turner and Stadtman, 1973; Sliwkowski and Stadtman, 1987; Zindel *et al.*, 1988).

Alanine represents the shortest amino acid with stereoisomers. Its moiety constitutes the side chain of the aromatic amino acids, and on its basis the different fates of the carbon atoms are demonstrated in Fig. 3. The simplest strategy for alanine degradation is the direct oxidation of alanine to pyruvate (Fig. 3A) by an NAD-dependent alanine dehydrogenase as present in some clostridia (Bogdahn and Kleiner, 1986). Most clostridia, however, will form pyruvate by a transamination reaction (Fig. 3B) involving 2-oxoglutarate and glutamate dehydrogenase (Bader *et al.*, 1982; Bogdahn and Kleiner, 1986). Pyruvate is generally oxidized (Fig. 3D) since the reduced ferredoxin generated can easily be reoxidized by hydrogenase.

3.6.4. Cysteine and Methionine

So far the decomposition of the sulfur-containing amino acids cysteine and methionine has not been studied in detail (Barker, 1961). Both are degraded by *C. propionicum* (Cardon and Barker, 1947) via an enoate, such as acrylate, as the characteristic intermediate of that organism. An involvement of cysteine desulfhydrase has been anticipated which would directly form pyruvate, sulfide, and ammonia from cysteine (Barker, 1961), whereas methionine is converted by a similar enzyme in *C. sporogenes* to 2-oxobutyrate, methyl sulfide, and ammonia (Wiesendanger and Nisman, 1953).

Some proteolytic clostridia form trace amounts of methane from methionine (Rimbault *et al.*, 1988).

3.6.5. Serine and Threonine

Serine can be fermented by many clostridial species (Barker, 1961). An iron-dependent serine dehydratase, as found for *C. acidiurici* (Carter and Sagers, 1972), transforms serine directly into pyruvate and NH_3. An alternative is the reversible aldol cleavage of serine into methylene tetrahydrofolate (THF) and glycine by serine hydroxymethyltransferase (Fig. 3H). It was supposed to be involved in acetate synthesis from CO_2 by *C. acidiurici* (Waber and Wood, 1979), but it is used only for biosynthetic purposes (Dürre and Andreesen, 1983). *Clostridium propionicum* forms acetate, CO_2, propionate, and NH_3 from serine (Cardon and Barker, 1947), thus indicating the involvement of an acrylate moiety during propionate formation.

Threonine having four carbon atoms but three functional groups can be handled in five different ways by the clostridial species. Deamination to 2-oxobutyrate and oxidative cleavage into CO_2, propionyl-CoA, and reduced ferredoxin represents the simplest way and is found in *C. tetanomorphum* (Tokushiga and Hayaishi, 1972). Both threonine and serine deaminase seem to have a common ancestor (Parsot, 1986). As a peculiarity of *C. propionicum,* part of the threonine is reduced to butyrate without any C-C rearrangements (Barker and Wiken, 1948). Oxidation of the hydroxyl group to 2-amino-3-oxobutyrate and the following cleavage to glycine and acetyl-CoA (Fig. 3I) has been reported for *C. sticklandii* (Golovchenko *et al.*, 1982). Threonine aldolase is an enzyme which primarily forms threonine from glycine and acetaldehyde (Fig. 3H) in *C. pasteurianum* (Thauer *et al.*, 1972). *Clostridium sporogenes* and related organisms excrete substantial amounts of 2-aminobutyrate from threonine (Mead, 1971). Formally, only the hydroxyl moiety is reduced by a system not yet studied.

3.6.6. Aspartate and Asparagine

Aspartate and its amide asparagine are not known as selective substrates for *Clostridium* spp., as is the case for members of the genus *Campylobacter* (Laanbroek *et al.*, 1978). Aspartate is utilized by *C. tetani* (Pickett, 1943), *C. tetanomorphum* (Woods and Clifton, 1937; Elsden and Hilton, 1979), *C. cochlearium,* and *C. difficile* (Mead, 1971). The degradation pathways have not been elucidated for these species, but can be anticipated as (1) transamination to oxaloacetate followed by decarboxylation to pyruvate, or (2) desaturative deamination (Fig. 3C) by aspartase forming fumarate. In *C. tetani* the latter can be reduced to succinate (Pickett, 1943) or hydrated to malate (Clifton, 1942) with subsequent oxidation to pyruvate directly or via oxaloacetate. *Clostridium sporogenes* can transform aspartate to alanine by decarboxylation (Meister *et al.*, 1951), whereas decarboxylation to β-alanine has not been observed for clostridial species (Barker, 1961).

3.6.7. Glutamate, Glutamine, and Histidine

Clostridium cochlearium represents the specialist for glutamate, glutamine, and histidine fermentation (Laanbroek *et al.*, 1979), although a variety of other species such as *C. tetanomorphum, C. tetani, C. malenominatum, C. limosum, C. cochlearium, C. sporosphaeroides*, and *C. symbiosum* ferment glutamate as sole substrate (Mead, 1971; Buckel, 1980; Buckel and Miller, 1987). Only the latter two species ferment glutamate by the hydroxyglutarate pathway (Buckel, 1980), whereas all the other clostridia use the cobamide-coenzyme-dependent pathway characterized in detail for *C. tetanomorphum* (Barker, 1961, 1981).

Histidine fermentation by *C. tetanomorphum* and *C. tetani* (Pickett, 1943) leads to glutamate. First it is desaturatively deaminated to urocanate (Wickremasinghe and Fry, 1954) to which water is added to give, after a rearrangement, imidazolone propionate. The imidazole ring is hydrolyzed to formiminoglutamate (Bochner and Savageau, 1979), which can be split to glutamate and formamide, as observed for *C. tetanomorphum* (Wachsman and Barker, 1955). Alternatively, the energy of the formimino moiety can be conserved by transfer to tetrahydrofolate to give formate and ATP via formyltetrahydrofolate synthetase (Barker, 1961). Histidine can also be decarboxylated to histamine under certain conditions (Recsei *et al.*, 1983).

Most clostridial species ferment glutamate via the cobamide-dependent pathway (Barker, 1961, 1978), which starts by a carbon rearrangement to give β-methylaspartate, which can be desaturatively deaminated to mesaconate (2-methylfumarate), followed by rehydration to citramalate (2-methyl-2-hydroxy-succinate). Cleavage of citramalate by the corresponding lyase yields pyruvate and acetate (*not* acetyl-CoA). Citramalate lyase exhibits relations to citrate lyase (Buckel and Bobi, 1976). This catabolic route is very effective since the intermediates represent no thermodynamic barrier (Buckel and Miller, 1987). The second pathway starts with the action of glutamate dehydrogenase, which is common in clostridia due to its involvement in transaminase reactions (Bogdahn and Kleiner, 1986). 2-Oxoglutarate is subsequently reduced to 2-hydroxyglutarate. After activation by CoA transferase, the iron-sulfur protein 2-hydroxyglutaryl-CoA dehydratase catalyzes the crucial step of dehydration of an α-hydroxy acid to yield glutaconyl-CoA (Schweiger *et al.*, 1987). A membrane-bound decarboxylase drives a sodium pump which may be involved in transport processes (Wohlfarth and Buckel, 1985). Crotonyl-CoA thus formed can only be disproportionated to acetyl-CoA and butyryl-CoA, as already outlined for the saccharolytic butyric acid-producing clostridia. A possible third pathway, via a reversed citric acid cycle, has not been found in fermentative bacteria (Buckel and Miller, 1987), since the intermediate formation of oxaloacetate and acetate represents a thermodynamic barrier. Such a barrier may be circumvented by anaerobic phototrophic bacteria via an ATP-dependent citrate lyase. A large group of clostridia such as *C. sporogenes, C. botulinum,* and *C. bifermentans* decarboxylate glutamate to 4(γ)-aminobutyrate, which is excreted (Mead, 1971; Elsden and Hilton, 1979). This compound (known as neurotransmitter) is fermented by *C. aminobutyricum* via transformation into 4(γ)-hydroxybutyrate, activation by CoA-transferase, dehydration to vinylacetyl-CoA, and isomerization of the double bond into crotonyl-CoA (Hardman and Stadtman, 1963), a well-known intermediate of clostridial metabolism.

3.6.8. Arginine, Ornithine, and Proline

Arginine, ornithine, and proline catabolisms are quite interrelated. Of the four pathways known for arginine breakdown, only the arginine deiminase pathway is found in clostridia (Cunin *et al.*, 1986). In hydrolytic reactions, first citrulline plus ammonia, and then ornithine plus carbamoyl phosphate are formed. The latter is transformed by carbamate kinase into bicarbonate, ammonia, and ATP. Thus, arginine fermentation to ornithine represents an ideal way for anaerobes to regenerate ATP by simple hydrolytic reactions avoiding redox reactions. Therefore, its breakdown is observed in many species, such as *C. sporogenes, C. sticklandii,* and *C. histolyticum.* The latter organism even excretes ornithine (Mead, 1971; Elsden and Hilton, 1979). Arginine can also be decarboxylated to agmatine and further hydrolyzed to urea and putrescine, the latter being a constituent of polyamines in clostridia (Kneifel *et al.*, 1986). Putrescine can also originate in *C. botulinum* by decarboxylation of ornithine (Mitruka and Costilow, 1967).

The catabolism of ornithine can be accomplished in two ways (as observed for glutamate). For *C. sticklandii* it serves as hydrogen donor in a Stickland reaction (Barker, 1981). The catabolism starts from D-ornithine generated by a racemase. By the ornithine mutase reaction (which involves cobamide coenzyme and pyridoxal phosphate), the amino group in position 5 (δ) migrates first to position 4 to avoid an otherwise spontaneous ring closure (to proline) by its oxidation to 2-amino-4-oxopentanoate, which no longer can form a stable ring system. The cleavage into alanine and acetyl-CoA proceeds by a pyridoxal phosphate- and coenzyme A-dependent thiolytic cleavage (Jeng *et al.*, 1974), thereby conserving energy. The enzyme is also present in *C. subterminale. Clostridium sporogenes* and others (Mead, 1971) follow the second strategy to ferment ornithine, using it as electron acceptor by forming 5(δ)-aminovalerate via D-proline. In this case, the amino group in position 2 is first attacked by ornithine cyclase (deaminating), which contains bound NAD. The enzyme is first reduced to form ammonia and 2-oxo-5-aminopentanoate, which spontaneously closes a ring to Δ^1-pyrroline-2-carboxylate as intermediate. The bound NADH is reoxidized by reducing the intermediate to L-proline (Muth and Costilow, 1974), which is racemized into the D form to avoid mixing with L-proline used for anabolism.

As an imino acid, proline can invariantly only be reduced to 5(δ)-aminovalerate by the group of *C. sporogenes, C. botulinum,* and *C. bifermentans,* but also by *C. sticklandii* (Mead, 1971; Elsden and Hilton, 1979). *Clostridium sporogenes* prefers proline over glycine as electron acceptor (Venugopalan, 1980), whereas glycine inhibits proline reduction in *C. sticklandii* (Schwartz *et al.,* 1979). Contrary to glycine, proline reduction to 5(δ)-aminovalerate (Seto, 1980) is not directly coupled with energy conservation; however, the formation of a proton gradient has been measured for *C. sporogenes* (Lovitt *et al.,* 1986). 5(δ)-Aminovalerate is further catabolized by *C. aminovalericum* via an internal Stickland-type reaction (Fig. 3G) to ammonia, acetate, and propionate as the oxidized products, and valerate as the reduced part (Hardman and Stadtman, 1960b). Most of the catabolic reactions are now understood (Barker *et al.,* 1987). Transamination, reduction, and CoA-transferase reactions form 5-hydroxyvaleryl-CoA, which is dehydrated. The double bond formed moves toward the carboxyl group to be reduced to valeryl-CoA or hydrated to 3-hydroxyvaleryl-CoA. This compound is oxidized to the corresponding oxo compound which is thiolytically cleaved by CoASH to acetyl-CoA and propionyl-CoA. Thus, its metabolism is quite similar to that of 4(γ)-aminobutyrate.

3.6.9. Lysine

Clostridium sticklandii and *C. subterminale* (SB$_4$) are typical lysine fermenters (Elsden and Hilton, 1979; Barker, 1981). There are two pathways, both yielding the same products (ammonia, acetate, and butyrate) by cleavage between carbon atoms 2 and 3 (which is well studied) or 4 and 5. Both pathways request the migration of the amino groups, which involves cobamide coenzyme-dependent enzymes, when the terminal amino group is transferred to position 5. This happens first in the case of D-lysine, whereas for L-lysine the amino group in the α position is moved in a pyridoxal phosphate-dependent reaction to β-lysine (3,6-diaminohexanoate), followed by the migration of the terminal amino group in a cobamide coenzyme-dependent reaction to position 5. These initiating rearrangements then allow an oxidation of the 3-amino group to 3-oxo-5-aminohexanoate, avoiding a spontaneous cyclization reaction (Fig. 3F). The enzymatic cleavage of this compound by acetyl-CoA is undertaken in a unique manner, involving concerted reactions leading to acetoacetate formation from acetyl-CoA (the latter becoming carbon atoms 3 and 4). The bound CoA moiety is involved in thiolytic cleavage at the 3-oxo position yielding L-3-aminobutyryl-CoA (made from the original carbon atoms 3–6 of lysine) (Yorifuji *et al.*, 1977), which is deaminated to crotonyl-CoA. This will be reduced to butyryl-CoA, which transfers its CoA moiety to acetoacetate which in turn can be thiolytically split into two acetyl-CoA. Again the energy conservation via phosphate acetyltransferase and acetate kinase is placed at the very end of the catabolism, thus ensuring carbon flow to the final products.

3.6.10. Pyrimidines and Purines

Pyrimidines and purines are part of nucleic acids and many coenzymes. Purines are also waste products of nitrogen metabolism excreted by many animals (Vogels and van der Drift, 1976). Therefore, clostridia specializing in purine fermentation can be easily isolated and have been extensively characterized (Rabinowitz, 1963; Vogels and van der Drift, 1976; Schiefer-Ullrich *et al.*, 1984; Berry *et al.*, 1987).

Clostridium oroticum seems to be the only *Clostridium* able to degrade the pyrimidine orotic acid (4-carboxyuracil) (Cato *et al.*, 1968), via dihydroorotate, *N*-carbamoylasparagine to ammonia, CO_2, and aspartate (Aleman and Handler, 1967; Pettigrew *et al.*, 1985). *Clostridium uracilicum* is no longer available; however, *C. glycolicum* was shown to act in a similar way on uracil (Mead *et al.*, 1979). In a survey of the proteolytic clostridia, *C. sporogenes* and certain types of *C. botulinum* were both shown to deaminate cytosine, and to reduce uracil and thymine to the dihydro forms (Hilton *et al.*, 1975). A hydrolytic cleavage to *N*-carbamoyl-β-alanine and further to ammonia, CO_2, and β-alanine (or β-aminoisobutyrate in the case of thymine) was not observed for *C. sporogenes*, but has been observed for *C. uracilicum* and many other organisms (Vogels and van der Drift, 1976). The general strategy adopted in the degradation of the heterocyclic aromatic pyrimidine ring consists of a reduction of the weak aromatic ring system. This can be accomplished even by NADPH and a flavoenzyme, because the aromaticity is further decreased in the hydroxylated form, which primarily exists in the oxo form. Thus, these lactams can be hydrolytically split. It is unclear whether energy can be conserved by transferring the *N*-carbamoyl moiety via carbamate kinase to ATP.

All three specialized purinolytic clostridia—*C. acidiurici*, *C. cylindrosporum*, and *C. purinolyticum*—are phenotypically closely related (Andreesen *et al.*, 1985). However, 16S rRNA studies indicate their distant relationships (see Chapter 1 of this book). The purine compounds adenine and guanine first must be deaminated to overcome their toxicity and/or insolubility (Dürre and Andreesen, 1983). Hypoxanthine and uric acid are directly converted to xanthine by xanthine dehydrogenase, a rather complex enzyme composed of selenium, molybdenum, iron-sulfur clusters, and FAD, which generally exhibits a broad substrate specifity (Wagner *et al.*, 1984). Xanthine is hydrolytically split in five consecutive reactions (Rabinowitz, 1963; Dürre and Andreesen, 1982b) via 4-ureido-5-imidazolecarboxylate to CO_2, ammonia, and 4-amino-5-imidazolecarboxylate; the latter to CO_2 and 4-aminoimidazole; the latter to ammonia and 4-imidazolone; and the latter finally to formiminoglycine. The formimino moiety is transferred to tetrahydrofolate to conserve energy via the formyltetrahydrofolate synthetase reaction. The formate is oxidized to CO_2 to reduce the glycine (formed before) directly to acetate and ammonia by the selenium-dependent glycine reductase and to conserve energy in this reductive deamination reaction (Dürre and Andreesen, 1982), which again constitutes the last step. Due to their restricted substrate utilization spectrum, the purinolytic *Clostridia* are mostly unable to synthesize *de novo* purine compounds, thus requiring thiamine and purine compounds for growth on the intermediate glycine (Schiefer-Ullrich *et al.*, 1984; Lebertz, 1984).

As a constituent of coenzymes, the pyridine nicotinic acid is a vitamin for many organisms. *Clostridium barkeri* has been isolated from enrichment culture on nicotinic acid and belongs taxonomically to a quite separate group of organisms (Chapter 1). Again the initial attack is done by a selenomolybdo-iron-sulfur flavoprotein, which introduces a hydroxyl group close to the nitrogen (Imhoff and Andreesen, 1979; Dilworth, 1983). After dissolving the aromaticity by reduction via ferredoxin, the adenosylcobalamin-dependent 2-methylenglutarate mutase constitutes a key enzyme which forms methylitaconate (Hartrampf and Buckel, 1986), which is isomerased to dimethylmaleate. The stereochemical course of the hydratase reaction to 2,3-dimethylmalate was studied as was the dimethylmalate lyase reaction (Eggerer, 1985), which yields the products propionate and pyruvate, from which the latter are further decomposed to give acetate, CO_2, energy, and the reduced ferredoxin to start the reduction of another pyridine ring system.

3.6.11. Amine Compounds

Amine compounds, such as ethanolamine, choline, betaine, creatinine, and creatine, have been positively tested as single substrates for a variety of clostridial species (Möller *et al.*, 1986). Ethanolamine is converted by a *Clostridium* sp. via the cobamide coenzyme-dependent ethanolamine deaminase into ammonia and acetaldehyde (Bradbeer, 1965), which can be dismutated to acetyl-CoA and ethanol. Trimethylamine was always formed from choline and betaine, indicating a reductive cleavage as observed for *C. sporogenes* where betaine only serves as electron acceptor (Naumann *et al.*, 1983). In contrast to members of the genus *Sporomusa* (Möller *et al.*, 1984), *Clostridium* spp. were unable to utilize dimethylglycine and sarcosine (*M-*

methylglycine) as sole substrates (Möller *et al.*, 1986). Clostridia generally degrade creatinine and creatine to *N*-methylhydantoin; only *C. sordellii* hydrolyzes it further to sarcosine (Möller *et al.*, 1986). Creatine, sarcosine, and betaine are only degraded to methylamine(s) and acetate by a recently described anaerobe if an electron donor and selenite is additionally supplied (Zindel *et al.*, 1988). Thus, the potential of the clostridial species might be even higher than anticipated. So far, no clostridial species has been shown to utilize the methylamine(s) generated in contrast to methanol (Schink, 1984a). However, some species of *Sporomusa* can utilize trimethylamine at least (Möller *et al.*, 1984; Hermann *et al.*, 1987).

3.6.12. Dinitrogen Fixation

Dinitrogen fixation is a property which has long been recognized in many clostridia, such as *C. pasteurianum*, despite their general role in the destruction of organic material (Winogradsky, 1895; Rosenblum and Wilson, 1950). Most of the species known are saccharolytic (Hammann and Ottow, 1976; Bogdahn and Kleiner, 1986). Although the enzymes of glutamate metabolism play a key role in NH_3 assimilation, *C. formicoaceticum* is able to ferment glutamate by the completely different enzyme set of the cobamide coenzyme-dependent pathway (Bogdahn *et al.*, 1983).

ACKNOWLEDGMENTS. Experimental work carried out in the laboratories of the authors was supported by Deutsche Forschungsgemeinschaft, Bundesministerium für Forschung und Technologie, and Fonds der Chemischen Industrie.

REFERENCES

Abraham, L. J., Wales, A. J., and Rood, J. L., 1985, World-wide distribution of the conjugative *Clostridium perfringens* tetracycline resistance plasmid, pCW3, *Plasmid* **14:**37–46.

Adams, M. W. W., and Mortenson, L. E., 1985, Mo reductases: Nitrate reductase and formate dehydrogenase, in: *Molybdenum Enzymes* (T. G. Spiro, ed.), John Wiley and Sons, New York, pp. 519–593.

Akedo, M., Cooney, C. L., and Sinskey, A. J., 1983, Direct demonstration of lactate-acrylate interconversion in *Clostridium propionicum*, *Bio/Technology* **1:**791–794.

Aleman, V., and Handler, P., 1967, Dihydroorotate dehydrogenase. I. General properties, *J. Biol. Chem.* **242:**4087–4096.

Andreesen, J. R., and Gottschalk, G., 1969, The occurrence of a modified Entner-Doudoroff pathway in *Clostridium aceticum*, *Arch. Mikrobiol.* **69:**160–170.

Andreesen, J. R., Gottschalk, G., and Schlegel, H. G., 1970, *Clostridium formicoaceticum* nov. spec. Isolation, description and distinction from *C. aceticum* and *C. thermoaceticum*, *Arch. Mikrobiol.* **72:**154–174.

Andreesen, J. R., Schaupp, A., Neurauter, C., Brown, A., and Ljungdahl, L. G., 1973, Fermentation of glucose, fructose and xylose by *Clostridium thermoaceticum*. Effects of metals on growth yield, enzymes and the synthesis of acetate from CO_2, *J. Bacteriol.* **114:**743–751.

Andreesen, J. R., El Ghazzawi, E., and Gottschalk, G., 1974, The effects of ferrous ions, tungstate and selenite on the level of formate dehydrogenase in *Clostridium formicoaceticum* and formate synthesis from CO_2 during pyruvate fermentation, *Arch. Microbiol.* **96:**103–118.

Andreesen, J. R., Zindel, U., and Dürre, P., 1985, *Clostridium cylindrosporum* (ex Barker and Beck 1942) nom. rev., *Int. J. Syst. Bacteriol.* **35:**206–208.

Antranikian, G., and Giffhorn, F., 1987, Citrate metabolism in anaerobic bacteria, *FEMS Microbiol. Rev.* **46:**175–198.

Antranikian, G., Friese, C., Quentmeier, A., Hippe, H., and Gottschalk, G., 1984, Distribution of the ability for citrate utilization amongst Clostridia, *Arch. Microbiol.* **138:**179–182.

Antranikian, G., Herzberg, C., and Gottschalk, G., 1987, Production of thermostable α-amylase, pullulanase, and α-glucosidase in continuous culture by a new *Clostridium* isolate, *Appl. Environ. Microbiol.* **53:**1668–1673.

Avrova, N. P., 1975, Synthesis of pectolytic enzymes by *Clostridium felsineum* and their hydrolysis of the pectin substances of flax straw, *Appl. Biochem. Microbiol.* **11:**736–741.

Bache, R., and Pfennig, N., 1981, Selective isolation of *Acetobacterium woodii* on methoxylated aromatic acids and determination of growth yields, *Arch. Microbiol.* **130:**255–261.

Bader, J., and Simon, H., 1983, ATP formation is coupled to the hydrogenation of 2-enoates in *Clostridium sporogenes*, *FEMS Microbiol. Lett.* **20:**171–175.

Bader, J., Rauschenbach, P., and Simon, H., 1982, On a hitherto unknown fermentation path of several amino acids by proteolytic clostridia, *FEBS Lett.* **140:**67–72.

Bahl, H., and Gottschalk, G., 1984, Parameters affecting solvent production by *Clostridium acetobutylicum* in continuous culture, *Biotechnol. Bioeng. Symp.* **14:**215–223.

Bahl, H., Gottwald, M., Kuhn, A., Rale, V., Andersch, W., and Gottschalk, G., 1986, Nutritional factors affecting the ratio of solvents produced by *Clostridium acetobutylicum*, *Appl. Environ. Microbiol.* **52:**169–172.

Balch, W. E., Fox, G. E., Magrum, L. J., Woese, C. R., and Wolfe, R. S., 1979, Methanogens: Reevaluation of a unique biological group, *Microbiol. Rev.* **43:**260–296.

Barker, H. A., 1961, Fermentations of nitrogenous organic compounds, in: *The Bacteria*, Vol. 2, *Metabolism* (I. C. Gunsalus and R. Y. Stanier, eds.), Academic Press, New York, pp. 151–207.

Barker, H. A., 1978, Explorations of bacterial metabolism, *Ann. Rev. Biochem.* **47:**1–33.

Barker, H. A., 1981, Amino acid degradation by anaerobic bacteria, *Ann. Rev. Biochem.* **50:**23–40.

Barker, H. A., and Wiken, T., 1948, The origin of butyric acid in the fermentation of threonine by *Clostridium propionicum*, *Arch. Biochem.* **17:**149–151.

Barker, H. A., D'Ari, L., and Kahn, J., 1987, Enzymatic reactions in the degradation of 5-aminovalerate by *Clostridium aminovalericum*, *J. Biol. Chem.* **262:**8994–9003.

Baumann, N. A., Hagen, P.-O., and Goldfine, H., 1965, Phospholipids of *Clostridium butyricum*. Studies on plasmalogen composition and biosynthesis, *J. Biol. Chem.* **240:**1559–1567.

Beesch, S. C., 1952, Acetone-butanol fermentation of sugars, *Eng. Proc. Devel.* **44:**1677–1682.

Bergère, J.-L., and Hermier, J., 1970, Spore properties of clostridia occurring in cheese, *J. Appl. Bacteriol.* **33:**167–179.

Berry, D. F., Francis, A. J., and Bollag, J. M., 1987, Microbial metabolism of homocyclic and heterocyclic aromatic compounds under anaerobic conditions, *Microbiol. Rev.* **51:**43–59.

Bhat, J. V., and Barker, H. A., 1948, Tracer studies on the role of acetic acid and carbon dioxide in the fermentation of lactate by *Clostridium lactoacetophilum*, *J. Bacteriol.* **56:**777–779.

Bochner, B. R., and Savageau, 1979, Inhibition of growth by imidazol(on)e propionic acid:

evidence in vivo for coordination of histidine catabolism with the catabolism of other amino acids, *Mol. Gen. Genet.* **168**:87–95.

Bogdahn, M., and Kleiner, D., 1986, Inorganic nitrogen metabolism in two cellulose-degrading clostridia, *Arch. Microbiol.* **145**:159–161.

Bogdahn, M., Andreesen, J. R., and Kleiner, D., 1983, Pathways and regulation of N_2, ammonium and glutamate assimilation by *Clostridium formicoaceticum*, *Arch. Microbiol.* **134**:167–169.

Booth, I. R., 1985, Regulation of cytoplasmic pH in bacteria, *Microbiol. Rev.* **49**:359–378.

Booth, I. R., and Mitchell, W. J., 1987, Sugar transport and metabolism in the clostridia, in: *Sugar Transport and Metabolism in Gram-Positive Bacteria* (J. Reizer and A. Peterkofsky, eds.), Ellis Horwood, Chichester, pp. 165–185.

Bornstein, B. T., and Barker, H. A., 1948, The nutrition of *Clostridium kluyveri*, *J. Bacteriol.* **55**:223–230.

Bradbeer, C., 1965, The clostridial fermentations of choline and ethanolamine. II. Requirement for a cobamide coenzyme by an ethanolamine deaminase, *J. Biol. Chem.* **240**:4645–4681.

Braun, M., Mayer, F., and Gottschalk, G., 1981, *Clostridium aceticum* (Wieringa), a microorganism producing acetic acid from molecular hydrogen and carbon dioxide, *Arch. Microbiol.* **128**:288–293.

Brefort, G., Magot, M., Ionesco, H., and Sebald, M., 1977, Characterization and transferability of *Clostridium perfringens* plasmids, *Plasmid* **1**:52–66.

Britz, M. L., and Wilkinson, R. G., 1982, Leucine dissimilation to isovaleric and isocaproic acids by cell suspensions of amino acid fermenting anaerobes: The Stickland reaction revisited, *Can. J. Microbiol.* **28**:291–300.

Bryant, M. P., 1972, Commentary on the Hungate technique for culture of anaerobic bacteria, *Am. J. Clin. Nutr.* **25**:1324–1328.

Buckel, W., 1980, Analysis of the fermentation pathways of Clostridia using double labelled glutamate, *Arch. Microbiol.* **127**:167–169.

Buckel, W., and Bobi, A., 1976, The enzyme complex citramalate lyase from *Clostridium tetanomorphum*, *Eur. J. Biochem.* **64**:255–262.

Buckel, W., and Miller, S. L., 1987, Equilibrium constants of several reactions involved in the fermentation of glutamate, *Eur. J. Biochem.* **164**:565–569.

Cameron, D. C., and Cooney, C. L., 1986, A novel fermentation: The production of R(−)-1,2-propanediol and acetol by *Clostridium thermosaccharolyticum*, *Bio/Technology* **4**:651–654.

Cardon, B. P., and Barker, H. A., 1947, Amino acid fermentations by *Clostridium propionicum* and *Diplococcus glycinophilus*, *Arch. Biochem.* **12**:165–180.

Carter, J. E., and Sagers, R. D., 1972, Ferrous ion-dependent L-serine dehydratase from *Clostridium acidi-urici*, *J. Bacteriol.* **109**:757–763.

Cato, E. P., Moore, W. E. C., and Holdeman, L. V., 1968, *Clostridium oroticum* comb. nov. amended description, *Int. J. Syst. Bacteriol.* **18**:9–13.

Cato, E. P., George, W. L., and Finegold, 1986, Genus *Clostridium*, in: *Bergey's Manual of Systematic Bacteriology*, Vol. 2 (P. H. A. Sreath, N. S. Mair, M. E. Sharpe, and J. G. Holt, eds.), Williams and Wilkins, Baltimore, pp. 1141–1200.

Clifton, C. E., 1942, The utilization of amino acids and related compounds by *Clostridium tetani*, *J. Bacteriol.* **44**:179–183.

Cornet, P., Tronik, D., Millet, J., and Aubert, J. P., 1983, Characterization of two *cel* (cellulose degradation) genes of *Clostridium thermocellum* coding for endoglucanases. *Bio/Technology* **1**:589–594.

Cummins, C. S., 1970, Cell wall composition in the classification of gram-positive anaerobes, *Int. J. Syst. Bacteriol.* **20**:413–419.

Cummins, C. S., and Johnson, J. L., 1971, Taxonomy of the clostridia: Wall composition and DNA homologies in *Clostridium butyricum* and other butyric-acid producing clostridia, *J. Gen. Microbiol.* **67**:33–46.

Cunin, R., Glamsdorff, N., Pierard, A., and Stalon, V., 1986, Biosynthesis and metabolism of arginine in bacteria, *Microbiol. Rev.* **50:**314–352.

D'Ari, L., and Barker, H. A., 1985, p-Cresol formation by cell-free extracts of *Clostridium difficile*, *Arch. Microbiol.* **143:**311–312.

Dilworth, G. L., 1983, Occurrence of molybdenum in the nicotinic acid hydroxylase from *Clostridium barkeri*, *Arch. Biochem. Biophys.* **221:**565–569.

Dimroth, P., 1987, Sodium ion transport decarboxylases and other aspects of sodium ion cycling in bacteria, *Microbiol. Rev.* **51:**320–340.

Dorn, M., Andreesen, J. R., and Gottschalk, G., 1978a, Fermentation of fumarate and L-malate by *Clostridium formicoaceticum, J. Bacteriol.* **133:**26–32.

Dorn, M., Andreesen, J. R., and Gottschalk, G., 1978b, Fumarate reductase of *Clostridium formicoaceticum.* A peripheral membrane protein, *Arch. Microbiol.* **119:**7–11.

Dürre, P., and Andreesen, J. R., 1982a, Selenium-dependent growth and glycine fermentation by *Clostridium purinolyticum, J. Gen. Microbiol.* **128:**1457–1466.

Dürre, P., and Andreesen, J. R., 1982b, Separation and quantitation of purines and their anaerobic and aerobic degradation products by high-pressure liquid chromatography, *Anal. Biochem.* **123:**32–40.

Dürre, P., and Andreesen, J. R., 1983, Purine and glycine metabolism by purinolytic clostridia, *J. Bacteriol.* **154:**192–199.

Dürre, P., Andersch, W., and Andreesen, J. R., 1981, Isolation and characterization of an adenine utilizing, anaerobic sporeformer, *Clostridium purinolyticum* sp. nov., *Int. J. Syst. Bacteriol.* **31:**184–194.

Dürre, P., Spahr, R., and Andreesen, J. R., 1983, Glycine fermentation via glycine reductase in *Peptococcus glycinophilus* and *Peptococcus magnus, Arch. Microbiol.* **134:**127–135.

Eggerer, H., 1985, Completion of the degradation scheme for nicotinic acid by *Clostridium barkeri, Curr. Top. Cell. Regul.* **26:**411–418.

Eichler, B., and Schink, B., 1985, Fermentation of primary alcohols and diols and pure culture of syntrophically alcohol-oxidizing anaerobes, *Arch. Microbiol.* **143:**60–66.

Elsden, S. R., and Hilton, M. G., 1978, Volatile acid production from threonine, valine, leucine and isoleucine by clostridia, *Arch. Microbiol.* **117:**165–172.

Elsden, S. R., and Hilton, M. G., 1979, Amino acid utilization patterns in clostridial taxonomy, *Arch. Microbiol.* **123:**137–141.

Elsden, S. R., Hilton, M. G., and Waller, J. M., 1976, The end products of the metabolism of aromatic amino acids by clostridia, *Arch. Microbiol.* **107:**283–288.

Elsden, S. R., Hilton, M. G., Parsley, K. R., and Self, R., 1980, The lipid fatty acids of proteolytic clostridia, *J. Gen. Microbiol.* **118:**115–123.

Finn, C. W., Silver, R. P., Habig, W. H., Hardegree, M. C., Zon, G., and Garon, C. F., 1984, The structural gene for tetanus neurotoxin is on a plasmid, *Science* **224:**881–884.

Freier, D., and Gottschalk, G., 1987, (L+)-lactate dehydrogenase of *Clostridium acetobutylicum* is activated by fructose-1,6-bisphosphate, *FEMS Microbiol. Lett.* **43:**229–233.

Fryer, T. F., and Mead, G. C., 1979, Development of a selective medium for the isolation of *Clostridium sporogenes* and related organisms, *J. Appl. Bacteriol.* **47:**425–431.

Garnier, T., and Cole, S. T., 1988, Complete nucleotide sequence and genetic organization of the bacteriocinogenic plasmid, pIP 404, from *Clostridium perfringens, Plasmid* **19:**134–150.

George, H. A., Johnson, J. L., Moore, E. C., Holdeman, L. V., and Chen, J. S., 1983, Acetone, isopropanol, and butanol production by *Clostridium beijerinckii* (syn. *Clostridium butylicum*) and *Clostridium aurantibutyricum, Appl. Environ. Microbiol.* **45:**1160–1163.

Giesel, H., and Simon, H., 1983, On the occurrence of enoate reductase and 2-oxo-carboxy-late reductase in clostridia and some observations on the amino acid fermentation by *Peptostreptococcus anaerobius, Arch. Microbiol.* **135:**51–57.

Giffhorn, S., 1980, Verwertung von Methanol, Ethanol und Lactat durch *Clostridium formicoaceticum,* Diplomarbeit, Universität Göttingen.

Golovchenko, N. P., Belokopytov, B. F., and Akimenko, V. K., 1982, Threonine catabolism in the bacterium *Clostridium sticklandii, Biochemistry* (USSR) **47**:969–974.

Gottschalk, G., 1986, *Bacterial Metabolism,* 2nd ed., Springer-Verlag, New York.

Gottschalk, G., and Andreesen, J. R., 1979, Energy metabolism in anaerobes, in: *International Review of Biochemistry,* Vol. 21, *Microbial Biochemistry* (J. R. Quayle, ed.), University Park Press, Baltimore, pp. 85–115.

Gottschalk, G., Andreseen, J. R., and Hippe, H., 1981, The genus *Clostridium* (nonmedical aspects), in: *The Prokaryotes* (M. P. Sarr, H. Stolp, H. G. Truper, A. Balows, and H. G. Schlegel, eds.), Springer-Verlag, Berlin, pp. 1767–1803.

Gottwald, M., and Gottschalk, G., 1985, The internal pH of *Clostridium acetobutylicum* and its effect on the shift from acid to solvent formation, *Arch. Microbiol.* **193**:42–46.

Gottwald, M., Hippe, H., and Gottschalk, G., 1984, Formation of n-butanol from D-glucose by strains of *Clostridium tetanomorphum* group, *Appl. Environ. Microbiol.* **48**(3):573–576.

Gregory, E. M., Moore, W. E. C., and Holdemann, L. V., 1978, Superoxide dismutase in anaerobes: Survey. *Appl. Environ. Microbiol.* **35**:988–991.

van Gylswyk, N. O., and van der Toorn, J. J. T. K., 1987, *Clostridium aerotolerans* sp. nov., a xylanolytic bacterium from corn stover and from the rumina of sheep fed corn stover, *Int. J. Syst. Bacteriol.* **37**:102–105.

Habermann, E., and Dreyer, F., 1986, Clostridial neurotoxins: Handling and action at the cellular and molecular level, *Curr. Top. Microbiol. Immunol.* **129**:93–180.

Hammann, R., and Ottow, J. C. G., 1976, Isolation and characterization of iron-reducing nitrogen-fixing saccharolytic clostridia from gley soils, *Soil Biol. Biochem.* **8**:357–364.

Hardman, J. K., and Stadtman, T. C., 1960a, Metabolism of ω-amino acids. I. Fermentation of γ-aminobutyric acid by *Clostridium aminobutyricum* n.sp., *J. Bacteriol.* **79**:544–548.

Hardman, J. K., and Stadtman, T. C., 1960b, Metabolism of ω-amino acids. II. Fermentation of δ-aminovaleric acid by *Clostridium aminovalericum* nov.sp., *J. Bacteriol.* **79**:549–552.

Hardman, J. K., and Stadtman, T. C., 1963, Metabolism of ω-amino acids. IV. γ-Aminobutyrate fermentation by cell-free extracts of *Clostridium aminobutyricum, J. Biol. Chem.* **238**:2088–2093.

Hartmanis, M. G. N., and Stadtman, T. C., 1986, Diol metabolism and diol dehydratase in *Clostridium glycolicum, Arch. Biochem. Biophys.* **245**:144–152.

Hartrampf, G., and Buckel, W., 1986, On the steric course of the adenosyl cobalamin-dependent 2-methyleneglutarate mutase reaction in *Clostridium barkeri, Eur. J. Biochem.* **156**:301–304.

Hefner, D. L., Squires, C. H., Evans, R. J., Kopp, B. J., and Yarus, M. J., 1984, Transformation of *Clostridium perfringens, J. Bacteriol.* **159**:460–464.

Hermann, M., Popoff, M. R., and Sebald, M., 1987, *Sporomusa paucivorans* sp. nov., a methylotrophic bacterium that forms acetic acid from hydrogen and carbon dioxide, *Int. J. System. Bacteriol.* **37**:93–101.

Hilton, M. G., Mead, G. C., and Elsden, S. R., 1975, The metabolism of pyrimidines by proteolytic clostridia, *Arch. Microbiol.* **102**:145–149.

Holdeman, L. V., Cato, E. P., and Moore, W. E. C., 1977, *Anaerobe laboratory manual,* 4th ed., Anaerobe Laboratory, Virginia Polytechnic Institute and State University, Blacksburg.

Holt, R. A., Cairns, A. J., and Morris, J. G., 1988, Production of butanol by *Clostridium puniceum* in batch and continuous culture, *Appl. Microbiol. Biotechnol.* **27**:319–324.

Hyun, H. H., and Zeikus, J. G., 1985, Simultaneous and enhanced production of thermostable amylases and ethanol from starch by cocultures of *Clostridium thermosulfurogenes* and *Clostridium thermohydrosulfuricum, Appl. Environ. Microbiol.* **49**:1174–1181.

Imhoff, D., and Andreesen, J. R., 1979, Nicotinic acid hydroxylase from *Clostridium barkeri:* Selenium-dependent formation of active enzyme, *FEMS Microbiol. Lett.* **5**:155–158.

Jain, M. K., and Zeikus, J. G., 1988, Taxonomic distinction of two new protein specific, hydrolytic anaerobes: Isolation and characterization of *Clostridium proteolyticum* sp. nov. and *Clostridium collagenovorans* sp. nov., *System. Appl. Microbiol.* **10**:134–141.

Jellet, J. J., Forrest, T. P., Macdonald, I. A., Merric, T. J., and Holdeman, L. V., 1980, Production of indole-3-propanoic acid and 3-(p-hydroxyphenyl)propanoic acid by *Clostridium sporogenes:* A convenient thin layer chromatography detection system, *Can. J. Microbiol.* **26**:448–453.

Jeng, I. M., Somack, R., and Barker, H. A., 1974, Ornithine degradation in *Clostridium sticklandii;* pyridoxal phosphate and coenzyme A dependent thiolytic cleavage of 2-amino-4-ketopentanoate to alanine and acetyl coenzyme A, *Biochemistry* **13**:2898–2903.

Johnson, J. L., 1973, Use of nucleic-acid homologies in the taxonomy of anaerobic bacteria, *Int. J. Syst. Bacteriol.* **23**:308–315.

Johnston, N. C., and Goldfine, H., 1988, Isolation and characterization of a novel four-chain ether lipid from *Clostridium butyricum:* The phosphatidylglycerol acetal of plasmenylethanolamine, *Biochim. Biophys. Acta* **961**:1–12.

Jungermann, K., Thauer, R. K., Leimenstoll, G., and Decker, K., 1973, Function of reduced pyridine nucleotide-ferredoxin oxidoreductases in saccharolytic Clostridia, *Biochim. Biophys. Acta* **305**:268–280.

Kaneuchi, C., Miyazato, T., Shinjo, T., and Mitsuoka, T., 1979, Taxonomic study of helically coiled, sporeforming anerobes isolated from the intestines of humans and other animals: *Clostridium cocleatum* sp. nov. and *Clostridium spiroforme* sp. nov., *Int. J. Syst. Bacteriol.* **29**:1–12.

Karlsson, S., Banhidi, Z. G., and Albertsson, A. C., 1988, Identification and characterization of alkali-tolerant clostridia isolated from biodeteriorated casein-containing building materials, *Appl. Microbiol. Biotechnol.* **28**:305–310.

Keanealy, W. R., and Waselefsky, D. M., 1985, Studies on the substrate range of *Clostridium kluyveri;* the use of propanol and succinate, *Arch. Microbiol.* **141**:187–194.

Kerby, R., and Zeikus, J. G., 1983, Growth of *Clostridium thermoaceticum* on H_2/CO_2 or CO as energy source, *Curr. Microbiol.* **8**:27–30.

Kneifel, H., Stetter, K., O., Andreesen, J. R., Wiegel, J., König, H., and Schoberth, S. M., 1986, Distribution of polyamines in representative species of archaebacteria, *System. Appl. Microbiol.* **7**:241–245.

Koch, R., Zablowski, P., and Antranikian, G., 1987, Highly active and thermostable amylases and pullulanases from various anaerobic thermophiles, *Appl. Microbiol. Biotechnol.* **27**:192–198.

Koser, S. A., 1968, *Vitamin Requirements of Bacteria and Yeasts*, Charles C Thomas, Springfield, IL.

Krumholz, L. R., and Bryant, M. P., 1985, *Clostridium pfennigii* sp. nov. uses methoxyl groups of monobenzenoids and produces butyrate, *Int. J. Syst. Bacteriol.* **35**:454–456.

Kubowitz, F., 1934, Über die Hemmung der Buttersäuregärung durch Kohlenoxyd, *Biochem. Z.* **274**:285–298.

Laanbroek, H. J., Lambers, J. T., De Vos, W. M., and Veldkamp, H., 1978, L-Aspartate fermentation by a free-living *Campylobacter* species, *Arch. Microbiol.* **117**:109–114.

Laanbroek, H. J., Smit, A. J., Klein Nulend, G., and Veldkamp, H., 1979, Competition for L-glutamate between specialised and versatile *Clostridium* species, *Arch. Microbiol.* **120**:61–66.

Lamed, R., Setter, E., Kenig, R., and Bayer, E. A., 1983, The cellulosome: A discrete cell surface organelle of *Clostridium thermocellum* which exhibits separate antigenic, cellulose-binding and various cellulolytic activities, *Biotechnol. Bioeng. Symp.* **13**:163–181.

Lebertz, H., 1984, Selenabhängiger Glycin-Stoffwechsel bei anaeroben Bakterien und vergleichende Untersuchungen zur Glycin-Reduktase und zur Glycin-Decarboxylase, PhD thesis, University of Göttingen.

Lebertz, H., and Andreesen, J. R., 1988, Glycine fermentation by *Clostridium histolyticum*, *Arch. Microbiol.* **150**:11–14.

Lee, C. K., Dürre, P., Hippe, H., and Gottschalk, G., 1987, Screening for plasmids in the genus *Clostridium*, *Arch. Microbiol.* **148:**107–114.

Leonhardt, U., and Andreesen, J. R., 1977, Some properties of formate dehydrogenase, accumulation and incorporation of [185]W-tungsten into proteins of *Clostridium formicoaceticum*, *Arch. Microbiol.* **115:**277–284.

Lin, Y., and Blaschek, H. P., 1984, Transformation of heat treated *Clostridium acetobutylicum* protoplasts with pUB110 plasmid DNA, *Appl. Environ. Microbiol.* **48:**737–742.

Loesche, W. J., 1969, Oxygen sensitivity of various anaerobic bacteria, *Appl. Microbiol.* **18:**723–727.

Lovitt, R. W., Kell, D. B., and Morris, J. G., 1986, Proline reduction by *Clostridium sporogenes* is coupled to vectorial proton ejection, *FEMS Microbiol. Lett.* **36:**269–273.

Lund, B. M., and Brocklehurst, T. F., 1978, Pectic enzymes of pigmented strains of *Clostridium*, *J. Gen. Microbiol.* **104:**59–66.

Lundie, L. L., and Drake, H. L., 1984, Development of a minimally defined medium for the acetogen *Clostridium thermoaceticum*, *J. Bacteriol.* **159:**700–703.

Machacek-Pitsch, C., Rauschenbach, P., and Simon, H., 1985, Observations on the elimination of water from 2-hydroxy acids in the metabolism of amino acids by *Clostridium sporogenes*, *Biol. Chem. Hoppe-Seyler* **366:**1057–1062.

MacMillan, J. D., and Vaughn, R. H., 1964, Purification and properties of polygalacturonic acid trans-eliminase produced by *Clostridium multifermentans*. *Biochemistry* **3:**564–572.

Madi, E., Antranikian, G., Ohmiya, K., and Gottschalk, G., 1987, Thermostable amylolytic enzymes from a new *Clostridium* isolate, *Appl. Environ. Microbiol.* **53:**1661–1667.

Marks, C. L., and Freese, E., 1987, Aspects of carbohydrate metabolism related to sporulation and germination, in: *Sugar Transport and Metabolism in Gram-Positive Bacteria* (J. Reizer and A. Peterkowsky, eds.) Ellis Horwood Ltd., Chicester, pp. 295–332.

Matteuzzi, D., Hollaus, F., and Biavati, B., 1978, Proposal of neotype for *Clostridium thermohydrosulfuricum* and the merging of *Clostridium tartarivorum* with *Clostridium thermosaccharolyticum*, *Int. J. Syst. Bacteriol.* **28:**528–531.

Mayer, F., 1988, Cellulosysis: ultrastructural aspects of bacterial systems, *Electron Microsc. Rev.* **1:**69–85.

Mayraud, D., and Bourgeau, G., 1982, Production of phenyl acetic acid by anaerobes, *J. Clin. Microbiol.* **16:**747–750.

Mead, G. C., 1971, The amino acid-fermenting Clostridia, *J. Gen. Microbiol.* **67:**47–56.

Mead, G. C., Adams, B. W., Hilton, M. G., and Lord, P. G., 1979, Isolation and characterization of uracil-degrading clostridia from soil, *J. Appl. Bacteriol.* **46:**465–472.

Meister, A., Sober, H. A., and Tice, S. V., 1951, Enzymatic decarboxylation of aspartic acid to α-alanine, *J. Biol. Chem.* **189:**577–590.

Mitruka, B. M., and Costilow, R. N., 1967, Arginine and ornithine catabolism by *Clostridium botulinum*, *J. Bacteriol.* **93:**295–301.

Möller, B., Oßmer, R., Howard, B. H., Gottschalk, G., and Hippe, H., 1984, Sporomusa, a new genus of Gram-negative anaerobic bacteria including *Sporomusa sphaeroides* spec. nov. and *Sporomusa ovata* sp. nov., *Arch. Microbiol.* **139:**388–396.

Möller, B., Hippe, H., and Gottschalk, G., 1986, Degradation of various amine compounds by mesophilic clostridia, *Arch. Microbiol.* **145:**85–90.

Morris, J. G., 1980, Oxygen tolerance/intolerance of anaerobic bacteria, in: *Anaerobes and Anaerobic Infections* (G. Gottschalk, N. Pfennig, and H. Werner, eds.), G. Fischer Verlag, Stuttgart, p. 7–15.

Moss, C. W., Lambert, M. A., and Goldsmith, D. J., 1970, Production of hydrocinnamic acid by clostridia, *Appl. Microbiol.* **19:**375–378.

Muldrow, L. L., Archibold, E. R., Nunez-Montiel, O. L., and Sheehey, R. J., 1982, Survey of the extrachromosomal gene pool of *Clostridium difficile*, *J. Clin. Microbiol.* **16:**137–640.

Muth, W. L., and Costilow, R. N., 1974, Ornithine cyclase (deaminating). III. Mechanism of the conversion of ornithine to proline, *J. Biol. Chem.* **249:**7463–7467.

Naumann, E., Hippe, H., and Gottschalk, G., 1983, Betaine: New oxidant in the Stickland reaction and methanogenesis from betaine and L-alanine by a *Clostridium sporogenes–Methanosarcina barkeri* coculture, *Appl. Environ. Microbiol.* **45:**474–483.

O'Brien, R. W., and Morris, J. G., 1971, O_2 and the growth and metabolism of *Clostridium acetobutylicum. J. Gen. Microbiol.* **68:**307–318.

Oulevey, J., Bahl, H., and Thiele, O. W., 1986, Novel alk-1-enyl ether lipids isolated from *Clostridium acetobutylicum, Arch. Microbiol.* **144:**166–168.

Oultram, J. D., and Young, M., 1985, Conjugal transfer of plasmid pAMβ1 from *Streptococcus lactis* and *Bacillus subtilis* to *Clostridium acetobutylicum, FEMS Microbiol. Lett.* **27:**129–134.

Pan-Hou, H. S. K., Hosono, M., and Imura, N., 1980, Plasmid-controlled mercury biotransformation by *Clostridium cochlearium* T-2, *Appl. Environ. Microbiol.* **40:**1007–1011.

Parsot, C., 1986, Evolution of biosynthetic pathways: a common ancestor for threonine synthase, threonine dehydratase and D-serine dehydratase, *EMBO. J.* **5:**3013–3019.

Perkins, W. E., 1965, Production of clostridial spores, *J. Appl. Bacteriol.* **28:**1–16.

Pettigrew, D. W., Bidigare, R. R., Mehta, B. J., Williams, M. I., and Sander, E. G., 1985, Dihydroorotase from *Clostridium oroticum*. Purification and reversible removal of essential zinc, *Biochem. J.* **230:**101–108.

Pickett, M. J., 1943, Studies on the metabolism of *Clostridium tetani, J. Biol. Chem.* **151:**203–209.

Pheil, C. G., and Ordal, Z. J., 1967, Sporulation of the thermophilic anaerobes, *Appl. Microbiol.* **15:**893–898.

Prazmowski, A., 1880, Untersuchung über die Entwicklungsgeschichte und Fermentwirkung einiger Bakterien-Arten. Inaug. Diss. Hugo Voigt, Leipzig, pp. 1–58.

Rabinowitz, J. C., 1963, Intermediates in purine breakdown, *Meth. Enzymol.* **6:**703–713.

Recsei, P. A., Moore, W. M., and Snell, E. E., 1983, Pyruvoyl-dependent histidine decarboxylase from *Clostridium perfringens* and *Lactobacillus buchneri, J. Biol. Chem.* **258:**439–444.

Reid, S. J., Allcock, E. R., Jones, D. T., and Woods, D. R., 1983, Transformation of *Clostridium acetobutylicum* protoplasts with bacteriophage DNA, *Appl. Environ. Microbiol.* **45:**305–307.

Rimbault, A., Niel, P., Virelizier, H., Darbord, J. C., and Leluan, G., 1988, L-Methionine, a precursor of trace methane in some proteolytic clostridia, *Appl. Environ. Microbiol.* **54:**1581–1586.

Roberts, T. A., 1967, Sporulation of mesophilic clostridia, *J. Appl. Bacteriol.* **30:**430–443.

Rogers, P., 1986, Genetics and biochemistry of *Clostridium* relevant to development of fermentation processes, *Adv. Appl. Microbiol.* **31:**1–60.

Rosen, B. P., and Silver, S., 1987, *Ion transport in prokaryotes,* Academic Press, San Diego.

Rosenblum, E. D., and Wilson, P. W., 1950, Molecular hydrogen and nitrogen fixation by *Clostridium, J. Bacteriol.* **59:**83–91.

Schantz, E. J., and Sugiyama, H., 1974, Toxic protein produced by *Clostridium botulinum, Agr. Food Chem.* **22:**26–33.

Schiefer-Ullrich, H., Wagner, R., Dürre, P., and Andreesen, J. R., 1984, Comparative studies on physiology and taxonomy of obligately purinolytic clostridia, *Arch. Microbiol.* **138:**345–353.

Schink, B., 1984a, *Clostridium magnum* sp. nov., a non-autotrophic homoacetogenic bacterium. *Arch. Microbiol.* **137:**250–255.

Schink, B., 1984b, Fermentation of tartrate enantiomers by anaerobic bacteria, and description of two new species of strict anaerobes, *Ruminococcus pasteurii* and *Ilyobacter tartaricus, Arch. Microbiol.* **139:**409–414.

Schink, B., and Zeikus, J. G., 1983, Characterisation of pectinolytic enzymes of *Clostridium thermosulfurogenes, FEMS Microbiol. Lett.* **17:**295–298.

Schink, B., Phelps, T. J., Eichler, B., and Zeikus, J. G., 1985, Comparison of ethanol degradation pathways in anoxic freshwater environments, *J. Gen. Microbiol.* **131:**651–660.

Schleifer, K. H., and Kandler, O., 1972, Peptidoglycan types of bacterial cell walls and their taxonomic implications, *Bacteriol. Rev.* **36:**407–477.

Schwartz, A. C., Quecke, W., and Brenschede, G., 1979, Inhibition by glycine of the catabolic reduction of proline in *Clostridium sticklandii:* Evidence on the regulation of amino acid reduction, *Z. Allg. Mikrobiol.* **19**:211–220.

Schwartz, W. H., Gräbnitz, F., and Staudenbauer, W. L., 1986, Properties of a *Clostridium thermocellum* endoglucanase produced in *Escherichia coli, Appl. Environ. Microbiol.* **51**:1293–1299.

Schweiger, G., Dutscho, R., and Buckel, W., 1987, Purification of 2-hydroxyglutaryl CoA-dehydratase from *Acidaminococcus fermentans.* An iron-sulfur protein, *Eur. J. Biochem.* **169**: 441–448.

Seto, B., 1980, The Stickland reaction, in: *Diversity in Bacterial Respiratory Systems,* Vol. 2 (C. J. Knowles ed.), CRC Press, Boca Raton, pp. 49–64.

Simon, H., Bader, J., Günther, H., Neumann, S., and Thanos, J., 1985, Chirale Verbindungen durch biokatalytische Reduktionen, *Angew. Chem.* **97**:541–555; *Angew. Chem., Int. Ed. Engl.* **24**:539–553.

Sleat, R., and Mah, R. A., 1985, *Clostridium populeti* sp. nov., a cellulolytic species from a woody-biomass digestor, *Int. J. Syst. Bacteriol.* **35**:160–163.

Sleat, R., Mah, R. A., and Robinson, R., 1984, Isolation and characterization of an anaerobic, cellulolytic bacterium, *Clostridium cellulovorans* sp. nov., *Appl. Environ. Microbiol.* **48**:88–93.

Sleytr, V. B., and Glauert, A. M., 1976, Ultrastructure of the cell walls of two closely related clostridia that possess different regular arrays of surface subunits, *J. Bacteriol.* **126**:869–882.

Sliwkowski, M. X., and Stadtman, T. C., 1987, Purification and immunological studies of selenoprotein A of the clostridial glycine reductase complex, *J. Biol. Chem.* **262**:4899–4904.

Stieb, M., and Schink, B., 1985, Anaerobic oxidation of fatty acids by *Clostridium bryantii* sp.nov., a sporeforming, obligately syntrophic bacterium, *Arch. Microbiol.* **140**:387–390.

Stieb, M., and Schink, B., 1986, Anaerobic degradation of isovalerate by a defined methanogenic coculture, *Arch. Microbiol.* **144**:291–295.

Stouthamer, A. H., 1979, The search for correlation between theoretical and experimental growth yields, in: *Microbial Biochemistry,* Intern. Rev. Biochem., Vol. 21 (J. R. Quayle, ed.), University Park Press, Baltimore, pp. 1–47.

Thauer, R. K., Kirchniawy, F. H., and Jungermann, K. A., 1972, Properties and function of the pyruvate-formate-lyase reaction in Clostridia, *Eur. J. Biochem.* **27**:282–290.

Thauer, R. K., Jungermann, K., and Decker, K., 1977, Energy conservation in chemotrophic anaerobic bacteria, *Bacteriol. Rev.* **41**:100–180.

Tokushigo, M., and Hayaishi, O., 1972, Threonine metabolism and its regulation in *Clostridium tetanomorphum, J. Biochem.* **72**:469–477.

Tran-Dinh, K., and Gottschalk, G., 1985, Formation of D(−)-1,2-propanediol and D(−)-lactate from glucose by *Clostridium sphenoides* under phosphate limitation, *Arch. Microbiol.* **142**:87–92.

Tschech, A., and Pfennig, N., 1984, Growth yield increase linked to caffeate reduction in *Acetobacterium woodii, Arch. Microbiol.* **137**:163–167.

Tschech, A., and Schink, B., 1985, Fermentative degradation of resorcinol and resorcylic acids, *Arch. Microbiol.* **143**:52–59.

Turner, D. C., and Stadtman, T. C., 1973, Purification of protein components of the clostridial glycine reductase system and characterization of protein A as a selenoprotein, *Arch. Biochem. Biophys.* **154**:366–381.

Venugopalan, V., 1980, Influence of growth conditions on glycine reductase of *Clostridium sporogenes, J. Bacteriol.* **141**:386–388.

Vogels, G. D., and van der Drift, C., 1976, Degradation of purines and pyrimidines by microorganisms, *Bacteriol. Rev.* **40**:403–468.

Vollherbst-Schneck, K., Sands, J. A., and Montenecourt, B. S., 1984, Effect of butanol on lipid composition and fluidity of *Clostridium acetobutylicum* ATCC 824, *Appl. Environ. Microbiol.* **47**:193–194.

Waber, J. L., and Wood, H. G., 1979, Mechanism of acetate synthesis from CO_2 by *Clostridium acidiurici, J. Bacteriol.* **140:**468–478.

Wachsman, J. T., and Barker, H. A., 1955, The accumulation of formamide during the fermentation of histidine by *Clostridium tetanomorphum, J. Bacteriol.* **69:**83–88.

Wagner, R., and Andreesen, J. R., 1987, Accumulation and incorporation of [185]W-tungsten into proteins of *Clostridium acidiurici* and *Clostridium cylindrosporum, Arch. Microbiol.* **147:** 295–299.

Wagner, R., Cammack, R., and Andreesen, J. R., 1984, Purification and characterization of xanthine dehydrogenase from *Clostridium acidiurici* grown in the presence of selenium, *Biochim. Biophys. Acta* **791:**63–74.

Weiss, N., Schleifer, K. H., and Kandler, O., 1981, The peptidoglycan types of Gram-positive anaerobic bacteria and their taxonomic implications, *Rev. Inst. Pasteur* (Lyon) **14:**3–12.

Whitmer, M. E., and Johnson, E. A., 1988, Development of improved defined media for *Clostridium botulinum* serotypes A, B, and E, *Appl. Environ. Microbiol.* **54:**753–759.

Wickremasinghe, R. L., and Fry, B. A., 1954, The formation of urocanic acid and glutamic acid in the fermentation of histidine by *Clostridium tetanomorphum, Biochem. J.* **58:**268–278.

Wiegel, J., and Ljungdahl, L. G., 1979, Ethanol as fermentation product of extreme thermophilic anaerobic bacteria, in: *Technische Mikrobiologie* (H. Dellweg, ed.), Verlag Versuchs- und Lehranstalt für Spiritusfabrikation und Fermentationstechnologie im Institut für Gärungsgewerbe und Biotechnologie, Berlin, pp. 117–127.

Wieringa, K. T., 1941, Über die Bildung von Essigsäure aus Kohlensäure und Wasserstoff durch anaerobe Bazillen, *Brennstoff-Chemie* **22:**161–164.

Wiesendanger, S., and Nisman, B., 1953, La L-méthionine démercapto-désaminase: Un nouvel enzyme à pyridoxal-phosphate, *Compt. Rend. Acad. Sci.* **237:**764–765.

Willis, A. T., 1977. *Anaerobic Bacteriology: Clinical and Laboratory Practice*, 3rd ed., Butterworths, London.

Winogradsky, S., 1895, Recherches sur l'assimilation de l'azote libre de l'atmosphère par les microbes, *Arch. Sci. Biol.* **3:**297–352.

Wohlfarth, G., and Buckel, W., 1985, A sodium ion gradient as energy source for *Peptostreptococcus asaccharolyticus, Arch. Microbiol.* **142:**128–135.

Wood, W. A., 1961, Fermentation of carbohydrates and related compounds, in: *The Bacteria*, Vol. 2, *Metabolism* (I. C. Gunsalus and R. Y. Stanier, eds.), Academic Press, New York, 1961, pp. 59–149.

Woods, D. D., and Clifton, C. E., 1937, Studies in the metabolism of the strict anaerobes (Genus *Clostridium*). VI. Hydrogen production and amino acid utilisation by *Clostridium tetanomorphum, Biochem. J.* **31:**1774–1788.

Yorifuji, T., Jeng, I. M., and Barker, H. A., 1977, Purification and properties of 3-keto-5-aminohexanoate cleaving enzyme from a lysine-fermenting *Clostridium, J. Biol. Chem.* **252:** 20–31.

Zindel, U., Freudenberg, W., Rieth, M., Andreesen, J. R., Schnell, J., and Widdel, F., 1988, *Eubacterium acidaminophilum* sp. nov., a versatile amino acid-degrading anaerobe producing or utilizing H_2 or formate, *Arch. Microbiol.* **150:**254–266.

Genetics of *Clostridium* 3

MICHAEL YOUNG, WALTER L. STAUDENBAUER,
and NIGEL P. MINTON

1. INTRODUCTION

The genus *Clostridium* does not represent a natural taxonomic grouping
(Chapter 1). Several of its remarkably diverse range of organisms are of
biotechnological interest because of the end products of their fermentative
metabolism (Chapters 4 and 5), the stereospecific reductions that they undertake (Chapter 6), and the potentially important enzymes that they elaborate (Chapter 7). Others are of considerable medical interest. For example, *C. botulinum* and *C. tetani* produce some of the most powerful toxins
known to man (Chapter 8). Genetic analysis of clostridia is still in its infancy,
but significant progress has been made in the last few years. Various elements of gene transfer technology have been developed in several species,
but it is not yet possible to provide a synopsis of techniques that can be
applied throughout the genus. This chapter reviews recent progress in the
genetic analysis of several species, especially *C. acetobutylicum.* Summaries of
important methods are included as appropriate.

2. CLASSICAL GENETIC ANALYSIS

An initial requirement for genetic analysis of any organism is the isolation of mutant strains. In clostridia, mutants defective in a variety of functions, including purine, pyrimidine, vitamin, and amino acid biosynthesis,
pathways of fermentative metabolism, and sporulation, have been isolated

MICHAEL YOUNG ● Department of Biological Sciences, University College of Wales,
Penglais, Aberystwyth, Dyfed SY23 3DA, Wales.   WALTER L. STAUDENBAUER
● Botany and Microbiology Institute, Department of Microbiology, Technical University of
Munich, D-8000 Munich 2, Federal Republic of Germany.   NIGEL P. MINTON ●
Molecular Genetics Group, Division of Biotechnology, Public Health Laboratory Service
Centre for Applied Microbiology and Research, Porton Down, Salisbury, Wiltshire SP4 0JG,
England.

and partially characterized (Robson *et al.*, 1974; Sebald and Costilow, 1975; Booth and Morris, 1982; Mendez and Gomez, 1982; Murray *et al.*, 1983; Bowring and Morris, 1985; see Rogers, 1986). In general, the most effective mutagens are those which act by a direct mutagenic mechanism, such as ethylmethanesulfonate and *N*-methyl-*N'*-nitro-*N*-nitrosoguanidine. Indirect mutagenic agents, such as UV light, which rely on subsequent misrepair of damaged DNA, have proved to be relatively ineffective (Walker, 1983; Bowring and Morris, 1984). This observation has led to the suggestion that clostridia are generally deficient in error-prone DNA repair processes (Walker, 1983; Bowring and Morris, 1985).

The study of mutant strains can provide valuable information on gene regulation and function, but further genetic analysis is possible only if mutations can be moved from one strain to another. Until a few years ago, procedures for undertaking gene transfer in clostridia did not exist; however, conjugal transfer of plasmids and transposons as well as phage transfection/plasmid transformation has now been documented in several species (see Tables I and II). Unfortunately, simple methods for the construction of a genetic map of the genome of any of the clostridia are still lacking. A major deficiency, as compared with their aerobic counterparts, the Bacilli, is the lack of a generalized transducing phage, which would prove invaluable in this context. In view of the absence of any simple means for undertaking classical genetic mapping, recombinant DNA technology has

Table I. Transformation of *Clostridium* spp.

Organism	Vector	Frequency[a]	Ref.
I. Wall-less cell transformation			
C. acetobutylicum	phage CA1	Not reported	Reid *et al.*, 1983
C. acetobutylicum	pUB110	2.8×10^1	Lin and Blaschek, 1984
C. acetobutylicum	phage HM3	5.4×10^4	Podvin *et al.*, 1988
C. acetobutylicum	pVA1, pVA677	1.0×10^5	Podvin *et al.*, 1988
C. perfringens	pJU124	0.7×10^2	Heefner *et al.*, 1984
C. perfringens	pJU12	1.6×10^2	Squires *et al.*, 1984
C. perfringens	pJU16	4.0×10^3	Mahony *et al.*, 1988
C. perfringens	pHR106	(1×10^{-6})	Roberts *et al.*, 1988
II. Intact cell transformation			
C. acetobutylicum	pMTL500E	2.9×10^3	Oultram *et al.*, 1988a
	pCB3	4.0×10^4	See Section 5.4
	pCL3	3.5×10^2	See Section 5.4
C. perfringens	pHR106	1.2×10^3	Allen and Blaschek, 1988
	pAMβ1	1.1×10^2	
C. thermohydrosulfuricum	pUB110, pGS13	(4×10^{-6})	Soutschek-Bauer *et al.*, 1985

[a]Frequencies are expressed as the number of transformants per μg of vector DNA, with the exception of the values in parentheses which represent the number of transformants per viable cell.

Table II. Regeneration of Clostridial Protoplasts

Organism	Ref.
C. acetobutylicum	Allcock et al., 1982
C. pasteurianum	Minton and Morris, 1983
C. tertium	Knowlton et al., 1984
C. saccharoperbutylacetonicum	Yoshino et al., 1984
C. saccharoperbutylacetonicum[a]	Reysset et al., 1987
C. perfringens	Stahl and Blaschek, 1985
C. thermohydrosulfuricum	Soutschek-Bauer et al., 1985

[a]It has been suggested (Reysset et al., 1987) that this strain be re-classified as C. acetobutylicum.

been employed to gain insights into the genetics of *Clostridium*. Many laboratories have been content to clone clostridial genes and physically to characterize them in *Escherichia coli*. Others, including our own, have strived to develop gene transfer systems using recombinant DNA methodology. Classical procedures such as transformation and conjugation have now been developed, but they have so far only been used to manipulate extrachromosomal genetic elements. Their application to the analysis of the bacterial chromosome is a subject for future research.

3. TRANSFORMATION

Development of transformation technology is of prime importance for genetic analysis of the clostridia. For many years the quest for reproducible procedures has focused on artificially induced wall-less cells as possible sources of "competent" bacteria. The advent of commercially available electroporation apparatus may now signal a change in emphasis to bacteria with their cell walls intact. Those clostridia in which transformation has been demonstrated are listed in Table I.

3.1. Protoplast/Autoplast Regeneration

The use of protoplasts for genetic analysis of gram-positive bacteria was pioneered during the late 1970s in various species of *Bacillus* and *Streptomyces* (Bibb et al., 1978; Chang and Cohen, 1979). Of central importance was the observation that L forms, or cells stripped of their walls in an osmotically stabilized medium (i.e., protoplasts or autoplasts), could be induced to take up exogenous DNA in the presence of polyethylene glycol (PEG). However, the technique can only be applied usefully to bacteria that can be induced to revert back to normal walled organisms after transformation. The potential of the method was widely recognized and numerous reports of the preparation and regeneration of clostridial protoplasts appeared in the 1980s (Table II). All of the procedures developed relied on the same essential principles. Cell walls were removed in a medium or buffer osmotically stabilized with sucrose; this occurred either naturally (autolysis) or was induced by the addition of lysozyme. Regeneration was subsequently achieved by plating on an isotonic solidified medium. Optimal concentrations of ionic constituents (i.e., Ca^{2+}, Mg^{2+}), agar, and osmotic stabilizer varied from one species to another.

It is often necessary to tailor protoplast production/regeneration procedures to the idiosyncrasies of the particular organism, or even strain, being studied; this can be a very time-consuming process. Unfortunately, the availability of clostridial protoplast regeneration procedures has rarely paved the way for demonstrable transformation. A case can be made for regarding the transformability of protoplasts as a reliable indicator of the effectiveness of any regeneration procedure. In this sense, the most successful method developed so far is that of Reysset *et al.* (1987) for *C. acetobutylicum.*

3.2. Transformation of *C. acetobutylicum*

There are two early reports suggesting that protoplasts of *C. acetobutylicum* can be transfected with phage DNA (Reid *et al.*, 1983) or transformed with the *Staphylococcus aureus* plasmid pUB110 (Lin and Blaschek, 1984). Recent work in Madeleine Sebald's laboratory at the Institut Pasteur (Reysset *et al.*, 1987) has culminated in the development of a good method for protoplast production and regeneration in *C. saccharoperbutylacetonicum* strain N1-4080, a Pro⁻ RifR derivative of strain N1-4 (Hongo, 1960). This organism, which is an extremely close relative of *C. acetobutylicum*, was chosen for several reasons. (1) It produces butanol over a wide range of pH values (Hongo, 1960). (2) Autoplasts are produced (Ogata *et al.*, 1975). (3) Much was known about autolysin production (Yoshino *et al.*, 1982). (4) Protoplasts had previously been shown to regenerate (Yoshino *et al.*, 1984). The last two features were important as there was evidence that autolysin production played a major role in the inhibition of cell wall regeneration. A highly efficient and reproducible procedure was devised which enabled 20–40% of protoplasts to revert to bacillary form (Reysset *et al.*, 1987). DNA from phage HM3, a lytic phage specific to strain N1-4, was initially employed for transformation experiments to avoid any problems that may have been associated with host restriction systems. Polyethylene glycol 4000-induced transfection of N1-4081 was achieved at frequencies of 5×10^4 plaque-forming units per μg DNA (Podvin *et al.*, 1988). In contrast with results that have been obtained with *C. perfringens* L forms (Heefner *et al.*, 1984; Mahony *et al.*, 1988), there was a linear relationship between the number of infectious centers and the DNA concentration (50–800 ng/ml). The method was applied to two

Streptococcus sanguis plasmids, pVA1 and pVA677 (Macrina *et al.*, 1980), which are pAMβ1 derivatives encoding resistance to erythromycin (Em^R). Between 10^4 and 10^5 transformants were obtained per μg DNA (Podvin *et al.*, 1988). A summary of the successful transformation procedure is given below.

Bacteria were grown in T69 defined medium (constituents, in grams per liter: KH_2PO_4, 0.5; ammonium acetate, 2.0; $MgSO_4\cdot7H_2O$, 0.3; $FeSO_4\cdot7H_2O$, 0.01; yeast extract, 1.0; casamino acids, 0.5; bactotryptone, 0.5; cysteine-HCl, 0.5; adjusted to pH 6.5 with NaOH and glucose added aseptically to 1%, wt./vol., final concentration). At a density of 1×10^8 cells/ml, preweighed, solid, sterile sucrose was added to a final concentration of 0.6 M and the cells converted to protoplasts by incubation for 1 hr at 34°C with 100 μg lysozyme and 20 μg penicillin/ml. Protoplasts were recovered by centrifugation, washed twice, and resuspended in protoplast buffer (T69 medium supplemented with 0.5 M xylose, 10 mM $MgCl_2$, and 25 mM $CaCl_2$). Plasmid DNA (50–800 ng) and PEG 4000 (35% wt./vol. final concentration) were added and the protoplasts incubated for 3 min at room temperature. The protoplast suspension was then diluted with low Ca^{2+} (1 mM $CaCl_2$) protoplast buffer containing choline (4 mg/ml), harvested by centrifugation, washed, and resuspended in the same buffer. Appropriate dilutions were added to T69X soft agar, which comprises T69 medium supplemented with $CaCl_2$ (1 mM), BSA (0.5%, wt./vol.), xylose (0.25M), choline (4 mg/ml), and agar (0.8%), and plated on T69X agar (2.5%, wt./vol.). After incubation for 20 hr at 34°C, a further T69X soft agar overlay containing Em (1 mg/ml) was added. Em^R transformants appeared after incubation for 4–6 days.

The key to the success of this protocol appears to have been the measures taken to limit autolysin activity. Principal among these were the isolation of a mutant (N1-4081) partially defective in autolysin production, use of penicillin in addition to lysozyme during protoplast formation, and inclusion of an autolysin inhibitor (Na polyanethole sulfonate or choline) in the medium. Other important features were the use of xylose (0.25 M) rather than sucrose (0.3 M) as osmotic stabilizer, and plating of protoplasts in a 0.8% (wt./vol.) soft agar overlay as opposed to direct plating on hard agar. It was essential to reduce the level of Ca^{2+} ions after transformation by centrifuging the protoplast suspensions, washing, and resuspending in protoplast buffer. Optimal concentrations of $CaCl_2$ for DNA uptake and protoplast regeneration were 25 and 1 mM, respectively (Reysset, personal communication).

3.3. Transformation of *C. acetobutylicum* by Electroporation

Until relatively recently, electroporation had been applied to the transformation/transfection only of eucaryotic cells. The procedure uses a high-voltage electric discharge through a cell suspension which is believed to induce transient "pores" in the cell membrane through which exogenous DNA can pass. Since the development of commercial electroporation apparatus (e.g., Gene Pulser℗, Bio-Rad Laboratories), the technique has been applied to the transformation of numerous gram-positive bacteria, e.g., *Streptomyces, Streptococcus, Lactobacillus,* and *Bacillus* (see Lucansky *et al.,* 1988). It has now been used successfully with *C. acetobutylicum* (Oultram *et al.,* 1988a).

The strain used was *C. acetobutylicum* NCIB 8052, while the plasmid employed was a derivative of pAMβ1 (pMTL500E; see Section 5.4.) encoding Em^R. Bacteria were grown to mid-exponential phase (OD$_{600}$ = 0.6) in 100 ml of 2 × YT broth supplemented with 0.5% (wt./vol.) glucose, harvested by centrifugation, washed once in 10 ml cooled (4°C) electroporation buffer (270 mM sucrose, 1 mM MgCl$_2$, 7 mM Na phosphate, pH 7.4), and resuspended in 5 ml cooled electroporation buffer. After holding the cell suspension on ice for 10 min, DNA (approx. 0.5 μg) was added to 0.8 ml of the cell suspension held in a Bio-Rad electroporation cuvette and the cells cooled on ice for an additional 8 min. An electric pulse of 0–2.5 kV (0–6.25 kV/cm in 0.4-cm cuvettes) was applied using a Gene Pulser℗ and the cells were again held on ice for 10 min before dilution (0.1 ml cell suspension in 0.8 ml 2 × YT + glucose) and incubation at 37°C for 1 hr. Em^R transformants appeared 16–24 hr after plating on selective medium. Manipulations were carried out in an anaerobic cabinet (Anaerobic Workstation Mark III, Don Whitley Ltd., Yorkshire) with minimal exposure to air during centrifugation (in sealed tubes) and electroporation. Transformation occurred most consistently at 5.0 kV/cm (giving time constants of 4.7–6.9 msec, and cell survival of 0.7–23% under the conditions used). Between 8 × 10^1 and 3 × 10^3 transformants were obtained per μg DNA.

The electroporation procedure outlined above provides a rapid and reliable means for transforming whole cells of *C. acetobutylicum* with plasmid DNA. The procedure does not require the production and regeneration of protoplasts and in this respect it should prove generally applicable to other *Clostridium* spp. Indeed, a similar protocol has already been successfully applied to whole cells of *C. perfringens* (Allen and Blaschek, 1988; Rood, personal communication).

3.4. Transformation of Other Clostridia

Two types of wall-less cells derived from *C. perfringens,* namely, L forms and autoplasts, have been transformed. The former are produced by growth in liquid medium containing an inhibitor of cell wall accretion, such as penicillin or D-cycloserine, whereas the latter form as a result of autolytic activity in medium or buffer containing an osmotic stabilizer. Both types of naked cell were first transformed by Heefner *et al.* (1984) using a *C. perfringens* conjugal plasmid, pJU124, encoding resistance to tetracycline (Tc^R). DNA uptake depended on the addition of PEG 8000 (30%) and Tc^R transformants arose at a frequency of about $10^2/\mu g$ DNA. Similar results were recently obtained using L forms from a different strain of *C. perfringens* (Mahony *et al.*, 1988; Roberts *et al.*, 1988). Both Heefner *et al.* (1984) and Mahony *et al.* (1988) obtained a nonlinear relationship between transformation frequency and DNA concentration. This may indicate that only a small proportion of the L form population was competent to be transformed. A number of *E. coli/C. perfringens* shuttle vectors have been constructed (see Section 5.5) and transformed into *C. perfringens* by the above procedure. However, these host–vector systems are of limited use currently because bacillary forms have not yet been regenerated from transformed L forms. Reversion of transformed autoplasts back to walled bacteria has been obtained by a somewhat cumbersome procedure (Heefner *et al.*, 1984), but transformability may prove to be strain-specific, since it was not demonstrated by Mahony *et al.* (1988).

Permeabilized cells of *C. thermohydrosulfuricum* have been transformed with plasmid pUB110 (encoding resistance to kanamycin, Km^R) and a derivative, pGS13, carrying a chloramphenicol resistance (Cm^R) marker (Soutschek-Bauer *et al.*, 1985) using the alkaline-Tris procedure of Fornari and Kaplan (1982). This procedure appears to be restricted to those bacteria, of which *C. thermohydrosulfuricum* is an example, possessing a paracrystalline proteinaceous surface layer in the cell wall. Southern hybridization experiments suggest that Km^R transformants arise by plasmid integration into the bacterial chromosome (W. Scholz and W. L. Staudenbauer, unpublished data).

4. CONJUGATION

The transfer of genetic information between strains of *C. perfringens* by a conjugal mechanism represents the earliest reported example of genetic exchange in the genus (see Brefort *et al.*, 1977). A summary of the types of conjugal transfer involving clostridial recipients and donors which have been demonstrated is outlined in Table III. A more detailed consideration of these processes is given below.

4.1. Indigenous Conjugative Plasmids

Of the indigenous conjugative plasmids found in *C. perfringens*, most attention has been given to a family of large plasmids conferring Tc^R which share substantial homology and are widely disseminated in different isolates of *C. perfringens* (Abraham *et al.*, 1985). Four representatives—pIP401 (53 kb, Tc^R, Cm^R), pJIR25 (52 kb, Tc^R, Cm^R), pJIR27 (50 kb, Tc^R, Cm^R), and pCW3 (47 kb, Tc^R)—have been studied in some detail (Sebald *et al.*,

Table III. Conjugal Gene Transfer

Recipient	Donor	Vector	Ref.
I. Plasmid-mediated			
C. perfringens	*C. perfringens*	Various indigenous R factors, e.g., pJIR62	Brefort *et al.*, 1977; Rood *et al.*, 1978; Rood, 1983; Abraham *et al.*, 1985
C. acetobutylicum	Various gram-positive species	Various streptococcal R factors, e.g., pAMβ1	Oultram and Young, 1985; Reysset *et al.*, 1985; Yu and Pearce, 1985
C. difficile	*C. perfringens*	pIP401	Ionesco, 1980
C. cochlearum	*C. cochlearum*	Uncharacterized plasmid	Pan-Hou *et al.*, 1980
II. Plasmid-free			
C. difficile	*C. difficile*	Tcl^r and Em^r	Ionesco, 1980; Smith *et al.*, 1981; Wüst and Hardegger, 1983; Hächler *et al.*, 1987
C. perfringens	*C. innocuum*	Em^r	Magot, 1983
C. acetobutylicum	*S. faecalis* and *B. subtilis*	Tn*916*, Tn*1545*	Davies *et al.*, 1988
C. tetani	*S. faecalis*	Tn*916*	Volk *et al.*, 1988
III. Mobilization of nonconjugative plasmids			
C. perfringens	*C. perfringens*	pIP404	Ionesco *et al.*, 1976
C. acetobutylicum	*B. subtilis, E. coli,* various streptococci	pAMβ1 cointegrates, pAM610 and plasmids carrying *oriT* of RK2	Yu and Pearce, 1985; Oultram *et al.*, 1987, 1988; Williams, Woolford, and Young, unpublished (see Section 4.4.2)

1975; Sebald and Brefort, 1975; Brefort *et al.*, 1977; Rood *et al.*, 1978; Magot, 1984; Abraham and Rood, 1985a,b). These plasmids are transferred to other strains of *C. perfringens* by a mechanism the details of which remain to be elucidated. A procedure for undertaking conjugal transfer of various genetic elements to clostridia is described below (see Section 4.2).

After transfer of plasmids pIP401, pJIR25, and pJIR27, some transcipients inherit only the Tc^R trait of the donor. These Cm^S transcipients harbor deleted plasmids (Brefort *et al.*, 1977; Abraham *et al.*, 1985) and they arise by the loss of a 6.2-kb Cm^R transposon (Tn*4451* in pIP401; Tn *4452* in pJIR27) during transfer (Abraham and Rood, 1987). Tn*4451* has been further characterized (Abraham and Rood, 1988). Sequencing of plasmids from which this element had deleted has shown that deletion occurs precisely. Tn*4451* has imperfect 12-bp terminal inverted repeat sequences that are similar to those of Tn*3*-like transposons, and there is some evidence that a 2-bp duplication of the target sequence is generated upon Tn*4451* insertion.

Conjugal transfer of a relatively small bacteriocinogenic plasmid, pIP404 (see 5.1), between strains of *C. perfringens,* and of a plasmid involved in mercury biotransformation in *C. cochlearum,* has also been reported (Table III).

4.2. Conjugal Streptococcal Plasmids

A number of conjugative streptococcal plasmids are transferable by filter mating to other genera of gram-positive bacteria, and the clostridia are no exception. The "best characterized" example is pAMβ1 (26.5 kb, Em^R), which has been transferred to *C. acetobutylicum* from a range of donors including *Streptococcus lactis, Streptococcus faecalis, Bacillus subtilis,* and other strains of *C. acetobutylicum* (Oultram and Young, 1985; Reysset and Sebald, 1985; Yu and Pearce, 1986). The highest transfer frequencies so far reported were obtained using *S. lactis* as donor (Oultram and Young, 1985). *Bacillus subtilis* seems to be a particularly inefficient donor of pAMβ1 and donor ability diminishes with repeated subculturing. Under these conditions, deleted forms of the plasmid tend to accumulate in this host (van der Lelie and Venema, 1987). Transfer of pAMβ1 to *C. pasteurianum* and *C. butyricum* has also been obtained (Richards *et al.,* 1988; Oultram, personal communication). Other conjugal streptococcal plasmids that have been transferred to *C. acetobutylicum* are pIP501, pJH4, and pVA797 (Reysset and Sebald, 1985; Yu and Pearce, 1986). The latter authors also introduced a derivative of pVA797 containing Tn*917* into *C. acetobutylicum,* but it is not known whether Tn*917* transposition occurs in this host.

Conjugal transfer of these plasmids depends on the establishment of close cell–cell contact and is therefore much more efficient on solid than in liquid medium. A simple filter-mating protocol that has been used to transfer pAMβ1 to *C. acetobutylicum* and can readily be adapted for transferring plasmids from *E. coli* (see Section 4.4.2) or conjugal transposons from streptococci (see Section 4.3) is described below.

Optimal transfer frequencies were obtained with bacteria growing exponentially. Many different media have been employed successfully: TYG (Brefort *et al.,* 1977) or CBM (O'Brien and Morris, 1971) for *C. acetobutylicum;* Brain Heart Infusion Broth (Difco or LabM) or Luria Broth for *B. subtilis, S. faecalis,* and *E. coli;* M17 medium (Terzaghi and Sandine, 1975) for *S. lactis.* Approximately equal numbers of donor and recipient bacteria were mixed and harvested by filtration through a 2.5-cm nitrocellulose membrane (Millipore) with a pore size of 0.45 microns. (Aerobically grown donors may be harvested by centrifugation and resuspended using the recipient culture before filtration.) Filters were placed, bacteria lowermost, on a nonselective medium such as Reinforced Clostridial Medium (Difco or

LabM) and incubated aerobically for 4–16 hr. At the end of the mating period bacteria were resuspended in holding buffer (25 mM potassium phosphate, 1 mM MgSO$_4$, pH 7.0) and appropriate dilutions plated on selective media enabling the numbers of recipient, donor, and transcipient bacteria to be determined.

4.3. Conjugative Transposons

The conjugative transposons are an interesting class of genetic elements of widespread occurrence in bacteria (Clewell and Gawron-Burke, 1986). They are often located in the bacterial chromosome and are transferred by a conjugal mechanism in the apparent absence of extrachromosomal DNA (Buu-Hoi and Horodniceanu, 1980; Shoemaker *et al.*, 1980). Naturally occurring examples of these elements have been found in *C. difficile* (Ionesco, 1980; Smith *et al.*, 1981; Wüst and Hardegger, 1983; Mächler *et al.*, 1987) and *C. innocuum* (Magot, 1983), but they have not been extensively investigated.

The best known examples are two elements, Tn*916* (16.4 kb, Tc^R) and Tn*1545* (25.3 kb, Km^R, Tc^R, MLS^R), found in Streptococci (Franke and Clewell, 1981; Gawron-Burke and Clewell, 1982; Courvalin and Carlier, 1986, 1987). They are being extensively characterized (Jones *et al.*, 1987a; Yamamoto *et al.*, 1987; Caillaud and Courvalin, 1987; Caillaud *et al.*, 1987; Senghas *et al.*, 1988). Analysis of Tn*916* using Tn*5* mutagenesis of the element cloned in *E. coli* has shown that functions required for intracellular transposition are a subset of those required for conjugative transposition. An excision step is believed to be essential for both types of transposition. Sequencing and/or heteroduplex analysis has shown that both Tn*916* and Tn*1545* lack terminal repeated sequences that are usually found in transposable elements. Neither transposon generates a duplication of the target DNA upon insertion and consensus target sites for insertion are very AT-rich ([A]AACTAAA[A] for Tn*916* and TTTNTN$_{3-5}$TAAAAA for Tn *1545*)—note that relatively few target sites have been sequenced to date.

Conjugal transfer of Tn*916* to a variety of *Streptococcus* species, *Staphylococcus aureus*, *Listeria monocytogenes*, and *B. subtilis* has been obtained (Gawron-Burke and Clewell, 1984; Jones *et al.*, 1987b; Kathariou *et al.*, 1987). Tn*1545* shows a similarly catholic host range (Courvalin and Carlier, 1986). Both elements have also been transferred to *C. acetobutylicum* (Davies *et al.*, 1988). The DNA of most clostridia has a very low GC content (Hill, 1966; Johnson and Francis, 1975). Since the consensus sites for insertion of Tn*916* and Tn*1545* appear to be very AT-rich (see above), it is likely that these elements will be useful genetic tools in the clostridia. Indeed, Tn *916* has been transferred to *C. tetani* where it is able to insert into multiple chromosomal sites (Volk *et al.*, 1988). Precise excision of both Tn*916* and Tn*1545* has been demonstrated in *E. coli* (Gawron-Burke and Clewell,

1984; Caillaud and Courvalin, 1987)—note that Tn*4451* and Tn*4452* from
C. perfringens also show this useful property (Abraham and Rood, 1987; see
above). Thus, a strategy for cloning clostridial genes inactivated by Tn
insertion can readily be envisaged (see Gawron-Burke and Clewell, 1984).

4.4. Conjugal Mobilization of Nonconjugative Plasmids

The above examples of conjugal transfer of plasmids and transposons
are of interest in their own right and deserve further extensive analysis.
Conjugal plasmids also provide the potential to introduce additional genetic material into clostridia by conjugal mobilization. This was already apparent in 1977, when some evidence was obtained suggesting that plasmid
pIP401 may occasionally mobilize another plasmid, pIP402 (63 kb, Em^R,
Cl^R), when both are present in the same donor strain of *C. perfringens*
(Brefort *et al.*, 1977). Plasmid pAMβ1 has been shown to mobilize a number of small nonconjugative plasmids from a variety of organisms
(Schaberg *et al.*, 1982; Lereclus *et al.*, 1983; Smith, 1985). The report of Yu
and Pearce (1986) that plasmids pAMβ1 and pVA797 mobilized a small
nonconjugative plasmid, pAM610, from various streptococci to *C. acetobutylicum* awaits confirmation (see Oultram *et al.*, 1987). Little is presently
known of the mechanism by which this "natural" mobilization of noncon-
jugal plasmids occurs.

Experimentally, two strategies have proved useful for mobilizing small
nonconjugative plasmids into *C. acetobutylicum*. They have been mobilized
in the form of cointegrates with large conjugal plasmids and also by the
provision of transfer functions *in trans*. These are considered in the following sections.

4.4.1. Formation of Cointegrates

Although plasmid pAMβ1 is inefficiently transferred from *B. subtilis* to
C. acetobutylicum, presumably as a result of its structural instability in the
former host (van der Lelie and Venema, 1987), it has been used successfully to mobilize small Rep$^-$ plasmids into *C. acetobutylicum* (Oultram *et al.*,
1987). The Rep$^-$ plasmids contained a segment of pAMβ1 (encoding the
Em^R gene). This enabled them to become established in a *B. subtilis* strain
harboring pAMβ1 as cointegrate molecules (cf. Smith and Clewell, 1984),
in which form they were transferred to and subsequently maintained in *C.
acetobutylicum*. This strategy was recently exploited both to obtain complementation of a leucine auxotroph of *C. acetobutylicum* by the corresponding
leucine gene isolated from *C. pasteurianum* (Oultram *et al.*, 1988b) and to
direct the expression in *C. acetobutylicum* of a pseudomonad *xylE* gene,
using a specially constructed expression cartridge based on the *C. pasteurianum* ferredoxin gene (see Minton *et al.*, 1988). Other derivatives of

pAMβ1, lacking a 5.1-kb *Eco*RI restriction fragment on which the replication origin is located (LeBlanc and Lee, 1984; see Section 5.3.1), have been constructed for use in mobilizing small Rep$^+$ gram-positive plasmids into *C. acetobutylicum* (Davies *et al.*, 1988).

4.4.2. Provision of Transfer Functions *in trans*

Trieu-Cuot *et al.* (1987) recently described a versatile method for transferring plasmids by conjugation directly from *E. coli* to a variety of gram-positive bacteria. It is based on a self-transferable IncP plasmid that can mobilize other plasmids harboring the origin of transfer (*oriT*) of the IncP plasmid RK2. The prototype vector that they constructed, pAT187, which comprises the replication origin of plasmid pAMβ1 (see above), a Km^R gene from Tn*1545* and *oriT* from RK2, in the backbone of plasmid pBR322, has been transferred to *C. acetobutylicum* (D. R. Williams, A. Woolford, and M. Young, unpublished work). A small DNA fragment encoding *oriT* has been incorporated into pMTL500E (see Section 5.4) and the resulting plasmid, pCTC1, with its readily selectable Em^R marker, is also efficiently transferred from *E. coli* to *C. acetobutylicum*. Two derivatives of pMTL20E, pMTL30 and pMTL31, containing *oriT* in different orientations outside the polylinker region have been constructed to facilitate the screening of replicons from different gram-positive plasmids for their ability to function in clostridia. Plasmid transfer from an *E. coli* donor with the *tra* genes of RK2 integrated into the chromosome has also been demonstrated. This novel method for the direct transfer of genetic information from *E. coli* to clostridia offers interesting possibilities for future research.

5. CLOSTRIDIAL PLASMIDS

5.1. Indigenous Plasmids

Plasmids are of widespread occurrence in clostridia but many are cryptic (reviewed by Rogers, 1986; Young *et al.*, 1986; Minton and Thompson, 1989). Several of the plasmids to which a function has been assigned are found in species of medical importance. For example, the genes encoding tetanus toxin and botulinum type G neurotoxin reside on extrachromosomal elements (Finn *et al.*, 1984; Eklund *et al.*, 1988). A number of plasmids isolated from *C. perfringens* encode bacteriocin production or resistance to antibiotics (see Section 4.1). The most extensively characterized clostridial plasmid is pIP404 (10.2 kb) from *C. perfringens*. This plasmid encodes bacteriocin BCN5. The complete nucleotide sequence of pIP404 has been determined. There are 10 open reading frames (ORFs), which account for about 77% of its coding capacity (Garnier and Cole, 1988a). Functions have been tentatively assigned to six of these ORFs. Apart from those involved in replication (see Section 5.2.3), these include the ORF encoding the 96,500 mol. wt. BCN5 protein, two ORFS (*uviAB*) implicated in bacteriocin immunity and/or secretion, and an ORF (*res*) encoding a resolvase (Garnier and Cole, 1986; 1988c; Garnier *et al.*, 1987a,b).

5.2. Replication Regions of Clostridial Plasmids

5.2.1. Isolation of Clostridial Replicons

Indigenous plasmids encoding selectable genetic markers have not been found in the biotechnologically important saccharolytic clostridia. Cloning vectors for use in these organisms have been constructed *in vitro*. An important step has been the physical characterization of the replication apparatus of plasmids able to replicate in *Clostridium*. Initial identification has been achieved by inserting restriction fragments into gram-positive replicon-cloning vectors. These are plasmids which carry antibiotic re-sistance genes that are expressed in gram-positive bacteria, such as *B. subtilis*, but which are able to replicate only in gram-negative bacteria, such as *E. coli*. Restriction fragments containing functional replicons promote establish-ment in a gram-positive host. Plasmids pMTL20/21C and pMTL20/21E (Fig. 1) are versatile vectors of this type that carry gram-positive genes encoding Cm^R and Em^R, respectively. The unique cloning sites of these plasmids reside within the *lacZ'* region, facilitating the detection of recombi-nants in *E. coli* as colorless colonies on agar medium supplemented with 5-bromo-4-chloro-3-indoyl-β-D-galactoside (X-Gal).

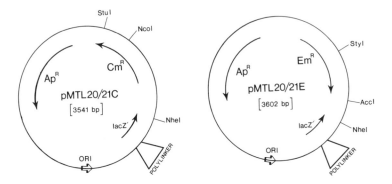

Figure 1. The pMTL gram-positive replicon-cloning vectors. The Ap^R gene and ColE1 rep-licon (ORI) function only in *E. coli*, while the Cm^R (from plasmid pC194) and the Em^R (from plasmid pAMβ1; Brehm *et al.*, 1986) genes express in both *E. coli* and gram-positive bacteria, e.g., *B. subtilis* and *C. acetobutylicum* (see Oultram *et al.*, 1988b). These plasmids will not repli-cate in a gram-positive bacterium unless an appropriate replication region is cloned into the polylinker. The restriction sites present within the extensive polylinker regions are as in pMTL20 and pMTL21 (Chambers *et al.*, 1988).

Since reproducible methods for transforming clostridia have been developed only very recently (Sections 3.2–3.4), clostridial replicons have hitherto been identified by their ability to function in *B. subtilis.* To date, restriction fragments from six different cryptic plasmids have been analyzed in this way, viz. pCB101 (6065 bp), pCB102 (8058 bp), and pCB103 (6.4 kb) from *C. butyricum* (Minton and Morris, 1981), pCP1 (4.05 kb) and pCP2 (6.1 kb) from *C. paraputrificum* (N. P. Minton and L. E. Clarke, unpublished data) and pIP404 (10,207 bp) from *C. perfringens* (Brefort *et al.,* 1977). Only fragments derived from pCB101 and pIP404 promoted autonomous replication in *B. subtilis* (Collins *et al.,* 1985; Luczak *et al.,* 1985; Garnier and Cole, 1988b).

5.2.2. The pCB101 Replicon

A 3.48-kb fragment of pCB101 appeared to function as an inefficient replicon in *B. subtilis,* since a recombinant plasmid, pRB1, carrying this fragment was only detectable using DNA-DNA hybridization; it was not visible in cleared lysates and was rapidly lost in the absence of selective pressure (Collins *et al.,* 1985). However, after repeated cycles of growth in the absence of selection (antibiotic) followed by the reimposition of selection (an increased concentration of antibiotic with each cycle), bacteria were obtained in which plasmid DNA was clearly visible. Interestingly, this was not due to a copy number mutation on the plasmid but was the consequence of a lesion in the *B. subtilis* chromosome (U. Vetter and W. L. Staudenbauer, unpublished data).

The pCB101 replicon has been further characterized by nucleotide sequencing and deletion analysis (J. K. Brehm, A. Pennock, M. Young, and N. P. Minton, unpublished data). A 2.5-kb subfragment which encompasses two ORFs, B and C, promoted autonomous replication in *B. subtilis.* Proteins equivalent in size to the predicted products of these ORFs (mol. wt. 43,000 and 27,000, respectively) were obtained using *in vitro* transcription/translation assays of appropriate template DNA. Manipulations (deletions/insertions) which interrupted the continuity of either ORF abolished replicative activity. Several independent lines of evidence suggest that the mechanism of pCB101 replication is similar to that of a family of plasmids isolated from aerobic gram-positive bacteria such as *Bacillus* and *Staphylococcus.* The protein encoded by ORF B exhibits significant homology with the replication proteins of the pC194/pUB110 family of plasmids. Furthermore, ORF B is preceded by a 25-bp sequence that has remarkable homology with the plus origin of these plasmids; in the case of pUB110, 23 out of 25 nucleotides are identical (see Minton and Oultram, 1988). Plasmids in this family replicate via a rolling circle mechanism, analogous to that of the isometric coliphage φX174 (Gruss *et al.,* 1987; Gros *et al.,* 1987). Single-stranded (ss) DNA is produced as a replication intermediate. Recent studies confirmed that plasmids replicating from the pCB101 replicon do indeed generate ssDNA in *B. subtilis* (J. K. Brehm and N. P. Minton, unpublished data).

5.2.3. The pIP404 Replicon

A 2.8-kb segment of the *C. perfringens* plasmid, pIP404, showed replicative activity in *B. subtilis* (Garnier and Cole, 1988b). This segment encoded two ORFs, denoted 4 and 5, together with a dA + dT-rich region comprising a family of direct repeats. There were two direct repeats of 8 and 10 bp, a 13-bp inverted repeat, and a dispersed tandem array of 6 copies of a 16-bp repeat and 16 copies of an 8-bp repeat. Similar highly structured regions of DNA are found at the replication origins of a number of gram-negative plasmids (F, RK2, R6K, and pSC101) and the *Bacillus stearothermophilus* plasmid pRAT11 (Vocke and Bastia, 1983; Scott, 1984; Imanaka *et al.*, 1986); they are thought to serve as recognition signals for assembly of the replisome complex.

Deletion analysis demonstrated that both the repeated region and ORF5 are essential for replication (Garnier and Cole, 1988b). ORF5 was therefore denoted *rep*. Its predicted product is a highly polar polypeptide (mol. wt. = 48,712). Deletions affecting the continuity of ORF4 led to an increase in plasmid copy number in *B. subtilis*. ORF4 was therefore denoted *cop*. The predicted 198 amino acid product of this ORF appeared to be membrane-bound, which is difficult to reconcile with its proposed function. A further interesting feature of the pIP404 replicon was the presence of a very strong promoter located some 30 bp 3' to the translational stop codon of the *rep* gene. Transcription from this promoter produced a 150-nucleotide RNA molecule that is complementary to the last 125 nucleotides of the *rep*, mRNA suggesting that it may act as an antisense RNA (Garnier and Cole, 1988b). The region of overlap encodes a domain of Rep containing a helix-turn-helix motif characteristic of DNA-binding proteins. However, antisense RNA molecules more usually have complementarity with the 5' end of the gene (RNA) that they regulate (Green *et al.*, 1986).

5.3. Other Replicons tlhat Function in Clostridia

The pAMβ1 Replicon

The nucleotide sequence of a 5101-bp *Eco*RI fragment carrying the previously identified (LeBlanc and Lee, 1984) pAMβ1 replicon has been determined (Swinfield *et al.*, 1989). It contains five major ORFs, the products of two of which (C and E) have been identified using *in vitro* transcription/translation assays. Although ORF C is not essential for replication, it is of interest because it encompasses an extensive region of directly repeated

DNA sequences, comprising two families of 9-bp repeats reiterated 51 times. As a result, the ORF C polypeptide (mol. wt. = 30,471) contains substantial tracts of repeated amino acid sequence (11 × V-D-P and 35 × T-E-P), which confers on it an anomalous electrophoretic mobility. The function of the ORF C product is not known. Very recent work has revealed the presence of genes encoding a topoisomerase and a resolvase in the plasmid sequences adjacent to this 5101-bp *Eco*RI fragment (T. J. Swinfield and N. P. Minton, unpublished data).

A 2589-bp segment encompassing ORF D and E (predicted mol. wt. of products 11,762 and 57,380, respectively) had replicative activity in *B. subtilis*. The product of ORF E is required for replication. The stop codon of ORF D is separated from the start codon of ORF E by 41 bp. Deletion of sequences 3' to the start of ORF D dramatically reduced the level of the ORF E polypeptide *in vitro,* suggesting that these two ORFs are cotranscribed. A region lying outside the minimal replicon appears to control plasmid copy number. Deletion of this region increased plasmid copy number and this was correlated with enhanced production of the ORF E polypeptide *in vitro.* No homology was found between the pAMβ1 replicon, neither at the DNA nor the protein level, and those of other gram-positive plasmids.

Plasmid pAMβ1 has proved particularly useful in the genetic manipulation of *C. acetobutylicum* (see Sections 4.2 and 4.4.1), which suggests that it could prove useful in vector development (see Section 5.4). In this regard, another particularly relevant feature of this plasmid is its mode of replication. In contrast with the pC194/pUB110 family, to which pCB101 belongs (see Section 5.2.2), this plasmid and derivatives containing its replicon produce no detectable ss DNA in *B. subtilis* (Jannière and Ehrlich, 1989). Single-stranded DNA is highly recombinogenic (Jannière and Ehrlich, 1987; Brunier *et al.*, 1988) and it would appear that a cardinal factor in the structural instability of most commonly used gram-positive cloning vectors is their rolling circle mode of replication. Vectors based on the pAMβ1 replicon exhibit 1000-fold greater structural stability in *B. subtilis* than those based on plasmids such as pC194 and pUB110 (Janniere and Ehrlich, 1989). The sequence data discussed above facilitated the construction of a number of *E. coli*/gram-positive shuttle vectors that have been shown to replicate in *C. acetobutylicum* following transformation by electroporation (see Section 3.3) or *oriT*-promoted conjugal mobilization (see Section 4.4.2). These are described in Section 5.4.

5.4. Shuttle Plasmids for Use in *C. acetobutylicum*

A variety of cloning vectors have been developed for use in *C. acetobutylicum*. The most versatile are those based on the pMTL20E backbone (see Section 5.2), into which various of the replicons that function in clostridia have been incorporated outside the polylinker and *lacZ'* regions. *E. coli* strains harboring recombinant plasmids created by cloning into the polylinker are detected by their white coloration on agar supplemented with X-Gal. Two vectors, pMTL500E (Oultram *et al.*, 1988a; Minton and Oultram, 1988) and pMTL501E (Swinfield *et al.*, 1989), are high copy number derivatives containing the pAMβ1 replicon, while a third version, pMTL502E (Swinfield *et al.*, 1989), also carries the pAMβ1 replicon but is maintained at a low copy number. An alternative high copy number vector, pCB3, carries the pCB101 replicon. All of these vectors have been introduced into *C. acetobutylicum* by electroporation. Interestingly, plasmid pCP3, constructed by inserting the whole of pCP1 (from *C. paraputrificum*) into pMTL20E, replicates at low copy number in *C. acetobutylicum* (T. J. Swinfield and N. P. Minton, unpublished data). pCP3 cannot be maintained in *B. subtilis* (see Section 5.2.1).

To date only pMTL500E has been used to introduce cloned heterologous genes into *C. acetobutylicum*. These were the *C. pasteurianum leuB* gene (Oultram *et al.*, 1988a) and the *C. thermocellum celA* gene (J. D. Oultram, W. L. Staudenbauer, and N. P. Minton, unpublished data). In the former case the transformed host was a Leu⁻ auxotroph of *C. acetobutylicum*, strain SBA9. Growth of cells of SAB9 transformed with the recombinant plasmid was no longer reliant on supplementation of the minimal medium with leucine. Acquisition of the *celA* gene was demonstrated by an *in situ* plate assay using Congo red (see Section 6.1.2).

By analogy with findings in *B. subtilis* (see Section 5.2.2), vectors based on the pCB101 replicon (such as pCB3) may show structural instability in *C. acetobutylicum*, making vectors based on the pAMβ1 replicon, such as pMTL500E to pMTL502E, more suitable for general use. This prediction has not yet been tested. Preliminary data on segregational stability indicate that pCB3 is maintained more stably than pMTL500E; after growth for 10 generations in the absence of selection (Em), there was 2% loss of the former plasmid and 54% loss of the latter. Similar segregational instability of another vector, pHV1432, carrying the pAMβ1 replicon has also been found in *B. subtilis* (Jannière and Ehrlich, 1989). Interestingly, stability is conferred in a related plasmid, pHV1431, by the presence of a 2-kb segment of adjacent pAMβ1 DNA. This segment has very recently been shown to encode a *res* gene similar to that identified by Garnier *et al.* (1988) on pIP404 (T. J. Swinfield and N. P. Minton, unpublished data). These *res* genes could play an active role in plasmid partitioning.

5.5. Shuttle Vectors for Use in *C. perfringens*

Squires *et al.* (1984) were the first to construct shuttle vectors for use in *E. coli* and *C. perfringens*, by combining small indigenous cryptic plasmids with pBR322 and then inserting a Tc^R gene from the conjugal R factor pCW3 into them. Many of the plasmids they constructed replicated and expressed the clostridial Tc^R gene in both hosts. Refined versions of these plasmids are now available (e.g., pHR106), into which multiple cloning sites have been incorporated, and the clostridial Tc^R gene replaced by a Cm^R gene from the *C. perfringens* plasmid pJIR62 (Roberts *et al.*, 1988).

Shuttle vectors, pTG36 and pTG67, based on pIP404 have also been constructed (Garnier and Cole, 1988b), but they have not yet been re-introduced into *C. perfringens*. Vectors based on this plasmid could prove valuable, since they are likely to be structurally stable. No detectable ss DNA is produced during pIP404 replication (Garnier and Cole, 1988b) and hence it is extremely unlikely that replication proceeds by a rolling-circle mechanism. Indeed, pIP404 shares no significant homology with gram-positive plasmids known to replicate by this mechanism, i.e., the pC194/pUB110 family (see Section 5.2.2).

6. ANALYSIS OF CLONED CLOSTRIDIAL GENES

6.1. Cloning of Clostridial Genes

In recent years, many genes have been isolated from both mesophilic and thermophilic clostridia (Table IV). These cloned genes encode a wide spectrum of proteins, such as antibiotic resistance determinants, ox-idoreductases, enzymes involved in solventogenesis, toxins, nitrogenase, and cellulolytic enzymes (for reviews, see Jones and Woods, 1986; Béguin *et al.*, 1987). They were all isolated from recombinant DNA libraries con-structed in *E. coli*. In spite of its phylogenetic unrelatedness to the clostridia, *E. coli* appears to be an excellent host for cloning clostridial genes. It not only offers the convenience of well-established cloning procedures, but has the further advantage that it can be grown both aerobically and anaerob-ically. In general, clostridial genes are stably maintained in *E. coli* and expressed at levels sufficient for their detection by appropriate screening procedures.

In most cases genes have been isolated from genomic DNA libraries and are considered as chromosomal. Exceptions are the tetanus toxin gene and the antibiotic resistance transposons of *C. perfringens* and *C. difficile*, which have been isolated from cloned plasmid DNA. Furthermore, several genes involved in bacteriocin production have been cloned and charac-terized from the *C. perfringens* plasmid pIP404 (Garnier and Cole, 1988a; see Section 5.2.3).

6.1.1. Cloning Protocols

Many clostridial gene banks have been constructed in multicopy plas-mid vectors such as pBR322 and the pUC series. A general cloning strategy

Table IV. Summary of Cloned Clostridial Genes

Organism	Genes cloned	Sequence[a]	Ref.
C. acetobutylicum	Alcohol dehydrogenase	−	Youngleson *et al.*, 1988
	Butyraldehyde dehydrogenase	−	Contag and Rogers, 1988
	Butyrate kinase	−	Cary *et al.*, 1988
	Endo-β-1,4-glucanase	+	Zappe *et al.*, 1988
	β-Glucosidase	−	Zappe *et al.*, 1986
	Glutamine synthetase (*glnA*)	+	Janssen *et al.*, 1988
	Phosphotransbutyrylase	−	Cary *et al.*, 1988
	Xylanase	−	Zappe *et al.*, 1987
C. acidiurici	Formyltetrahydrofolate synthetase	−	Whitehead and Rabino-witz, 1986
C. butyricum	Hydrogenase	−	Karube *et al.*, 1983
	β-Isopropylmalate dehydrogenase	−	Ishii *et al.*, 1983
	2 Chloramphenicol acetyltransfer-ases	−	Dubbert *et al.*, 1988
C. cellulolyticum	2 Endo-β-1,4-glucanases	−	Faure *et al.*, 1988
C. difficile	Chloramphenicol acetyltransferase	−	Wren *et al.*, 1988
	Enterotoxin A	−	Wren *et al.*, 1987
C. pasteurianum	Ferredoxin	+	Graves *et al.*, 1985
	Galactokinase	±	Daldal and Applebaum, 1985
	Mo-pterin-binding protein (*mop*)	+	Hinton and Freyer, 1986
	Nitrogenase Fe protein (*nifH*)	+	Wang *et al.*, 1988
	Nitrogenase MoFe protein (*nifDK*)	±	Wang *et al.*, 1987
C. perfringens	Chloramphenicol acetyltransferase	−	Abraham and Rood, 1987
	Sialidase	+	Roggentin *et al.*, 1988
	Tetracycline resistance determi-nant	−	Abraham *et al.*, 1988
	Bacteriocin (pIP404 *bcn*)	+	Garnier and Cole, 1986
	Bacteriocin immunity/secretion (pIP404 *uviAB*)	+	Garnier and Cole, 1988a
	Recombinase (pIP404 *res*)	+	Garnier *et al.*, 1987b
C. stercorarium	Endo-β-1,4-glucanase	+	Schwarz *et al.*, 1988c
	β-Glucosidase	−	Schwarz *et al.*, 1988c
	2 Xylanases	−	Schwarz *et al.*, 1988c
	β-Xylosidase	−	Schwarz *et al.*, 1988c
C. tetani	Tetanus toxin	+	Eisel *et al.*, 1986
			Fairweather *et al.*, 1986
C. thermoaceticum	Formyltetrahydrofolate synthetase	−	Lovell *et al.*, 1988
	Leucine dehydrogenase	−	Shimoi *et al.*, 1987
C. thermocellum	Endo-β-1,4-glucanase (*celA*)	+	Béguin *et al.*, 1985
			Schwarz *et al.*, 1986
	Endo-β-1,4-glucanase (*celB*)	+	Grépinet and Béguin, 1986

(continued)

Table IV. (*Continued*)

Organism	Genes cloned	Sequence[a]	Ref.
			Béguin *et al.*, 1983
	Endo-β-1,4-glucanase (*celC*)	+	Pétré *et al.*, 1986,
			Schwarz *et al.*, 1988a
	Endo-β-1,4-glucanase (*celD*)	+	Joliff *et al.*, 1986a
	Endo-β-1,4-glucanase (*celE*)	+	Hall *et al.*, 1988
C. thermocellum	Endo-β-1,3-glucanase (*licA*)	−	Schwarz *et al.*, 1988b
	Endo-β-1,3-1,4-glucanase (*licB*)	−	Schwarz *et al.*, 1985
	β-Glucosidase (*bglA*)	+	Gräbnitz and Stauden-
			bauer, 1988
	β-Glucosidase (*bglB*)	+	Romaniec *et al.*, 1987b
			Gräbnitz and Stauden-
			bauer, 1988
			Kadam *et al.*, 1988
	Xylanase (*xynZ*)	+	Hazlewood *et al.*, 1988

[a] +,Complete sequence; ±, partial sequence.

involves the following steps: (1) partial digestion of genomic DNA with a restriction enzyme such as *Sau*3A, (2) size fractionation of linearized DNA by preparative electrophoresis and recovery of 4- to 10-kb fragments by electroelution, (3) ligation of passenger DNA into suitably cleaved (e.g., *Bam*HI) and dephosphorylated vector DNA, (4) transformation of CaCl$_2$-treated *E. coli* HB101 or JM109 cells with selection for ampicillin resistant transformants, and (5) identification of recombinant clones by loss of tetracycline resistance (pBR322) or formation of colorless colonies in the presence of X-Gal (pUC vectors).

Under these conditions 70–90% of clones carry inserts with an average size of 4.5–5.0 kb. Employing the Clark–Carbon equation and assuming a clostridial genome size equivalent to that of *E. coli* (4500 kb), it can be calculated that 3000–5000 clones are required to have a 99% probability of isolating an average-sized gene. The required size of the gene bank can be reduced to about 1000 clones if the insert size is increased to 20–40 kb by employing phage λ or cosmid vectors. *Clostridium thermocellum* and *C. stercorarium* gene banks have been constructed using such vectors (Cornet *et al.*, 1983a; Schwarz *et al.*, 1985, 1988c). In addition to the larger insert size, λ vectors offer experimental advantages over plasmid vectors. Exoenzymes, which are not secreted by *E. coli*, can readily be detected in phage lysates. Furthermore, screening protocols employing antibodies or hybridization probes are more conveniently carried out with phage plaques than with colonies. *Clostridium perfringens* toxin A was cloned in *E. coli* using the replacement vector λ EMBL3 (Wren *et al.*, 1987). The expression vector λ gt11 has recently been employed for cloning fragments of clostridial toxins as β-galactosidase fusion proteins (Hanna *et al.*, 1988; Muldrow *et al.*, 1988).

The enzyme *Sau*3A is commonly employed for restriction of genomic DNA because, in contrast to other isoschizomers that recognize the GATC sequence, cleavage is not inhibited by adenosine methylation. However, Garnier and Cole (1986) observed that *Sau*3A was unable to cut *C. perfringens* DNA. This was attributed to a restriction–modification system acting on the GATC sequence and presumably methylating the cytosine. *Clostridium perfringens* DNA could be digested with the enzyme *Mbo*I, which only cleaves GATC if the adenosine in this sequence is unmethylated. Restriction endonucleases have been detected in several other commonly used laboratory strains of clostridia (see Roberts, 1987; Richards *et al.*, 1988).

To avoid the systematic exclusion of sequences from gene banks resulting from an unfortunate distribution of restriction sites, it is advisable to construct additional banks employing different enzymes. Those with G + C-rich recognition sequences cut clostridial DNA only rarely (owing to its low G + C content) and should therefore be avoided.

6.1.2. Screening Strategies

In most cases, screening of clostridial gene banks has relied on expression of the heterologous genes in *E. coli*. Depending on the gene of interest, recombinants have been identified either by complementation of defective host genes or by screening for a novel phenotype. Several clostridial genes involved in amino acid metabolism have been isolated by complementation of *E. coli* auxotrophs (Ishii *et al.*, 1983; Cornet *et al.*, 1983a; Efstathiou and Truffaut, 1986; Zappe *et al.*, 1986; Oultram *et al.*, 1988a). The *C. butyricum* hydrogenase gene was identified by cloning into a Hyd⁻ background (Karube *et al.*, 1983). Similarly, the galactokinase gene of *C. pasteurianum* was cloned by complementation of an *E. coli galK* mutant (Daldal and Applebaum, 1985). The glutamine synthetase gene of *C. acetobutylicum* enabled *E. coli glnA* mutants to utilize ammonium sulfate as sole nitrogen source (Usdin *et al.*, 1986).

Appropriate *E. coli* mutants have also been employed for the isolation of *C. acetobutylicum* genes involved in the acetone-butanol fermentation. Youngleson *et al.* (1988) reported the cloning of an NADP-dependent butanol-ethanol dehydrogenase by screening a *C. acetobutylicum* gene bank for sensitivity to sublethal concentrations of allyl alcohol. The recombinant gene was shown to complement an *E. coli adh* mutant deficient in NAD-dependent alcohol dehydrogenase. Recently, the butyraldehyde dehydrogenase gene of *C. acetobutylicum* was isolated by complementation of an *E. coli* aldehyde dehydrogenase-deficient mutant (Contag and Rogers, 1988).

Both the phosphotransbutyrylase and the butyrate kinase genes were cloned on a recombinant plasmid, which conferred on *E. coli* acetoacetyl-CoA transferase mutants the ability to grow on butyrate as sole carbon source (Cary *et al.*, 1988).

Endo-β-glucanases and xylanase of *C. thermocellum* and other cellulolytic clostridia are easily detectable on plates containing the appropriate substrate by the Congo red assay. Upon staining with Congo red, positive clones are surrounded by a yellow hydrolysis zone on a red background. By screening for the degradation of carboxymethylcellulose (CMC), 15 distinct endo-β-1,4-glucanase genes have been identified in genomic libraries of *C. thermocellum* (Millet *et al.*, 1985; Romaniec *et al.*, 1987a; Hazlewood *et al.*, 1988). Recombinant clones producing β-glucosidases or β-cellobiosidase were revealed by fluorescence under UV irradiation, using the fluorogenic substrate methylumbelliferyl-β-glucoside (MU-β-glucoside) or MU-β-cellobioside, respectively (Gräbnitz and Staudenbauer, 1988). Likewise, clones expressing *C. perfringens* sialidase have been detected by screening for hydrolysis of MU-α-*N*-acetylneuraminic acid (Roggentin *et al.*, 1988).

In the absence of a readily discernible phenotype, positive clones can be detected by either immunological screening or colony hybridization. The latter method has the advantage that it does not depend on expression of the clostridial gene in the heterologous host. However, at least a partial amino acid sequence of the gene product is required to design an appropriate oligonucleotide to be used as hybridization probe. Generally, mixed oligonucleotides of 17–18 bases in length representing a 32- or 64-fold degeneracy are employed. Instead of screening a genomic library generated by cloning random DNA fragments, defined fragments resulting from complete digestion with suitable enzymes are cloned. The suitability of different enzymes can be assessed by Southern hybridization of genomic digests with the oligonucleotide probe.

Both immunological screening and hybridization with DNA probes were employed for isolating clostridial genes encoding formyltetrahydrofolate (FTHF) synthetase. Recombinants harboring the FTHF synthetase gene of *C. acidiurici* were identified by screening with antibody (Whitehead and Rabinowitz, 1986). The corresponding gene from *C. thermoaceticum* was detected by colony hybridization with a heptadecanucleotide pool (Lovell *et al.*, 1988). Screening with synthetic oligonucleotide probes was also used for identifying recombinant clones carrying the *C. pasteurianum* ferredoxin gene (Graves *et al.*, 1985) and for cloning the tetanus toxin gene (Eisel *et al.*, 1986; Fairweather and Lyness, 1986). In contrast, the detection of cloned nitrogenase genes from *C. pasteurianum* was accomplished by hybridization with DNA fragments carrying *Klebsiella pneumoniae nif* genes (Chen *et al.*, 1986). Lack of extensive sequence similarity between the *C. pasteurianum* and *K. pneumoniae* genes due to a marked difference in G + C content was compensated for by applying hybridization conditions of low stringency.

The clostridial origin of cloned DNA fragments has been routinely confirmed by Southern hybridization. In some instances, multiple bands have been observed, suggesting the presence of several homologous genes. For example, the molybdenum-pterin-binding protein of *C. pasteurianum* is encoded by a multigene family comprising three distinct genes (Hinton *et al.*, 1987) and five *nifH*-like sequences have been detected in this organism (Wang *et al.*, 1988). However, in the case of clones carrying *C. thermocellum* DNA, multiple bands of hybridization indicated the presence of insertion sequences, which can occur at various loci in the genome (Hazlewood *et al.*, 1988; cf. Gomez *et al.*, 1981). These observations suggest that clostridial genomes may display considerable plasticity due to the presence of transposable elements.

6.2. Gene Structure and Expression

6.2.1. Codon Usage

A striking characteristic of clostridial DNA is its low G + C content, which ranges from about 26% (*C. acetobutylicum, C. pasteurianum*) to 38% (*C. thermocellum*). This feature facilitates enzymatic sequencing of cloned DNA fragments by the dideoxy chain termination method. Interestingly, the G + C content of coding regions is consistently higher than that of noncoding intergenic regions. Thus, in the *nif* region of *C. pasteurianum*, the G + C content of the ORFs is 26% whereas that of the noncoding sequences is only 17% (Chen *et al.*, 1986). Similarly, the tetanus toxin gene has a G + C content of 28% whereas that of the flanking noncoding regions is only 19% (Eisel *et al.*, 1986).

The extremely low G + C ratio of *C. acetobutylicum* and *C. pasteurianum* DNA affects codon usage, which is strongly biased toward codons in which A and U predominate (Table V). A nearly identical pattern of codon utilization has been reported for the ORFs of the *C. perfringens* plasmid pIP404 (Garnier and Cole, 1988a). Biased codon usage is most prominent in amino acids with four to six synonymous codons. Among the six arginine codons, AGA predominates, whereas the CGX family is rarely used. Similarly, among the six leucine codons, UUA is used preferentially. There is a strong prejudice toward the use of A and U at the degenerate third position of all codons. Consequently, there are striking differences between mesophilic clostridia and *E. coli* in codon utilization for the amino acids arginine, leucine, threonine, proline, glycine, and isoleucine. This may influence translation efficiency in heterologous hosts (see Section 6.2.4).

Inspection of Table V also reveals a clear difference in the codon usage pattern between the mesophilic clostridia and *C. thermocellum*. Codon usage in *C. thermocellum* is less biased and more similar to that in *B. subtilis*. Less than 60% of the codons employed end in A or U (as compared with more than 80% for *C. acetobutylicum* and *C. pasteurianum*). A distinctive

Table V. Codon Usage of Clostridial Genes[a]

Amino acid	Codon	C. acetobutylicum (%)	C. pasteurianum (%)	C. thermocellum (%)	B. subtilis (%)	E. coli (%)
Arg	CGU	8.3	7.4	12.2	25.2	58.1
	C	0.0	3.7	6.1	17.5	35.0
	A	0.0	5.6	1.8	9.1	2.3
	G	0.0	0.0	4.9	11.1	3.2
	AGA	91.7	86.5	51.8	27.7	1.2
	G	0.0	0.0	23.2	9.4	0.3
Leu	CUU	8.2	30.4	31.4	26.1	8.6
	C	3.3	3.6	5.4	9.8	6.6
	A	6.5	10.7	3.0	6.3	1.8
	G	3.3	1.8	17.5	21.8	69.1
	UUA	72.1	46.4	12.1	22.4	5.8
	G	6.5	7.1	30.5	13.6	8.2
Ser	UCU	30.8	24.7	13.7	24.4	26.5
	C	1.5	10.8	13.0	12.0	25.6
	A	35.4	28.0	24.8	18.7	8.3
	G	3.1	1.1	11.5	10.0	11.4
	AGU	23.1	21.5	15.9	10.7	6.5
	C	6.2	14.0	21.1	24.1	21.6
Thr	ACU	49.3	47.6	25.2	14.8	23.8
	C	2.9	10.7	29.5	14.1	50.6
	A	47.8	41.7	32.4	43.3	5.9
	G	0.0	0.0	12.9	27.9	19.7
Pro	CCU	32.4	30.2	41.1	33.6	9.0
	C	0.0	1.9	13.7	9.8	6.0
	A	67.6	66.0	9.7	19.1	19.9
	G	0.0	1.9	35.4	37.5	65.1
Ala	GCU	52.2	49.5	21.1	27.5	27.9
	C	2.9	4.8	17.3	20.0	18.8
	A	43.5	41.9	42.6	27.1	22.9
	G	1.5	3.8	19.0	25.4	30.5
Gly	GGU	22.6	43.9	22.9	25.4	47.8
	C	14.5	4.7	22.6	29.6	40.8
	A	61.3	48.7	46.5	31.5	4.6
	G	1.6	2.7	8.1	13.5	6.8
Val	GUU	47.1	57.0	35.9	31.4	37.5
	C	5.9	4.1	10.1	24.7	12.9
	A	45.1	35.5	28.6	24.5	22.9
	G	2.0	3.3	25.4	19.4	26.8
Ile	AUU	44.6	25.4	36.9	50.0	37.3
	C	8.9	18.5	15.9	39.4	62.2
	A	46.4	56.2	47.2	10.6	0.5

Table V. (*Continued*)

Amino acid	Codon	C. acetobutylicum (%)	C. pasteurianum (%)	C. thermocellum (%)	B. subtilis (%)	E. coli (%)
Lys	AAA	76.6	75.2	66.8	75.4	76.7
	G	23.4	24.8	33.2	24.6	23.3
Asn	AAU	84.9	62.7	50.2	53.1	24.2
	C	15.2	37.3	49.8	46.9	75.8
Gln	CAA	78.6	79.1	36.9	54.2	26.6
	G	21.4	20.9	63.1	45.8	73.4
His	CAU	92.9	71.4	65.0	68.6	38.9
	C	7.1	28.6	35.0	31.4	61.1
Glu	GAA	89.4	91.2	67.9	69.5	73.4
	G	10.6	8.8	32.1	30.5	26.6
Asp	GAU	83.6	73.0	55.3	63.8	51.0
	C	16.4	27.0	44.7	36.2	49.0
Tyr	UAU	74.4	80.0	65.6	61.8	40.6
	C	25.6	20.0	34.5	38.2	59.4
Cys	UGU	70.0	88.9	51.4	45.7	42.0
	C	30.0	11.1	48.7	54.3	58.0
Phe	UUU	84.6	44.2	73.9	64.0	43.5
	C	15.4	55.8	26.1	36.0	56.5
Met	AUG	100.0	100.0	100.0	100.0	100.0
Trp	UGG	100.0	100.0	100.0	100.0	100.0

[a]Data for clostridia are compiled from the published sequences representing a total of 893 (*C. acetobutylicum*), 1559 (*C. pasteurianum*), and 4048 (*C. thermocellum*) codons. Codon usage data for *E. coli* and *B. subtilis* are taken from Ogasawara (1985).

feature of *C. thermocellum* genes is the frequent use of the codons AGG (arginine), UUG (leucine), and CCG (proline), which are rarely found in mesophilic clostridia.

The ORFs of most clostridial genes that have been sequenced so far start with the initiation codon AUG. However, the alternate initiation codon GUG is used quite frequently and is found at the start of the reading frames of the *celA* and *celE* genes of *C. thermocellum* as well as the *nifD* gene of *C. pasteurianum*. The only ORF so far identified that starts with the unusual initiation codon UUG is that of an endoglucanase gene from *C. acetobutylicum* (Zappe *et al.*, 1988).

Clostridial ORFs terminate most frequently with the ochre stop codon

UAA. Exceptions are the *celA* and the *uviAB* genes, which terminate with the amber codon UAG, and the *celC* gene, which terminates with the opal codon UGA.

6.2.2. Ribosome Binding Sites

Besides the initiation codon, ribosome binding sites comprise an upstream purine-rich sequence that is complementary to the 3'-terminal sequence of 16S rRNA (3'-UCUUUCCUCCACU-5' in *B. subtilis* versus 3'-AUUCCUCCACU-5' in *E. coli*) as was first noted by Shine and Dalgarno (1974). Three parameters influence the efficiency of ribosome binding: (1) the free energy change consequent upon mRNA-rRNA interaction, (2) the distance between the Shine–Dalgarno sequence and the initiator codon, and (3) the extent to which the Shine–Dalgarno site is masked by secondary structure.

A survey of putative ribosomal binding sites identified in clostridial genes is presented in Table VI. All 15 mRNAs display regions of extensive complementarity to the 3' terminus of *B. subtilis* 16S RNA, followed by a possible initiation codon and an ORF. In most cases the initiation codon has been identified directly by amino acid sequencing. The calculated free energies of Shine–Dalgarno pairing (Salser, 1977) vary from -15 to -21 kcal/mole as compared to -9 kcal/mole for the prototype *E. coli* binding sequence AGGA. Ribosome binding sites of gram-positive mRNAs are characteristically able to form highly stable complexes with 16S rRNA, whereas the potential for base pairing by *E. coli* Shine–Dalgarno sites varies over a wide range (McLaughlin *et al.*, 1981). The strong Shine–Dalgarno complementarities of gram-positive genes might compensate for the use of alternative initiator codons (GUG and UUG) which presumably bind with low efficiency to fMet-tRNA.

The spacing between the Shine–Dalgarno sequence and the initiator codon is critical (Hager and Rabinowitz, 1985). In *E. coli* the optimal distance is 7 ± 2 nucleotides. Spacings in clostridia range from 6 to 10 nucleotides. The mean spacing of the 15 sites in Table VI is 8 nucleotides, as is observed for *E. coli* ribosome binding sites. An exception is the *celD* gene of *C. thermocellum* with a spacing of 13 nucleotides. However, this gene contains a second (albeit weaker) Shine–Dalgarno site situated 6 nucleotides upstream from the initiation codon.

6.2.3. Transcription Signals

Identification of transcription initiation sites by mRNA mapping has been reported for the ferredoxin mRNA (Graves and Rabinowitz, 1986), the *celA* message (Béguin *et al.*, 1986), and several pIP404 transcripts (Gar-

Table VI. Nucleotide Sequences of Putative Clostridial Ribosome Biding Sites

Organism	Gene	Sequence	Free energy (kcal/mol)[a]	Spacing
C. acetobutylicum	Endoglucanase	UUUAUAAUAGGGGGUAUUAACUUGUUU	-16.4	7
	glnA	AUGUAAAGGGGGAGUUGUAAAAUGGCA	-16.7	9
C. pasteurianum	Ferredoxin	UUUUAAGGAGGUGUAUUUUUCAUGGCA	-21.3	10
	mop	UAAAACUAGGAGGAAUUAAUUAUGAGU	-15.9	8
	nifD	UUUUGAUGAGGGGUGAAUUUCGUGAGC	-20.0	8
	nifH1	AAUUUUUAGGAGGAAUGUUUAAUGAGA	-16.2	8
	nifK	AGUUGUAGGAGGGGAAGCGUAAUGUUA	-17.9	9
C. perfringens	pIP404 bcn	AAGAAAAGAGAGGUUUUAAAAUGGCA	-17.6	8
	pIP404 uviA	AGUAAUAGGGGUAGAGUGCGUAUGAGU	-17.2	10
	pIP404 uviB	AUGGAUUGGAGGUUAGCUAAAAUGGAU	-15.9	9
	Sialidase	UUUAUGGAGGAGAUUAUAUUUAUGUGU	-15.8	11
C. tetani	tet	UUUAAUUAGGAGAUGAUACGUAUGCCA	-15.2	8
C. thermocellum	celA	UUUAAAAAGGAGGAAAAAAAAGUGAAG	-18.3	8
	celB	UUGUUUAGGAGGAAAAAUGCAAUGAAA	-15.9	9
	celC	CAAUAUUUCAGGAGGAAAAAAUGGUG	-15.9	6
	celD	AAAGGGGGAUAAAGGUAAAAAAUGAGU	-16.7 (-9.3)	13 (6)
	celE	UUUUGUAAAGGAGAGGGUAAUGUGAAA	-17.5	6
	bglA	AUCUUAAAGGGUGUGGUAAACAUGUCA	-17.5	8
	bglB	CAAAGCAUGAGGAGGAUAGAAAUGGCG	-15.9	6

[a]Free energies were calculated using an algorithm based on the data of Salser (1977). Spacing is measured from the first base to the right of the purine run (GGAGG or its equivalent). Sequences are aligned at the initiation codon and purine runs and initiation codons are underlined. References are given in Table I.

nier and Cole, 1988a). The deduced promoter sequences are presented in Table VII. Also included is the promoter for the small RNA1 of pIP404, which may function as an antisense RNA in plasmid replication control (Garnier and Cole, 1988b; see Section 5.2.3). It should be noted that the pIP404 promoters P1–P5 differ from the constitutively expressed clostridial promoters by their UV inducibility.

The powerful ferredoxin and RNA1 promoters strongly resemble the *E. coli* and *B. subtilis* −10 (TATAAT) and −35 (TTGACA) consensus sequences. This is corroborated by the finding that the ferredoxin promoter is recognized by the major forms of *E. coli* (E.σ-70) and *B. subtilis* (E.σ-43) RNA polymerase *in vitro* (Graves and Rabinowitz, 1986).

Comparison of the ferredoxin promoter with other gram-positive promoters recognized by *B. subtilis* RNA polymerase (E.σ-43) indicated that some sequences outside the canonical −10 and −35 consensus sequences

Table VII. Clostridial Promoter Sequences

Organism	Gene	Nucleotide sequence[a]
		−45 −35 −15 −10
C. pasteurianum	Ferredoxin	ATAAATTACACT TTTAAA -AAGTTTAAAAACATGA TACAAT AAGTTATG
C. perfringens	pIP404 bcn P1	TTATAAATTTAG TTTACA AAATTGAAGTCAAATTA CTTTTTAT ATTATG
	pIP404 bcn P2	AAAAAAAAAAAA TTATAA -ATTTAGTTTACAAAAT TGAAGT CAAATTACT
	pIP404 bcn P3	TTAATTTTTAGG TTTACA -TTTTTAAAACTAAACT CTTTTTAT TTATT
	pIP404 uviAB P4	TGTTTGGGTTTA TTGACT TATTTATGAAAAAGTTG TAAAAT TAATACGAACATAT
	pIP404 uviAB P5	AAAAAATAGATA TTTACA AAATAGACTAAAAAAG CTTTTTAT ATAGTATAAGCTTTT
	pIP404 res	CAGTATCAAAAA TCCACA TTTTTGATACATTATTT TTTTGT ACAGAAAAA
	pIP404 RNA1	TTTATTTAAAGT TTGAAA AAAATTTTTTTATATTA TATAAT CTTTGAAG
C. thermocellum	celA P1	AATGTTTTTGTA TAAACA TGACAAAATAAATATGA TATAAT GATTGTA
	celA P2	GTTATTGGTTTG GTAAA ------TGTTTTTGGGTA ACGATAT TTATTTT

[a]Sequences are aligned by the −35 and −10 regions. The 3′ nucleotide listed is the major site of initiation.

are conserved. Graves and Rabinowitz (1986) therefore proposed that an "extended" promoter recognition sequence is required by gram-positive bacteria. Interestingly, comparison of the region immediately upstream of the -10 consensus reveals a conserved sequence (APyATNA) between the ferredoxin and RNA1 promoters. This sequence is also present in the -18 to -13 region of the *celA* P1 promoter but is absent in the UV-induced pIP404 promoters. It remains to be seen whether this sequence represents an "extended" promoter element recognized by the major form of clostridial RNA polymerase.

Putative promoter sequences have been identified upstream from ribosome binding sites of several clostridial genes. However, promoter identification by sequence inspection is fraught with difficulty, which is only compounded in gram-positive bacteria by the presence of a multiplicity of minor RNA polymerase holoenzymes with different promoter specificities (Doi and Wang, 1986). Thus, the major *celA* promoter (designated P2 in Table VII) comprises sequences resembling the -35 (CTAAA) and -10 (CCGATAT) consensus sequences of *B. subtilis* σ-28 promoters. In general, homologies between clostridial noncoding regions and consensus sequences recognized by various σ factors from *E. coli* and *B. subtilis* are too weak and too numerous to be of any predictive value; promoters are best identified experimentally, by footprinting (Galas and Schmitz, 1978), for example.

Dyad symmetry elements followed by oligo(U) tracts serve as ρ factor-independent terminators in *E. coli* and *B. subtilis*. A palindromic structure of this nature is located just downstream from the coding region of the ferredoxin gene. Graves and Rabinowitz (1986) showed that the 3′ termini of ferredoxin mRNAs map within the dyad symmetry element. This terminator, when transcribed into RNA, has the potential to form a hairpin with a stem structure at least 5 nucleotides long and a minimum free energy of -8 kcal/mole. Although similar regions of dyad symmetry have been observed downstream from the ORFs of several clostridial genes, biochemical identification has only been obtained with the *C. thermocellum celA* gene (Béguin *et al.*, 1986), the *bcn* and *uviAB* genes of pIP404 (Garnier and Cole, 1986, 1988c), and the pIP404 RNA 1 transcript (Garnier and Cole, 1988b). In spite of the general lack of G + C residues within these transcription terminator sequences, the corresponding RNA transcripts can form stable stem-loop structures preceding or including a stretch of U residues. However, the free energy of RNA hairpin formation is not directly related to terminator strength, since termination of transcription may already occur within the dyad element, as shown for the *C. pasteurianum* ferredoxin gene (Graves and Rabinowitz, 1986).

Bacterial genes are often transcribed in groups resulting in the formation of polycistronic mRNAs. Surprisingly, most clostridial genes analyzed so far are apparently translated from monocistronic transcripts. Sequence analysis revealed an operon structure (*nifH-nifD-nifK*) for the nitrogenase genes of *C. pasteurianum* (Chen *et al.*, 1986) and for the *uviAB* genes of pIP404 (Garnier and Cole, 1988a). A feature of the *C. pasteurianum nif* operon is a 1 base pair overlap between the UAA terminator codon of the *nifD* cistron and the AUG initiator codon of the *nifK* cistron (Wang *et al.*, 1987), suggesting that these cistrons are translationally coupled.

Clustering of cellulolytic genes has been observed in several cases. Two endoglucanase genes from *C. cellulolyticum* (Faure *et al.*, 1988) as well as an endo-β-glucanase and an β-glucosidase gene from *C. acetobutylicum* (Zappe *et al.*, 1986) have been cloned on single restriction fragments. Furthermore, the *celC* and *licA* genes were found to be located in close proximity on the *C. thermocellum* chromosome (Schimming *et al.*, 1988). The latter two genes apparently constitute a transcription unit involved in the degradation of mixed-linkage β-1,3-1,4-glucans. An ORF encoding a putative endo-glucanase has been identified upstream from the *C. thermocellum celE* gene (Hall *et al.*, 1988).

6.2.4. Heterologous Gene Expression

The available evidence suggests that clostridial genes are efficiently transcribed in *E. coli*. Gene expression is generally independent of the orientation of the cloned gene within the vector DNA and this has been taken as evidence for the functioning of clostridial transcription signals in the heterologous host (e.g., Squires *et al.*, 1984). However, promoters apparently recognized by additional (minor?) clostridial σ-factors, such as the major promoter (P1) of the *C. thermocellum celA* gene, are not recognized by *E. coli* RNA polymerase. Analysis of *celA* and ferredoxin transcripts has shown that the *E. coli* enzyme efficiently uses additional promoters that are only weakly recognized by the normal host enzyme. Transcription of clostridial genes in *E. coli* may therefore frequently result from initiation at fortuitous promoter sequences present in the A + T-rich intergenic regions.

Inefficient translation can constitute a barrier to the expression of clostridial genes in heterologous hosts. Ribosome binding sites are probably recognized correctly, but the strongly biased codon usage of mesophilic clostridia might severely limit heterologous translation. For example, *E. coli* cells harboring the *C. perfringens bcn* gene produced no detectable bac-teriocin, even when the transcriptional and translational signals were re-placed by those of *lacZ* (Garnier and Cole, 1986). The inefficient ex-pression of tetanus toxin fragments in *E. coli* has also been attributed to unfavorable condon usage (Eisel *et al.*, 1986). However, not all genes from mesophilic clostridia are poorly expressed in *E. coli* due to limited tRNA availability. For example, the alcohol dehydrogenase and endoglucanase genes of *C. acetobutylicum* and the galactokinase gene of *C. pasteurianum* were expressed efficiently. Whereas the production of endoglucanase in *C. acetobutylicum* requires induction by molasses, this enzyme was constitutively synthesized in *E. coli* (Zappe *et al.*, 1986). Interestingly, the overexpressed galactokinase gene was found to be unstable due to frequent inactivation by acquisition of insertion elements (Daldal and Applebaum, 1985).

Probably as a result of their less biased codon usage (see Section 6.2.1), genes from thermophilic clostridia are quite efficiently translated in *E. coli*. Several enzymes from *C. thermoaceticum* (leucine dehydrogenase, FTHF synthetase) and *C. thermocellum* (endoglucanases, β-glucosidases) have been overproduced in *E. coli*. High-level expression may result from increased dosage of genes on multicopy plasmid vectors and further enhancement can be obtained by placing the gene under the control of an efficient *E. coli* promoter. For example, overproduction of the *C. thermocellum celD* gene product was accomplished by translational fusion of the N-terminal end of the endoglucanase with the N terminus of *E. coli* β-galactosidase (Joliff *et al.*, 1986a). Intracellular accumulation of this fusion protein resulted in the precipitation of cytoplasmic granules.

In order to avoid debilitation of the host arising from excessive production of potentially deleterious proteins, the use of controllable expression vectors is advisable. Expression systems comprising the powerful leftward promoter P_L of bacteriophage λ, regulated by the heat-sensitive λ repressor *cI857*, were especially suitable for the controlled production of proteins from cloned thermophilic clostridial genes (Schwarz *et al.*, 1987). Under optimal conditions the recombinant clostridial protein can amount to 10–15% of the total cellular protein. The thermostability of the recombinant proteins might possibly be exploited for rapid purification by thermal denaturation (precipitation) of the thermolabile host proteins.

6.2.5. Localization Signals

A common feature of bacterial protein export is the requirement for an N-terminal leader (signal) sequence, which is removed upon translocation of the preprotein across the cell membrane (Michaelis and Beckwith, 1982). Leader peptides are 15–40 amino acid residues in length and consist of three parts: (1) a positively charged N-terminal region with one or more basic amino acids, (2) a hydrophobic core region, and (3) a C-terminal part, of which the last three amino acids constitute the signal peptidase recognition site. Signal sequences from gram-positive bacteria tend to be longer and more basic at the N terminus than those from gram-negative bacteria.

The putative leader peptides of several clostridial endoglucanases have been deduced from the nucleotide sequence of the encoded gene. These include the endoglucanase of *C. acetobutylicum* and those encoded by *celA*, *celB*, *celC*, *celD*, and *celE* of *C. thermocellum* (Table IV). Comparative analysis indicates that their basic N-terminal regions differ in charge as well as in length. Likewise, the hydrophobic portions show no obvious homologies. It should be noted that only in the case of the *celA* product has the postulated cleavage site actually been confirmed by sequencing of the extracellular protein (Béguin *et al.*, 1985). The leader sequence of *celC* is atypical and apparently nonfunctional in *E. coli*. The signal cleavage site of the *celE* enzyme could not be identified due to further proteolytic processing (Hall *et al.*, 1988).

Localization experiments have indicated that clostridial endoglucanases expressed in *E. coli* are distributed between the cytoplasmic and periplasmic compartments (Cornet *et al.*, 1983b; Schwarz *et al.*, 1987). Little or no endoglucanase activity is found in the culture medium. This deficiency reflects the lack of a true secretory system in *E. coli*. The appearance of hydrolysis zones on CMC plates cannot be considered as an indication of endoglucanase secretion; presumably it results from enzyme release by cell lysis.

Defective export of heterologous exoenzymes in *E. coli* can be expected to interfere with membrane function. Overproduction of the *celA* gene product led to a rapid loss of cell viability (Schwarz *et al.*, 1987). In contrast, this enzyme was efficiently secreted by *B. subtilis* (Soutschek-Bauer and Staudenbauer, 1987). No loss of enzyme activity due to proteolytic degradation was observed during prolonged incubation of culture supernatants. *Bacillus subtilis* may therefore be a preferred host for the production of clostridial exoenzymes. An additional consideration in favor of *B. subtilis* is the codon usage of this organism, which is less biased than that of *E. coli* and therefore closer to that found in *Clostridium*. Indeed, Garnier and Cole (1988a) found that pIP404 genes, which are not translated in *E. coli*, are apparently expressed at moderate levels in *B. subtilis*.

7. CONCLUDING REMARKS

Genetic analysis in clostridia is at a very interesting stage. Two gene transfer procedures have now been developed that should prove widely applicable throughout the genus, namely, electroporation (Section 3.4) and conjugal plasmid mobilization (Section 4.4.3). Conjugative transposons afford a means for cloning genes inactivated by transposon insertion (Section 4.3). The way is now open to using these and other genetic tools to analyze problems of fundamental biological interest in clostridia, such as regulation of metabolite flow through branched fermentation pathways, and the molecular bases of oxygen toxicity and endospore formation, to name but three. There is much to be done.

ACKNOWLEDGMENTS. The work carried out in the authors' laboratories was funded within the framework of contracts BAP-0044-UK, BAP-0045-D, and BAP-0046-UK of the Biotechnology Action Programme of the Commission of the European Communities. MY gratefully acknowledges financial support from the SERC Biotechnology Directorate.

REFERENCES

Abraham, L. J., and Rood, J. I., 1985a, Molecular analysis of transferable tetracycline resistance plasmids from *Clostridium perfringens*, *J. Bacteriol.* **161:**636–640.

Abraham, L. J., and Rood, J. I., 1985b, Cloning and analysis of the *Clostridium perfringens* tetracycline resistance plasmid, pCW3, *Plasmid* **13:**155–162.

Abraham, L. J. and Rood, J. I., 1987, Identification of Tn*4451* and Tn*4452*, chloramphenicol resistance transposons from *Clostridium perfringens*, *J. Bacteriol.* **169:**1579–1584.

Abraham, L. J., and Rood, J. I., 1988, The *Clostridium perfringens* chloramphenicol resistance transposon Tn*4451* excises precisely in *Escherichia coli*, *Plasmid* **19:**164–168.

Abraham, L. J., Wales, A. J. and Rood, J. I., 1985, World-wide distribution of the conjugative *Clostridium perfringens* tetracycline resistance plasmid, pCW3, *Plasmid* **14:**37–46.

Abraham, L. J., Berryman, D. I., and Rood, J. I., 1988, Hybridization analysis of the class P tetracycline resistance determinant from the *Clostridium perfringens* R-plasmid, pCW3, *Plasmid* **19:**113–120.

Allcock, E. R., Reid, S. J., Jones, D. T., and Woods, D. R., 1982, *Clostridium acetobutylicum* protoplast formation and regeneration, *Appl. Env. Microbiol.* **43:**719–721.

Allen, S. P., and Blaschek, H. P., 1988, Electroporation-induced transformation of intact cells of *Clostridium perfringens*, *Appl. Env. Microbiol.* **54:**2322–2324.

Béguin, P., Cornet, P., and Millet, J., 1983, Identification of the endoglucanase encoded by the *celB* gene of *Clostridium thermocellum*, *Biochimie* **65:**495–500.

Béguin, P., Cornet, P., and Aubert, J. P., 1985, Sequence of a cellulase gene of the thermophilic bacterium *Clostridium thermocellum*, *J. Bacteriol.* **162:**102–105.

Béguin, P., Rocancourt, M., Chebrou, M. C., and Aubert, J. P., 1986, Mapping of mRNA encoding endoglucanase A from *Clostridium thermocellum*, *Mol. Gen. Genet.* **202:**251–254.

Béguin, P., Millet, J., and Aubert, J. P., 1987, The cloned *cel* (cellulose degradation) genes of *Clostridium thermocellum* and their products, *Microbiol. Sciences* **4:**277–280.

Bibb, M. J., Ward, J. M., and Hopwood, D. A., 1978, Transformation of plasmid DNA into *Streptomyces* at high frequency, *Nature* **274:**398–400.

Booth, I. R., and Morris, J. G., 1982, Carbohydrate transport in *Clostridium pasteurianum*, *Biosci. Rep.* **2:**47–53.

Bowring, S. N., and Morris, J. G., 1985, Mutagenesis of *Clostridium acetobutylicum*, *J. Appl. Bacteriol.* **58:**577–584.

Bréfort, G., Magot, M., Ionesco, H., and Sebald, M., 1977, Characterization and transferability of *Clostridium perfringens* plasmids, *Plasmid* **1:**52–66.

Brehm, J. K., Salmond, G. P. C., and Minton, N. P., 1987, Sequence of the adenine methylase gene of the *Streptococcus faecalis* plasmid pAMβ1, *Nucl. Acids Res.* **15:**3177.

Brunier, D., Michel, B., and Ehrlich, S. D., 1988, Copy choice illegitimate DNA recombination, *Cell* **52:**883–892.

Buu-Hoi, A., and Horodniceanu, T., 1980, Conjugative transfer of multiple antibiotic resistance markers in *Streptococcus pneumoniae*, *J. Bacteriol.* **143:**313–320.

Caillaud, F., and Courvalin, P., 1987, Nucleotide sequence of the ends of the conjugative shuttle transposon Tn*1545*, *Mol. Gen. Genet.* **209:**110–115.

Caillaud, F., Carlier, C., and Courvalin, P., 1987, Physical analysis of the conjugative shuttle transposon Tn*1545*, *Plasmid* **17:**58–60.

Cary, J. W., Petersen, D. J., and Bennett, G. N., 1988, Cloning and expression of *Clostridium acetobutylicum* phosphotransbutyrylase and butyrate kinase in *Escherichia coli, J. Bacteriol.* **170:**4613–4618.

Chambers, S. P., Prior, S. E., Barstow, D. A. and Minton, N. P., 1988, The pMTL *nic⁻* cloning vectors. I. Improved pUC polylinker regions to facilitate the use of sonicated DNA for nucleotide sequencing, *Gene* **68:**139–149.

Chang, S., and Cohen, S. N., 1979, High frequency transformation of *Bacillus subtilis* protoplasts by plasmid DNA, *Mol. Gen. Genet.* **168:**111–115.

Chen, K. C. K., Chen, J. S., and Johnson, J. L., 1986, Structural features of multiple *nifH*-like sequences and very biased codon usage in nitrogenase genes of *Clostridium pasteurianum, J. Bacteriol.* **166:**162–172.

Clewell, D. B., and Gawron-Burke, C., 1986, Conjugative transposons and the dissemination of antibiotic resistance in streptococci, *Ann. Rev. Microbiol.* **40:**635–659.

Collins, M. E., Oultram, J. D., and Young, M., 1985, Identification of restriction fragments from two cryptic *Clostridium butyricum* plasmids that promote the establishment of a replication-defective plasmid in *Bacillus subtilis, J. Gen. Microbiol.* **131:**2097–2105.

Contag, P. R., and Rogers, P., 1988, The cloning and expression of the *Clostridium acetobutylicum* B643 butyraldehyde dehydrogenase by complementation of an *Escherichia coli* aldehyde dehydrogenase negative mutant, ASM Annual Meeting Abstracts, 1988, p. 169.

Cornet, P., Tronik, D., Millet, J., and Aubert, J. P., 1983a, Cloning and expression in *Escherichia coli* of *Clostridium thermocellum* genes coding for amino acid synthesis and cellulose hydrolysis, *FEMS Microbiol. Lett.* **16:**137–141.

Cornet, P., Millet, J., Béguin, P., and Aubert, J. P., 1983b, Characterization of two *cel* (cellulose degradation) genes of *Clostridium thermocellum* coding for endoglucanases, *Bio/Technology* **1:**589–594.

Courvalin, P., and Carlier, C., 1986, Transposable multiple antibiotic resistance in *Streptococcus pneumoniae, Mol. Gen. Genet.* **205:**291–297.

Courvalin, P., and Carlier, C., 1987, Tn*1545:* a conjugative shuttle transposon, *Mol. Gen. Genet.* **206:**259–264.

Daldal, F., and Applebaum, J., 1985, Cloning and expression of *Clostridium pasteurianum* galactokinase gene in *Escherichia coli* K-12 and nucleotide sequence analysis of a region affecting the amount of the enzyme, *J. Mol. Biol.* **186:**533–544.

Davies, A., Oultram, J. D., Pennock, A., Williams, D. R., Richards, D. F., Minton, N. P., and Young, M., 1988, Conjugal gene transfer in *Clostridium acetobutylicum*, in: *Genetics and Biotechnology of Bacilli*, Vol. 2 (A. T. Ganesan and J. A. Hoch, eds.), Academic Press, London, pp. 391–395.

Doi, R. H., and Wang, L. F., 1986, Multiple prokaryotic ribonucleic acid polymerase sigma factors, *Microbiol. Rev.* **50:**227–243.

Dubbert, W., Luczak, H., and Staudenbauer, W. L., 1988, Cloning of two chloramphenicol acetyltransferase genes from *Clostridium butyricum* and their expression in *Escherichia coli* and *Bacillus subtilis, Mol. Gen. Genet.* **214:**328–332.

Efstathiou, I., and Truffaut, N., 1986, Cloning of *Clostridium acetobutylicum* genes and their expression in *Escherichia coli* and *Bacillus subtilis, Mol. Gen. Genet.* **204:**317–321.

Eisel, U., Jarausch, W., Goretzki, K., Henschen, A., Engels, J., Weller, U., Hudel, M., Habermann, E., and Niemann, H., 1986, Tetanus toxin: primary structure, expression in *Escherichia coli,* and homology with botulinum toxins, *EMBO J.* **5:**2495–2502.

Eklund, M. W., Poysky, F. T., Mseitif, L. M., and Strom, M. S., 1988, Evidence for plasmid-mediated toxin and bacteriocin production in *Clostridium botulinum* type G, *Appl. Env. Microbiol.* **54:**1405–1408.

Fairweather, N. F., and Lyness, V. A., 1986, The complete nucleotide sequence of the tetanus toxin, *Nucl. Acid. Res.* **14:**7809–7812.

Fairweather, N. F., Lyness, V. A., Pickard, D. J., Allen, G., and Thomson, R. O., 1986, Cloning, nucleotide sequencing, and expression of tetanus toxin fragment C in *Escherichia coli, J. Bacteriol.* **165:**21–27.

Faure, E., Bagnara, C., Belaich, A., and Belaich, J. P., 1988, Cloning and expression of two cellulase genes of *Clostridium cellulolyticum* in *Escherichia coli, Gene* **65**:51–58.

Finn Jr, C. W., Silver, R. P., Habig, W. H., Hardegree, M. C., Zon, G., and Garon, C. F., 1984, The structural gene for tetanus neurotoxin is on a plasmid, *Science* **224**:881–884.

Fornari, C. S., and Kaplan, S., 1982, Genetic transformation of *Rhodopseudomonas sphaeroides* by plasmid DNA, *J. Bacteriol.* **154**:1513–1515.

Franke, A. E., and Clewell, D. B., 1981, Evidence for a chromosome-borne resistance transposon (Tn*916*) in *Streptococcus faecalis* that is capable of "conjugal" transfer in the absence of a conjugative plasmid, *J. Bacteriol.* **145**:494–502.

Galas, D., and Schmitz, A., 1978, DNAase footprinting: a simple method for the detection of protein-DNA binding specificity, *Nucl. Acids Res.* **5**:3157–3170.

Garnier, T., and Cole, S. T., 1986, Characterization of a bacteriocinogenic plasmid from *Clostridium perfringens* and molecular genetic analysis of the bacteriocin-encoding gene, *J. Bacteriol.* **168**:1189–1196.

Garnier, T., and Cole, S. T., 1988a, Complete nucleotide sequence and genetic organization of the bacteriocinogenic plasmid, pIP404, from *Clostridium perfringens, Plasmid* **19**:134–150.

Garnier, T., and Cole, S. T., 1988b, Identification and molecular genetic analysis of replication functions of the bacteriocinogenic plasmid pIP404 from *Clostridium perfringens, Plasmid* **19**:151–160.

Garnier, T., and Cole, S. T., 1988c, Studies of UV-inducible promoters from *Clostridium perfringens* in vivo and in vitro, *Mol. Microbiol.* **2**(5):607–614.

Garnier, T., Le Grice, S. F. J., and Cole, S. T., 1987a, Characterization of the promoters for two UV-inducible transcriptional units carried by plasmid pIP404 from *Clostridium perfringens*, in: *Genetics and Biotechnology of the Bacilli*, Vol. 2 (A. T. Ganesan and J. A. Hoch, eds.) Academic Press, London, pp. 211–214.

Garnier, T., Saurin, W., and Cole, S. T., 1987b, Molecular characterization of the resolvase gene, *res*, carried by a multicopy plasmid from *Clostridium perfringens:* Common evolutionary origin for prokaryotic site-specific recombinases, *Mol. Microbiol.* **1**:371–376.

Gawron-Burke, C., and Clewell, D. B., 1982, A transposon in *Streptococcus faecalis* with fertility properties, *Nature* **300**:281–284.

Gawron-Burke, C. and Clewell, D. B., 1984, Regeneration of insertionally inactivated streptococcal DNA fragments after excision of transposon Tn*916* in *Escherichia coli:* Strategy for targeting and cloning of genes from Gram-positive bacteria, *J. Bacteriol.* **159**:214–221.

Gräbnitz, F., and Staudenbauer, W. L., 1988, Characterization of two β-glucosidase genes from *Clostridium thermocellum, Biotechnol. Lett.* **10**:73–78.

Graves, M. C., and Rabinowitz, J. C., 1986, In vivo and in vitro transcription of the *Clostridium pasteurianum* ferredoxin gene. Evidence for "extended" promoter elements in gram-positive organisms, *J. Biol. Chem.* **261**:11409–11415.

Graves, M. C., Mullenbach, G. T., and Rabinowitz, J. C., 1985, Cloning and nucleotide sequence of the *Clostridium pasteurianum* ferredoxin gene, *Proc. Natl. Acad. Sci. USA* **82:**1653–1657.

Green, P. J., Pines, O., and Inouye, M., 1986, The role of antisense RNA in gene regulation, *Ann. Rev. Biochem.* **55**:569–597.

Grépinet, O., and Béguin, P., 1986, Sequence of the cellulase gene of *Clostridium thermocellum* coding for endoglucanase B, *Nucl. Acid. Res.* **14**:1791–1799.

Gros, M. F., te Riele, H., and Ehrlich, S. D., 1987, Rolling circle replication of single-stranded DNA plasmid pC194, *EMBO J.* **6**:3863–3869.

Gruss, A., Ross, H. F., and Novick, R. P., 1987, Functional analysis of a palindromic sequence required for normal replication of several staphylococcal plasmids, *Proc. Natl. Acad. Sci. USA* **84**:2165–2169.

Hächler, H., Kayser, F. H., and Berger-Bächi, B., 1987, Homology of a transferable tetracycline resistance determinant of *Clostridium difficile* with *Streptococcus* (*Enterococcus*) *faecalis* transposon Tn *916, Antimicrob. Ag. Chemother.* **31:**1033–1038.

Hager, P. W., and Rabinowitz, J. C., 1985, Translational specificity in *Bacillus subtilis*, in: *The Molecular Biology of the Bacilli*, Vol. 2 (D. Dubnau, ed.), Academic Press, New York, pp. 1–32.

Hall, J., Hazlewood, G. P., Barker, P. J., and Gilbert, H. J., 1988, Conserved reiterated domains in *Clostridium thermocellum* endoglucanases are not essential for catalytic activity, *Gene* **69:**29–38.

Hanna, P. C., Wnek, A. P., and McClane, B. A., 1988, Cloning of *Clostridium perfringens* type A enterotoxin gene fragment, ASM Annual Meeting Abstracts, 1988, p. 36.

Hazlewood, G. P., Romaniec, M. P. M., Davidson, K., Grépinet, O., Béguin, P., Millet, J., Raynaud, O., and Aubert, J. P., 1988, A catalogue of *Clostridium thermocellum* endoglucanase, β-glucosidase and xylanase genes cloned in *Escherichia coli, FEMS Microbiol. Lett.* **51:**231–236.

Heefner, D. L., Squires, C. H., Evans, R. J., Kopp, B. J., and Yarus, M. J., 1984, Transformation of *Clostridium perfringens, J. Bacteriol.* **159:**460–464.

Hinton, S. M., and Freyer, G., 1986, Cloning, expression and sequencing the molybdenum-pterin binding protein (*mop*) gene of *Clostridium pasteurianum* in *Escherichia coli, Nucl. Acids Res.* **14:**9371–9380.

Hinton, S. M., Slaughter, C., Eisner, W., and Fisher, T., 1987, The molybdenum-pterin binding protein is encoded by a multigene family in *Clostridium pasteurianum, Gene* **54:**211–219.

Hill, L. R., 1966, An index to deoxyribonucleic acid base compositions of bacterial species, *J. Gen. Microbiol.* **44:**419–437.

Hongo, M., 1960, US Patent 2, 945, 786.

Imanaka, T., Ishikawa, H., and Aiba, S., 1986, Complete nucleotide sequence of the low copy number plasmid pRAT11 and replication control by the RepA protein in *Bacillus subtilis, Mol. Gen. Genet.* **205:**90–96.

Ionesco, H., 1980, Transfert de la résistance à la tétracycline chez *Clostridium difficile, Ann. Microbiol. Inst. Pasteur* **131:**171–179.

Ishii, K., Kudo, T., Honda, H., and Horikoshi, K., 1983, Molecular cloning of β-isopropylmalate dehydrogenase gene from *Clostridium butyricum* M588, *Agric. Biol. Chem.* **47:**2313–2317.

Jannière, L., and Ehrlich, S. D., 1987, Recombination between short repeated sequences is more frequent in plasmids than in the chromosome of *Bacillus subtilis, Mol. Gen. Genet.* **210:**116–121.

Jannière, L., and Ehrlich, S. D., 1989, Structurally stable DNA cloning vectors, *Gene* (in press).

Janssen, P. J., Jones, W. A., Jones, D. T., and Woods, D. R., 1988, Molecular analysis and regulation of the *glnA* gene of the Gram-positive anaerobe *Clostridium acetobutylicum, J. Bacteriol.* **170:**400–408.

Johnson, J. L., and Francis, B. S., 1975, Taxonomy of the clostridia: ribosomal ribonucleic acid homologies among the species, *J. Gen. Microbiol.* **88:**229–244.

Joliff, G., Béguin, P., Juy, M., Millet, J., Ryter, A., Poljak, R., and Aubert, J. P., 1986a, Isolation, crystallization and properties of a new cellulase of *Clostridium thermocellum* overproduced in *Escherichia coli, Bio/Technology* **4:**896–900.

Joliff, G., Béguin, P., and Aubert, J. P., 1986b, Nucleotide sequence of the cellulase gene *celD* encoding endoglucanase D of *Clostridium thermocellum, Nucleic Acids Res.* **14:**8605–8613.

Jones, D. T., and Woods, D. R., 1986, Gene transfer, recombination and gene cloning in *Clostridium acetobutylicum, Microbiol. Sci.* **3:**19–22.

Jones, J. M., Gawron-Burke, C., Flannagan, S. E., Yamamoto, M., Senghas, E., and Clewell, D. B., 1987a, Structural and genetic studies of the conjugative transposon Tn*916*, in: *Staphylococcal Genetics* (J. J. Ferretti and R. Curtiss III, eds.,) ASM, Washington, pp. 54–60.

Jones, J. M., Yost, S. C., and Pattee, P. A., 1987b, Transfer of the conjugal tetracycline resistance transposon Tn916 from *Streptococcus faecalis* to *Staphylococcus aureus* and identification of some insertion sites in the staphylococcal chromosome, *J. Bacteriol.* **169**:2121–2131.

Kadam, S., Demain, A. L., Millet, J., Béguin, P., and Aubert, J. P., 1988, Molecular cloning of a gene for a thermostable β-glucosidase from *Clostridium thermocellum* into *Escherichia coli*, *Enzyme Microb. Technol.* **10**:9–13.

Karube, I., Urano, N., Yamada, T., Hirochika, H., and Sakaguchi, K., 1983, Cloning and expression of the hydrogenase gene from *Clostridium butyricum* in *Escherichia coli*, *FEBS Lett.* **158**:119–122.

Kathariou, S., Metz, P., Hof, H., and Goebel, W., 1987, Tn916-induced mutations in the hemolysin determinant affecting virulence of *Listeria monocytogenes*, *J. Bacteriol.* **169**:1291–1297.

Knowlton, S., Ferchak, J. D., and Alexander, J. K., 1984, Protoplast regeneration in *Clostridium tertium:* Isolation of derivatives with high frequency regeneration, *Appl. Env. Microbiol.* **48**:1246–1247.

LeBlanc, D. J., and Lee, L. N., 1984, Physical and genetic analyses of streptococcal plasmid pAMβ1 and cloning of its replication region, *J. Bacteriol.* **157**:445–453.

Lereclus, D., Menou, G., and Lecadet, M.-M., 1983, Isolation of a DNA sequence related to several plasmids from *Bacillus thuringiensis* after a mating involving the *Streptococcus faecalis* plasmid pAMβ1, *Mol. Gen. Genet.* **191**:307–313.

Lin, Y., and Blaschek, H. P., 1984, Transformation of heat-treated *Clostridium acetobutylicum* with pUB110 plasmid DNA, *Appl. Env. Microbiol.* **48**:737–742.

Lovell, C. R., Przybyla, A., and Ljungdahl, L. G., 1988, Cloning and expression in *Escherichia coli* of the *Clostridium thermoaceticum* gene encoding thermostable formyltetrahydrofolate synthetase, *Arch. Microbiol.* **149**:280–285.

Lucansky, J. B., Muriana, P. M., and Klaenhammer, T. R., 1988, Application of electroporation for transfer of plasmid DNA to *Lactobacillus, Lactococcus, Leuconostoc, Listeria, Pediococcus, Bacillus, Staphylococcus, Enterococcus* and *Propionibacterium, Mol. Microbiol.* **2**:637–646.

Luczak, H., Schwarzmoser, H., and Staudenbauer, W. L., 1985, Construction of *Clostridium butyricum* plasmids and transfer to *Bacillus subtilis*, *Appl. Microbiol. Biotechnol.* **23**:114–122.

McLaughlin, J. R., Murray, C. L., and Rabinowitz, J. C., 1981, Unique features in the ribosome binding site sequence of the gram-positive *Staphylococcus aureus* β-lactamase gene, *J. Biol. Chem.* **256**:11283–11291.

Macrina, F. L., Keeler, C. L., Jones, K. R. and Wood, P. H., 1980, Molecular characterization of unique deletion mutants of the streptococcal plasmid, pAMβ1, *Plasmid* **4**:8–16.

Magot, M., 1983, Transfer of antibiotic resistances from *Clostridium innocuum* to *Clostridium perfringens* in the absence of detectable plasmid DNA, *FEMS Microbiol. Lett.* **18**:149–151.

Magot, M., 1984, Physical characterization of the *Clostridium perfringens* tetracycline-chloramphenicol resistance plasmid pIP401, *Ann. Microbiol. Inst. Pasteur* **135**:269–282.

Mahony, D. E., Mader, J. A., and Dubel, J. R., 1988, Transformation of *Clostridium perfringens* L forms with shuttle plasmid DNA, *Appl. Env. Microbiol.* **54**:264–267.

Mendez, B. S., and Gomez, R. F., 1982, Isolation of *Clostridium thermocellum* auxotrophs, *Appl. Env. Microbiol.* **43**:495–496.

Michaelis, S., and Beckwith, J., 1982, Mechanism of incorporation of cell envelope proteins in *Escherichia coli*, *Ann. Rev. Microbiol.* **36**:435–465.

Millet, J., Pétré, D., Béguin, P., Raynaud, O., and Aubert, J. P., 1985, Cloning of ten distinct DNA fragments of *Clostridium thermocellum* coding for cellulases, *FEMS Microbiol. Lett.* **29**:145–149.

Minton, N. P., and Morris, J. G., 1981, Isolation and partial characterization of three cryptic plasmids from strains of *Clostridium butyricum*, *J. Gen. Microbiol.* **127**:325–331.

Minton, N. P., and Morris, J. G., 1983, Regeneration of protoplasts of *Clostridium pasteurianum* ATCC 6013, *J. Bacteriol.* **155**:432–434.

Minton, N. P., and Oultram, J. D., 1988, Host:vector systems for gene cloning in *Clostridium*, *Microbiol. Sci.* **5**:310–315.

Minton, N. P., Brehm, J. K., Oultram, J. D., Swinfield, T. J., and Thompson, D. E., 1988, Construction of plasmid vector systems for gene transfer in *Clostridium acetobutylicum*, in: *Anaerobes Today* (J. M. Hardie and S. P. Boriello, eds.), John Wiley and Sons, Chichester, pp. 125–134.

Minton, N. P., and Thompson, D. E. 1989, Genetics of anaerobes, in: *Anaerobes in Human Disease* (B. I. Duerden and B. S. Drassar, eds.), Edward Arnold Ltd, London (in press).

Muldrow, L. L., Ibeanu, G. C., Lee, N. I., Bose, N. K., and Johnson, J., 1988, Molecular cloning of *Clostridium difficile* toxin B gene fragments in *Escherichia coli*, ASM Annual Meeting Abstracts, 1988, p. 37.

Murray, W. D., Wemyss, K. B., and Khan, A. W., 1983, Increased ethanol production and tolerance by a pyruvate-negative mutant of *Clostridium saccharolyticum*, *Eur. J. Appl. Microbiol. Biotechnol.* **18**:71–74.

O'Brien, R. W., and Morris, J. G., 1971, Oxygen and the growth and metabolism of *Clostridium acetobutylicum*, *J. Gen. Microbiol.* **68**:307–318.

Ogasawara, N., 1985, Markedly unbiased codon usage in *Bacillus subtilis*, *Gene* **40**:145–150.

Ogata, S., Choi, K. H., and Hongo, M., 1975, Sucrose-induced autolysis and development of protoplast-like cells of *Clostridium saccharoperbutylacetonicum*, *Agric. Biol. Chem.* **39**:1247–1254.

Oultram, J. D., and Young, M., 1985, Conjugal transfer of plasmid pAMβ1 from *Streptococcus lactis* and *Bacillus subtilis* to *Clostridium acetobutylicum*, *FEMS Microbiol. Lett.* **27**:129–134.

Oultram, J. D., Davies, A., and Young, M., 1987, Conjugal transfer of a small plasmid from *Bacillus subtilis* to *Clostridium acetobutylicum* by cointegrate formation with plasmid pAMβ1, *FEMS Microbiol. Lett.* **42**:113–119.

Oultram, J. D., Loughlin, M., Swinfield, T. J., Brehm, J. K., Thompson, D. E., and Minton, N. P., 1988a, Introduction of plasmids into whole cells of *Clostridium acetobutylicum* by electroporation *FEMS Microbiol. Lett.* **56**:83–88.

Oultram, J. D., Peck, H., Brehm, J. K., Thompson, D., Swinfield, T. J., and Minton, N. P., 1988b, Introduction of genes for leucine biosynthesis from *Clostridium pasteurianum* into *Clostridium acetobutylicum*, *Mol. Gen. Genet.* **214**:177–179.

Pan-Hou, H. S. K., Hosono, M., and Imura, N., 1980, Plasmid-controlled mercury biotransformation by *Clostridium cochlearum* T-2, *Appl. Env. Microbiol.* **40**:1007–1011.

Pétré, D., Millet, J., Longin, R., Béguin, P., Girard, M., and Aubert, J.-P., 1986, Purification of the endoglucanase C of *Clostridium thermocellum* produced in *Escherichia coli*, *Biochimie*, **68**:687–695.

Podvin, L., Reysset, G., Hubert, J., and Sebald, M., 1988, Recent developments in the genetics of *Clostridium acetobutylicum*, in: *Anaerobes Today* (J. M. Hardie and S. P. Borriello, eds.), John Wiley and Sons, Chichester, pp. 135–140.

Reid, S. J., Allcock, E. R., Jones, D. T., and Woods, D. R., 1983, Transformation of *Clostridium acetobutylicum* protoplasts with bacteriophage DNA, *Appl. Env. Microbiol.* **45**:305–307.

Reysset, G., and Sebald, M., 1985, Conjugal transfer of plasmid-mediated antibiotic resistance from streptococci to *Clostridium acetobutylicum*, *Ann Microbiol. Inst. Pasteur* **136**:275–282.

Reysset, G., Hubert, J., Podvin, L., and Sebald, M., 1987, Protoplast formation and regeneration of *Clostridium acetobutylicum* strain N1-4080, *J. Gen. Microbiol.* **133**:2595–2600.

Richards, D. F., Linnett, P. E., Oultram, J. D., and Young, M., 1988, Restriction endonucleases in *Clostridium pasteurianum* ATCC 6013 and *C. thermohydrosulfuricum* DSM 568, *J. Gen. Microbiol.* **134**:3151–3157.

Roberts, R. J., 1987, Restriction enzymes and their isoschizomers, *Nucl. Acids Res.* **15**(suppl):r189–r217.

Roberts, I., Holmes, W. M., and Hylemon, P. B., 1988, Development of a new shuttle plasmid system for *Escherichia coli* and *Clostridium perfringens, Appl. Env. Microbiol.* **54**:268–270.

Robson, R. L., Robson, R. M., and Morris, J. G., 1974, The biosynthesis of granulose by *Clostridium pasteurianum, Biochem. J.* **144**:503–511.

Rogers, P., 1986, Genetics and biochemistry of *Clostridium* relevant to development of fermentation processes, *Adv. Appl. Microbiol.* **31**:1–60.

Roggentin, P., Rothe, B., Lottspeich, F., and Schauer, R., 1988, Cloning and sequencing of a *Clostridium perfringens* sialidase gene, *FEBS Lett.* **238**:31–34.

Romaniec, M. P. M., Clarke, N. G., and Hazlewood, G. P., 1987a, Molecular cloning of *Clostridium thermocellum* DNA and the expression of further novel endo-β-1,4-glucanase genes in *Escherichia coli, J. Gen. Microbiol.* **133**:1297–1307.

Romaniec, M. P. M., Davidson, K., and Hazlewood, G. P., 1987b, Cloning and expression in *Escherichia coli* of *Clostridium thermocellum* DNA encoding β-glucosidase activity, *Enzyme Microb. Technol.* **9**:474–478.

Rood, J. I., 1983, Transferable tetracycline resistance in *Clostridium perfringens* strains of porcine origin, *Canad. J. Microbiol.* **29**:1241–1246.

Rood, J. I., Scott, V. N. and Duncan, C. L., 1978, Identification of a transferable tetracycline resistance plasmid (pCW3) from *Clostridium perfringens, Plasmid* **1**:563–570.

Schaberg, D. R., Clewell, D. B., and Glatzer, L., 1982, Conjugative transfer of R-plasmids from *Streptococcus faecalis* to *Straphylococcus aureus, Antimicrob. Ag. Chemother.* **22**:204–207.

Salser, W., 1977, Secondary structure prediction of RNA, *Cold Spring Harbor Symp. Quant. Biol.* **42**:985–995.

Scott, J. R., 1984, Regulation of plasmid replication, *Microbiol. Rev.* **48**:1–23.

Schimming, S., Schwarz, W. H., and Staudenbauer, W. L., 1988, Clustering of *Clostridium thermocellum* genes involved in β-glucan degradation, *FEMS Microbiol. Lett.* (submitted).

Schwarz, W. H., Bronnenmeier, K., and Staudenbauer, W. L., 1985, Molecular cloning of *Clostridium thermocellum* genes involved in β-glucan degradation in bacteriophage lambda, *Biotechnol. Lett.* **7**:859–864.

Schwarz, W. H., Gräbnitz, F., and Staudenbauer, W. L., 1986, Properties of a *Clostridium thermocellum* endoglucanase produced in *Escherichia coli, Appl. Env. Microbiol.* **51**:1293–1299.

Schwarz, W. H., Schimming, S., and Staudenbauer, W. L., 1987, High-level expression of *Clostridium thermocellum* cellulase genes in *Escherichia coli, Appl. Microbiol. Biotechnol.* **27**:50–56.

Schwarz, W. H., Schimming, S., Rücknagel, K. P., Burgschwaiger, S., Kreil, G., and Staudenbauer, W. L., 1988a, Nucleotide sequence of the *celC* gene encoding endoglucanase C of *Clostridium thermocellum, Gene* **63**:23–30.

Schwarz, W. H., Schimming, S., and Staudenbauer, W. L., 1988b, Isolation of a *Clostridium thermocellum* gene encoding a thermostable β-1,3-glucanase (laminarinase), *Biotechnol. Lett.* **10**:225–230.

Schwarz, W. H., Jauris, S., Kouba, M., and Staudenbauer, W. L., 1988c, Molecular cloning and expression in *Escherichia coli* of *Clostridium stercorarium* genes involved in cellulose degradation, *Biotechnol. Lett.* (submitted).

Sebald, M. and Brefort, G., 1975, Transfert du plasmide tétracycline-chloramphénicol chez *Clostridium perfringens, C. R. Acad. Sci. Paris Ser. D.* **281**: 317–319.

Sebald, M., and Costilow, R. N., 1975, Minimal growth requirements for *Clostridium perfringens* and isolation of auxotrophic mutants, *Appl. Microbiol.* **29**:1–16.

Sebald, M., Bouanchaud, D., and Bieth, G., 1975, Nature plasmidique de la résistance à plusieurs antibiotiques chez *C. perfringens* type A, souche 659. *C. R. Acad. Sci. Paris Ser. D.* **280**:2401–2404.

Senghas, E., Jones, J. M., Yamamoto, M., Gawron-Burke, C., and Clewell, D. B., 1988, Genetic organization of the bacterial conjugative transposon Tn*916, J. Bacteriol.* **170**:245–259.

Shimoi, H., Nagata, S., Esaki, N., Tanaka, H., and Soda, K., 1987, Leucine dehydrogenase of a thermophilic anaerobe, *Clostridium thermoaceticum:* Gene cloning, purification and characterization, *Agric. Biol. Chem.* **51**:3375.

Shine, J., and Dalgarno, L., 1974, The 3'-terminal sequence of *Escherichia coli* 16S ribosomal RNA: complementarity to nonsense triplets and ribosome binding sites, *Proc. Natl. Acad. Sci. USA* **71**:1342–1346.

Shoemaker, N. B., Smith, M. D., and Guild, W. R., 1980, DNase-resistant transfer of chromosomal *cat* and *tet* insertions by filter mating in *Pneumococcus, Plasmid* **3**:80–87.

Smith, M. D., 1985, Transformation and fusion of *Streptococcus faecalis* protoplasts, *J. Bacteriol.* **162**:92–97.

Smith, M. D., and Clewell, D. B., 1984, Return of *Streptococcus faecalis* DNA cloned in *Escherichia coli* to its original host via transformation of *Streptococcus sanguis* followed by conjugative mobilization, *J. Bacteriol.* **160**:1109–1114.

Smith, C. J., Markowitz, S. M., and Macrina, F. L., 1981, Transferable tetracycline resistance in *Clostridium difficile, Antimicrob. Ag. Chemother.* **19**:997–1003.

Soutschek-Bauer, E., and Staudenbauer, W. L., 1987, Synthesis and secretion of a heat-stable carboxymethylcellulase from *Clostridium thermocellum* in *Bacillus subtilis* and *Bacillus stearothermophilus, Mol. Gen. Genet.* **208**:537–541.

Soutschek-Bauer, E., Hartl, L., and Staudenbauer, W. L., 1985, Transformation of *Clostridium thermohydrosulfuricum* DSM 568 with plasmid DNA, *Biotechnol. Lett.* **7**:705–710.

Squires, C. H., Heefner, D. L., Evans, R. J., Kopp, B. J., and Yarus, M. J., 1984, Shuttle plasmids for *Escherichia coli* and *Clostridium perfringens, J. Bacteriol.* **159**:465–471.

Stal, M. H., and Blaschek, H. P., 1985, Protoplast formation and cell wall regeneration in *Clostridium perfringens, Appl. Env. Microbiol.,* **50**:1097–1099.

Swinfield, T. J., Oultram, J. D., Thompson, D. E., Brehm, J. K., and Minton, N. P., 1989, Physical characterization of the replication region of the *Streptococcus faecalis* plasmid pAMβ1, *Gene* (in press).

Terzaghi, B. E., and Sandine, W. E., 1975, Improved medium for lactic streptococci and their bacteriophages, *Appl. Microbiol.* **29**:807–813.

Trieu-Cuot, P., Carlier, C., Martin, P., and Courvalin, P., 1987, Plasmid transfer by conjugation from *Escherichia coli* to gram-positive bacteria, *FEMS Microbiol. Lett.* **48**:289–294.

Usdin, K. P., Zappe, H., Jones, D. T., and Woods, D. R., 1986, Cloning, expression, and purification of glutamine synthetase from *Clostridium acetobutylicum. Appl. Env. Microbiol.* **52**:413–419.

van der Lelie, D., and Venema, G., 1987, *Bacillus subtilis* generates a major specific deletion in pAMβ1, *Appl. Env. Microbiol.* **53**:2458–2463.

Vocke, C., and Bastia, D., 1983, DNA-protein interaction at the origin of DNA replication of the plasmid pSC101, *Cell* **35**:495–502.

Volk, W. A., Bizzini, B., Jones, K. R., and Macrina, F. L., 1988, Inter- and intrageneric transfer of Tn*916* between *Streptococcus faecalis* and *Clostridium tetani, Plasmid* **19**:255–259.

von Heijne, G., 1986, A new method for predicting signal sequence cleavage sites, *Nucl. Acids Res.* **14**:4683–4690.

Walker, G. C., 1983, Genetic strategies in strain design for fermentations, in: *Basic Biology of New Developments in Biotechnology* (A. Hollander, A. I. Laskin, and P. Rogers, eds.), Plenum Press, New York, pp. 349–376.

Wang, S. Z., Chen, J. S., and Johnson, J. L., 1987, Nucleotide and deduced amino acid sequences of *nifD* encoding the A 2 n-subunit of nitrogenase MoFe protein of *Clostridium pasteurianum, Nucl. Acids Res.* **15**:3935.

Wang, S. Z., Chen, J. S., and Johnson, J. L., 1988, The presence of five *nifH*-like sequences in *Clostridium pasteurianum:* Sequence divergence and transcription properties, *Nucl. Acids Res.* **16**:439–454.

Whitehead, T. R., and Rabinowitz, J. C., 1986, Cloning and expression in *Escherichia coli* of the gene for 10-formyltetrahydrofolate synthetase from *Clostridium acidi-urici, J. Bacteriol.* **167**:205–209.

Wren, B.W., Clayton, C. L., Mullany, P. P., and Tabaqchali, S., 1987, Molecular cloning and expression of *Clostridium difficile* toxin A in *Escherichia coli, FEBS Lett.* **225**:82–86.

Wren, B. W., Mullany, P., Clayton, C., and Tabaqchali, S., 1988, Molecular cloning and genetic analysis of a chloramphenicol acetyltransferase determinant from *Clostridium difficile, Antimicrob. Ag. Chemother.* **32**:1213–1217.

Wüst, J., and Hardegger, U., 1983, Transferable resistance to clindamycin, erythromycin, and tetracycline in *Clostridium difficile, Antimicrob. Ag. Chemother.* **23**:784–786.

Yamamoto, M., Jones, J. M., Senghas, E., Gawron-Burke, C., and Clewell, D. B., 1987, Generation of Tn5 insertions in streptococcal conjugative transposon Tn*916, Appl. Env. Microbiol.* **53**:1069–1072.

Yoshino, S., Ogata, S., and Hayashida, S., 1982, Some properties of autolysin of *Clostridium saccharoperbutylacetonicum, Agric. Biol. Chem.* **46**:1243–1248.

Yoshino, S., Ogata, S., and Hayashida, S., 1984, Regeneration of protoplasts of *Clostridium saccharoperbutylacetonicum, Agric. Biol. Chem.* **48**:249–250.

Young, M., Collins, M. E., Oultram, J. D., and Pennock, A., 1986, Genetic exchange and prospects for cloning in clostridia, in: *Bacillus Molecular Genetics and Biotechnology Applications* (A. T. Ganesan and J. A. Hoch, eds.), Academic Press, London, pp. 259–281.

Youngleson, J. S., Santangelo, J. D., Jones, D. T., and Woods, D. R., 1988, Cloning and expression of *Clostridium acetobutylicum* alcohol dehydrogenase gene in *Escherichia coli, Appl. Environ. Microbiol.* **54**:676–682.

Yu, P.-L., and Pearce, L. E., 1986, Conjugal transfer of streptococcal antibiotic resistance plasmids into *Clostridium acetobutylicum, Biotechnol. Lett.* **8**:469–474.

Zappe, H., Jones, D. T., and Woods, D. R., 1986, Cloning and expression of *Clostridium acetobutylicum* endoglucanase, cellobiase and amino acid biosynthesis genes in *Escherichia coli, J. Gen. Microbiol.* **132**:1367–1372.

Zappe, H., Jones, D. T., and Woods, D. R., 1987, Cloning and expression of a xylanase gene from *Clostridium acetobutylicum* P262 in *Escherichia coli, Appl. Microbiol. Biotechnol.* **27**:57–63.

Zappe, H., Jones, W. A., Jones, D. T., and Woods, D. R., 1988, Structure of an endo-β-1,4-glucanase gene from *Clostridium acetobutylicum* P262 showing homology with endoglucanase genes from *Bacillus* spp., *Appl. Env. Microbiol.* **54**:1289–1292.

Solvent Production

4

DAVID T. JONES and DAVID R. WOODS

1. INTRODUCTION

Fermentation processes using anaerobic microorganisms provide a potential route for the conversion of plant biomass and wastes from agriculture and industry to chemical feedstocks and fuels (Wiegel, 1980; Zeikus, 1980; Rogers, 1984). However, only a few industrial fermentation processes exist which utilize single species of anaerobic microorganisms for the production of acids and alcohols. The need to develop novel processes which are efficient and less energy-intensive, and which can compete economically with processes employing chemical synthesis, represents a major challenge for the biotechnology industry. Of the existing industrial fermentation processes, the production of bioethanol has achieved the greatest success and attracted the most attention. The only large-scale industrial fermentation utilizing anaerobic bacteria which has made a significant contribution to the production of chemical feedstocks is the acetone/butanol/ethanol fermentation (ABE fermentation) using *Clostridium acetobutylicum* strains (Jones and Woods, 1986). The ABE fermentation was the major route used for the production of these solvents during the first part of the century and up until the early 1960s was able to compete successfully with synthetic processes.

The ability of many species of *Clostridium* to produce significant yields of ethanol and butanol has continued to focus interest on these species as potential industrial fermentation organisms (Wiegel, 1980; Zeikus, 1980, 1985; Zeikus *et al.*, 1981; Rogers, 1984; Zeikus, 1985). Most saccharolytic species of *Clostridium* are able to catabolize a wide range of hexose and pentose sugars to produce a variety of acids and neutral solvents. Many species are also able to hydrolyze polysaccharides such as starch, pectin, inulin, and xylan and a number of species are able to degrade hemi-

DAVID T. JONES and DAVID R. WOODS ● Microbiology Department, University of Capetown, Rondebosch 7700, Cape Town, South Africa. *Present address for D. T. J.*: Department of Microbiology, University of Otago, Dunedin, New Zealand.

cellulose and crystalline cellulose (Gottschalk *et al.*, 1981; Zeikus, 1985). Thermophilic species which are capable of producing high yields of ethanol have been identified as having potential in industrial fermentation processes as the operation of fermentation processes at elevated temperatures provides a number of advantages (Wiegel, 1980; Zeikus *et al.*, 1981).

At present, anaerobic fermentation processes for the production of fuels and chemicals suffer from a number of serious limitations including low yields, low productivity, and low final product concentrations. Unless some of these limitations can be overcome, it is unlikely that the fermentation route will become competitive.

2. SOLVENT-PRODUCING CLOSTRIDIA

2.1. Ethanol-Producing Species

Many facultative and obligate anaerobic bacteria produce varying amounts of ethanol as a subsidiary end product of fermentation. Only a few species are capable of fermenting hexose sugars stereometrically to produce 2 moles of ethanol and 2 moles of CO_2 as their major end product. Ethanol may be produced via a number of different biochemical pathways. A few species such as *Zymomonas mobilis*, *Sarcina ventriculi*, and *Erwinia amylovorons* utilize pyruvate decarboxylase to convert pyruvate generated by means of the Embden–Meyerhof or Entner–Doudoroff pathways to CO_2 and acetaldehyde, which is then reduced to ethanol. The heterofermentative lactic acid bacterium *Leuconostoc mesenteroides* produces ethanol by the reduction of acetyl phosphate, which is generated by glucose metabolism via the pentose phosphate pathway. A number of different groups of bacteria including clostridia produce ethanol via the reduction of acetyl-CoA which is generated by the cleavage of pyruvate produced during glycolysis.

The production of varying amounts of ethanol along with CO_2, H_2, acetate, and lactate as end products is widespread among both the mesophilic and thermophilic species of saccharolytic *Clostridium* (Holdeman *et al.*, 1977; Corry, 1978; Gottschalk *et al.*, 1981). At least 30 species of *Clostridium* have been reported to produce ethanol in amounts varying from traces to close to the theoretical maximum of 2 moles of ethanol per mole of glucose fermented. In most species, both the nature and concentration of the fermentation substrate and the fermentation conditions, such as pH and temperature, can drastically alter the amount of ethanol produced. Yields of ethanol ranging from 1.7 to 1.9 moles/mole of hexose fermented have been reported for a number of mesophilic species such as *C. spo-*

rogenes, C. sordelii, and *C. indolis* when grown in complex media containing glucose (Corry, 1978). *Clostridium sphenoides* and *C. sordeli* have also been reported to produce significant amounts of ethanol when grown in complex media in the absence of added sugars. More recently, *C. saccharolyticum,* which has the ability to utilize a wide variety of carbohydrates when grown in complex media, was reported to produce yields of up to 1.8 moles of ethanol/mole of glucose (Murray and Khan, 1983a). Acidic conditions and an increased partial pressure of H_2 have been reported to enhance ethanol production in this species (Murray and Khan, 1983b). Some mesophilic cellulolytic species of *Clostridium* such as *C. cellobioparum* (Hungate, 1944), *C. cellulolyticum* (Petitdemange *et al.,* 1984), and *C. chartatbidium* (Kelly *et al.,* 1987) produce limited amounts of ethanol.

Amino acids can also serve as substrate for ethanol production, but yields tend to be very low with a yield of 0.57 moles/mole of serine by *C. botulinum* being amongst the highest (Corry, 1978). *Clostridium sporogenes,* which is both a saccharolytic and proteolytic species, produces high yields of ethanol from glucose, but when amino acid substrates are fermented by means of the Stickland reaction, little or no ethanol is produced (Holdeman *et al.,* 1977; Gottschalk *et al.,* 1981). Low yields of ethanol are also produced by some clostridia during growth on various di- and tricarboxylic acids, such as fumarate, malate, and citrate (Walther *et al.,* 1977; Antranikian *et al.,* 1984).

Wiegel (1980) pointed out that a high proportion of the known anaerobic thermophiles and extreme thermophiles produce ethanol from sugars as the major fermentation product and he suggested that the reason for the widespread production of this end product could be that it is a volatile neutral compound which would evaporate readily at elevated temperatures, thereby minimizing the problems caused by end product accumulation.

Among the saccharolytic thermophilic clostridia which produce ethanol as an end product, four species—*C. thermosaccharolyticum, C. thermohydrosulfuricum, C. thermosulfurogenes,* and *C. thermocellum*—have been the most intensively studied. During vegetative growth, *C. thermosaccharolyticum* produces mainly CO_2, H_2, acetate, and butyrate with some lactate being produced under certain conditions. However, when nongrowing cells are held in continuous dilution culture or during the shift to endospore formation, this organism undergoes a shift to ethanol production (Hsu and Ordal, 1970; Landuyt *et al.,* 1983). A mutant defective in acetate production which produced higher yields of ethanol than the parental strain has been isolated (Rothstein, 1986).

Clostridium thermohydrosulfuricum was identified as one of the thermophilic anaerobes capable of producing the highest yields of ethanol (Zeikus, 1979; Zeikus *et al.,* 1980; Ng *et al.,* 1981; Lovitt *et al.,* 1984). This bacterium appears to be widespread in nature and has been isolated from a variety of thermophilic marine and terrestrial environments (Wiegel *et al.*, 1979; Zeikus, 1979; Zeikus *et al.*, 1980). *Clostridium thermohydrosulfuricum* is able to ferment a wide range of carbohydrates including starch, maltose, cellobiose, fructose, sucrose, lactose, mannose, and xylose to produce varying amounts of CO_2, ethanol, acetate, lactate, and H_2 (Wiegel *et al.*, 1979; Zeikus *et al.*, 1980; Parkkinen, 1986). Optimal growth occurs between 67 and 70°C with a doubling time of 70–90 min (Wiegel, 1980). The yield of ethanol produced varies from 0.5 to 1.8 moles/mole of glucose, depending on the strain and culture conditions. Ethanol production is influenced by pH, temperature, substrate composition and concentration (Hyun and Zeikus, 1985; Parkkinen, 1986).

Clostridium thermosulfurogenes is a recently isolated thermophilic species which exhibits a similar substrate range and produces end products similar to those of *C. thermosulfuricum* (Schink and Zeikus, 1983). This species produces elemental sulfur from thiosulfate which is deposited on the cell surface during growth. When grown on pectin it produces methanol and isopropanol in addition to other end products.

Clostridium thermocellum is able to grow on cellulose as a sole carbon source and is widely distributed in decaying organic material, where it is often found in a stable association with other anaerobes, including ethanologens and methanogens (Wiegel, 1980). Crystalline cellulose is degraded by a complex of extracellular cellulases to produce soluble sugars (Lamed *et al.*, 1983; Bayer *et al.*, 1985; Hon-nami *et al.*, 1986). Although cellobiose is the principle sugar utilized, many strains grow well on glucose but the utilization of other sugars is limited (Patni and Alexander, 1971; Ng *et al.*, 1977; Ng and Zeikus, 1982; Brener and Johnson, 1984). The end products produced by *C. thermocellum* are similar to those produced by *C. thermohydrosulfuricum*, but the ratio in which they are produced differs significantly and the yields of ethanol produced are low (Ng and Zeikus, 1982). Although *C. thermocellum* has the potential for the direct conversion of cellulose to ethanol, the slow growth rate coupled with the limited production and tolerance to ethanol has limited the development of a practical fermentation process.

2.2. Butanol-Producing Species

Among the clostridia the ability to produce butanol along with varying amounts of acetone, isopropanol, and ethanol is restricted to saccharolytic mesophilic species which are able to produce butyric acid as an end product (Holdeman *et al.*, 1977; Gottschalk *et al.*, 1981). In most species, the production of solvents only occurs late in the fermentation cycle following a shift from the pathways leading to acetate and butyrate production. The production of butanol and ethanol is usually associated with the uptake and reutilization of acids and the production of acetone or isopropanol. However, the ability to produce solvents is not a stable trait, and strains which have the ability to produce significant amounts of solvents when first isolated frequently undergo degenerative changes resulting in the loss of ability to produce solvents (McCoy and Fred, 1941; Kutzenok and Aschner, 1952; Finn and Nowrey, 1958; Hartmanis et al., 1986; Adler and Crow, 1987). The ability to produce solvents is also influenced by the type and concentration of substrate, the pH and buffering capacity of the culture medium, and the culture conditions. The cultivation of cells under high partial pressures of H_2 also causes a shift to solvent production in some species such as in C. roseum and C. rubrum (Hugo et al., 1972). Due to the variation in the amount and the ratio of the neutral end products, solvent production cannot be used as a reliable taxonomic criterion.

A number of species which produce butyric acid as their major fermentation product, such as C. butyricum, C. cadaveros, C. cochlearium, C. felsineum, C. pasteurianum, C. roseum, C. rubrum, C. sporogenes, and C. tryobutyricum, have been reported to produce limited amounts of solvents under the appropriate growth conditions. Strains of clostridia which are capable of producing significant amounts of butanol and other neutral products fall into at least four distinct groups based on their DNA homology and other characteristics, and include C. acetobutylicum, C. beijerinkii, C. aurantibutyricum, and C. tetanomorphum (Cummins and Johnson, 1971; George et al., 1983; Gottwald et al., 1984; Chen and Hiu, 1986). Strains of C. acetobutylicum which characteristically produce butanol, acetone, and ethanol in the ratio of 6 : 3 : 1 have been used extensively for the industrial production of solvents and have been the most intensively studied (Jones and Woods, 1986). In the past, numerous other species names were given to patent strains used for the industrial production of solvents, and the haphazard and arbitrary way in which the nomenclature was applied led to considerable confusion regarding the classification of these strains.

Strains now classified as C. beijerinkii, which includes most strains previously classified as C. butylicum, constitute a second group of solvent producers which do not show DNA homology with the C. acetobutylicum group. This group contains both high and low solvent-producing strains which produce either acetone or isopropanol in addition to butanol (Chen and Hiu, 1986). The C. aurantibutyricum group also contains butanol-producing strains which produce both acetone and isopropanol (George et al., 1983). A strain which produced butanol and ethanol but no acetone or isopropanol has been assigned to the C. tetanomorphum group (Gottwald et al., 1984). At present, many ethanol- and butanol-producing strains of Clostridium remain poorly classified and there is still no accepted standard classification for the clostridia group as a whole (Gottschalk et al., 1981).

3. BIOCHEMISTRY AND REGULATION OF SOLVENT PRODUCTION

3.1. Fermentation Strategies and Pathways

All saccharolytic solvent-producing clostridia which have been studied use the fructose biphosphate pathway (Embden–Myerhof pathway) for the metabolism of hexose sugars. During glycolysis, 1 mole of hexose produces 2 moles of pyruvate with the net production of 2ATP and 2NADH. In addition to utilizing hexose sugars, many species are capable of fermenting pentose sugars via the hexose monophosphate pathway (Warburg–Dickens pathway), using the transaldolase and transketolase reactions. The fermentation of 3 moles of pentose results in the production of 2 moles of fructose-6-phosphate and 1 mole of glyceraldehyde-3-phosphate with the net production of 5 moles of ATP and 5 moles of NADH.

In all solvent-producing clostridia, the pyruvate generated by glycolysis can be cleaved into CO_2 and acetyl-CoA by means of pyruvate ferredoxin oxidoreductase, resulting in the production of reduced ferredoxin. The reduced ferredoxin can then be regenerated by the action of hydrogenase, resulting in the production of H_2. Under the appropriate conditions most saccharolytic clostridia are able to utilize an alternative pathway to metabolize pyruvate to lactate by means of lactic dehydrogenase. The production of formate by some species of clostridia indicates that these organisms are also able to cleave pyruvate to acetyl-CoA and formate by means of pyruvate-formate lyase.

Due to the small number of ATP molecules generated during fermentation, the growth of anaerobic bacteria utilizing this mechanism for energy generation can be considered to be limited by the rate of the energy-yielding reactions. Obligate anaerobes appear to be optimally adapted to grow efficiently under these energy-stressed conditions. The fermentation strategies used, and the type of end products produced, can be considered to be adaptations to allow effective growth under a particular set of environmental conditions. Depending on the pathways used, the fermentation of a mole of glucose can generate between 1 and 4 moles of ATP. Those clostridia which ferment glucose to produce lactate or ethanol and CO_2 have a net production of 2 moles of ATP. The use of acetate kinase to produce acetate results in the generation of an additional 2 moles of ATP/mole of glucose metabolized, and the use of butyrate kinase to produce butyrate results in the generation of an extra mole of ATP. However, the ability to divert acetyl-CoA to regenerate ATP by means of acetate and butyrate production is directly linked to the ability of the cell to evolve H_2. In the absence of H_2 production, the fermentation substrate must serve as both H_2 donor and H_2 acceptor. An intermediate product of substrate degradation, such as pyruvate or acetyl-CoA, acts as an acceptor for H_2

from $NADH_2$ and the reduced end products such as butyrate, lactate, ethanol, or butanol are then excreted by the cell.

To divert all of the acetyl-CoA to acetate, the NADH produced during glycolysis must be regenerated by the transfer of H_2 to ferredoxin, which can then be released as H_2 by hydrogenase. However, the equilibrium of this reaction is unfavorable as it involves a change to a more negative redox potential of H_2 and can only occur if the partial pressure of H_2 is kept very low. In nature this reaction occurs only in species which exist in symbiotic association with H_2-consuming species, which permits interspecific hydrogen transfer. This allows all of the acetyl-CoA to be converted to acetate, generating a net total of 4 moles of ATP. Free-living species are unable to dispose of all their excess reducing power in this way and must produce some reduced end products in the form of acids or alcohols, which results in a reduction in the amount of ATP which can be generated. This means that a direct relationship exists between the amount of H_2 produced and the amount of ATP which is regenerated. The saccharolytic clostridia utilize branched pathways whereby the reduction of acetate, which results in ATP production but no additional consumption of reducing power, is linked to the production of either ethanol or butyrate, which allows the disposal of the excess reducing equivalents. Many species of *Clostridium* which employ the branched acetate/butyrate pathway are able to switch to using the butanol pathway when the concentration of acid end products becomes inhibitory. The routes of carbon and electron flow are directly dependent on the amount of H_2 which can be produced (see Table I). As the amount of H_2 produced decreases, the cell must direct more of the

Table I. Comparison of Energy Generation and the Production and Disposal of Reducing Power by Various Catabolic Pathways Utilized by the Solvent-Producing Clostridia (moles/mole glucose)

Pathway	Net amount of ATP produced	Net amount of reducing equivalents produced as NAD(P)H or reduced ferredoxin	Net amount of reducing equivalents disposed of by the pathway
Glycolysis via EMP	2	2	0
Pyruvate cleavage	0	2	0
Hydrogen production	0	0	0–4
Acetate production	2	0	0
Butyrate production	1	0	2
Lactate production	0	0	2
Ethanol production	0	0	2
Butanol production	0	0	4

carbon flow to produce reduced end products such as butyrate, ethanol, or butanol. When H_2 production is blocked, many species switch to lactate production, which can quantitatively consume the NADH produced during the metabolism of glucose to pyruvate.

3.2. Regulation of Ethanol Production

Most ethanol-producing clostridia form similar end products. However, the proportion in which they are produced is dependent on the particular species or strain and the culture conditions (Wiegel, 1980; Zeikus et al., 1981; Zeikus, 1985). The nature and the concentration of the carbon source, the nitrogen source, and other nutrients have been shown to affect ethanol production in some species. Growth conditions such as temperature, pH, the amount of agitation, and the partial pressure of H_2 in the culture vessel may also have a marked effect. The yield and proportion of end products is dependent on the specific activity and regulation of the enzymes of the different biochemical pathways. The key enzymes involved in ethanol formation are shown in Figure 1, a number of which are subject to inhibition or activation by end products or metabolic intermediates.

Four oxidoreductases, which interact with ferredoxin, play an important role in controlling the electron flow, which determines the proportion of reduced end products produced by the cell (Fig. 1). Pyruvate ferredoxin oxidoreductase is responsible for the generation of reduced ferredoxin during the cleavage of pyruvate. In many species of *Clostridium*, the primary role of NADH–ferredoxin oxidoreductase, which requires acetyl-CoA as an activator (Jungermann et al., 1973), is the regeneration of NAD^+ by the production of reduced ferredoxin. This reaction involves a change to a more negative redox potential from E'_0 -320 mV to E'_0 -420 mV, and is unfavorable for the generation of reduced ferredoxin and the subsequent evolution of H_2. It only appears to take place in cells growing under optimal conditions. In most species the reverse reaction appears to be limited and is strongly inhibited by NADH (Jungermann et al., 1973).

Since most clostridia appear to lack the enzymes required for the oxidation of glucose-6-phosphate to generate NADPH, it has been assumed that the major role of NADPH–ferredoxin oxidoreductase is for the production of NADPH for biosynthesis. However, the presence of NADPH-specific alcohol dehydrogenases in many species suggests that this enzyme may play a role in the diversion of reducing power to form ethanol when the cell is unable to produce H_2. Ferredoxin–H_2 oxidoreductase (hydrogenase), which is responsible for the regeneration of reduced ferredoxins, has not been well characterized in the ethanol-producing species, and varies in specific activity and regulatory properties.

In *C. thermosaccharolyticum*, which is capable of producing ethanol yields

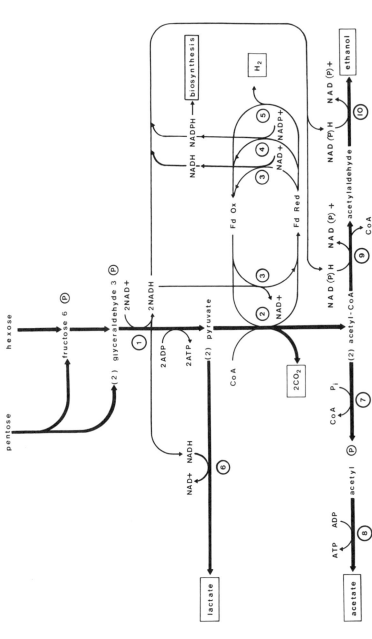

Figure 1. Catabolic pathways used by the ethanol-producing clostridia. Heavy arrows indicate the direction of carbon flow and light arrows the direction of electron flow. Enzymes are indicated by numbers as follows: (1) glyceraldehyde-3-phosphate dehydrogenase; (2) pyruvate–ferredoxin oxidoreductase; (3) NADH–ferredoxin oxidoreductase; (4) NADPH–ferredoxin oxidoreductase; (5) hydrogenase; (6) lactate dehydrogenase; (7) phosphate acetyltransferase (phosphotransacetylase); (8) acetate kinase; (9) acetaldehyde dehydrogenase; (10) ethanol dehydrogenase.

as high as 1.6–1.8 moles, very little H_2 and acetate are produced and increases in the partial pressure of H_2 inhibit growth (Zeikus, 1979; Zeikus *et al.*, 1980; Ng *et al.*, 1981; Lovitt *et al.*, 1984; Parkkinen, 1986). The bacterium is able to transfer significant quantities of electrons from reduced ferredoxin produced by pyruvate cleavage to generate both NADPH and NADH via the action of ferredoxin NAD(P)H oxidoreductase. *Clostridium thermohydrosulfuricum* contains both NADH- and NADPH-linked acetaldehyde and ethanol dehydrogenase (Lamed and Zeikus, 1980; Zeikus, 1985). The presence of a reversible NADPH-linked alcohol ketone/aldehyde oxidoreductase results in end product inhibition at relatively low concentrations of ethanol. Increases in both the partial pressure of H_2 and ethanol accumulation can readily reverse electron flow causing an increase in the pool of reduced ferredoxin and the accumulation of high levels of NAD(P)H, which finally results in the cessation of carbon flow (Zeikus, 1985).

On the other hand, most strains of *C. thermocellum* exhibit relatively high hydrogenase activity (Table II). They are able to dispose of a substantial amount of reducing power as H_2, allowing a large proportion of the acetyl-CoA to be converted to acetate and the generation of more ATP. In *C. thermocellum,* an increase in the partial pressure of H_2 has little effect on growth and H_2 production (Lamed and Zeikus, 1980; Zeikus, 1985). The NADH–ferredoxin oxidoreductase activity in *C. thermocellum* is very low and operates only in the catabolic direction (Lamed and Zeikus, 1980). NADPH–ferredoxin oxidoreductase activity operates in the reverse direction but probably only has a biosynthetic role, as *C. thermocellum* has been reported to contain only an NADH-specific ethanol dehydrogenase, which is undirectional and which is inhibited by ethanol and NAD (Zeikus, 1985). This suggests that this organism is not able to redirect the reducing power generated during pyruvate cleavage to produce ethanol, limiting ethanol yields to less than 1 mole.

Most ethanol-producing species of clostridia growing under optimal conditions convert the major portion of the pyruvate to acetyl CoA with only minor amounts being reduced to lactate. However, under conditions where the cleavage of pyruvate is limited by the rate of ferredoxin re-

Table II. Comparison of Fermentation Products of *C. thermocellum* and *C. thermohydrosulfuricum*[a] (mmoles/100 mmoles hexose)

Species	CO_2	H_2	Acetate	Lactate	Ethanol
C. thermocellum LQRI	148	122	71	10	67
C. thermohydrosulfuricum 39E	207	11	11	2	194

[a]From Ben-Bassat *et al.*, 1980.

generation, substantial amounts of lactate may be produced (Ng and Zeikus, 1982; Germain *et al.*, 1986). In both ethanol- and butanol-producing species, fructose-1,6-diphosphate has been identified as the activator of lactate dehydrogenase (Germain *et al.*, 1986; Turunen *et al.*, 1987). Conditions which result in an increase in the pol of NADH lead to the inhibition of glyceraldehyde-3-phosphate dehydrogenase activity, resulting in an accumulation of fructose-1,6-diphosphate. Activation of lactate dehydrogenase permits the diversion of pyruvate to lactate with a corresponding decrease in ethanol and acetate production (Germain *et al.*, 1986; Turunen *et al.*, 1987). The production of lactate by *C. thermohydrosulfuricum* from glucose was increased under carbon- or nitrogen-limited conditions (Ng and Zeikus, 1982). Lactate production was also increased in xylose fermentations by increasing the dilution rate or by increasing nitrogen levels.

3.3. Regulation of Butanol Production

The ability to produce butanol appears to be restricted to species which are able to produce butyrate as a fermentation product. The majority of solvent-producing strains produce CO_2, H_2, acetate, and butyrate during the initial growth phase in batch culture. The shift from acid production to solvent production only occurs during the second phase of the fermentation and is accompanied by a decrease in H_2 production and growth rate, as cells enter the stationary phase. The formation of butanol by strains of *C. acetobutylicum*, *C. aurantibutyricum*, and *C. beijerinckii* is accompanied by the formation of acetone or isopropanol, which is linked to the uptake and reutilization of acids. The formation of butanol accompanied by the production of H_2 and acetate has, however, been reported to occur during exponential growth in some strains of *C. tetanomorphum* (Gottwald *et al.*, 1984). In these strains, butanol formation was not accompanied by the formation of acetone or isopropanol.

The biochemical pathways used for acid and solvent production in *C. acetobutylicum* have been well characterized (Davies, 1943; Doelle, 1975; Gottschalk, 1979; Andersch *et al.*, 1983; Hartmanis and Gatenbek, 1984; Hartmanis *et al.*, 1984; Ballongue *et al.*, 1986; Hartmanis, 1987; Dürre *et al.*, 1987; Matta-El-Ammouri *et al.*, 1987) (Fig. 2).

In *C. acetobutylicum*, the switch from acid production to solvent production is characterized by a decrease in activity in all of the acid pathway enzymes with the exception of butyrate kinase, and the induction of the solvent pathway enzymes, which involves a shift in both carbon and electron flow (Hartmanis *et al.*, 1984; Ballongue *et al.*, 1985, 1986; Dürre *et al.*, 1987). During the acidogenic phase, the cell is able to regenerate the reduced ferredoxin produced during pyruvate cleavage by the production of H_2. Part of the reducing equivalent in the form of NADH which is produced during glycolysis can also be disposed of through the activity of

116   DAVID T. JONES and DAVID R. WOODS

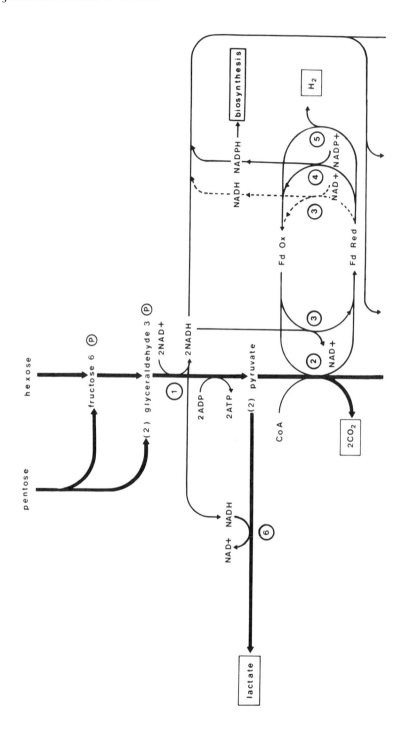

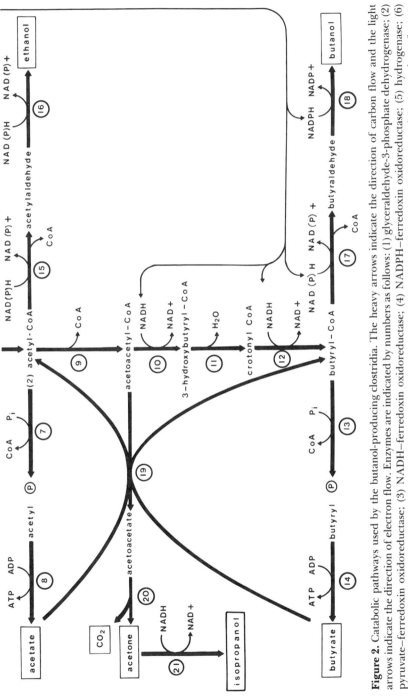

Figure 2. Catabolic pathways used by the butanol-producing clostridia. The heavy arrows indicate the direction of carbon flow and the light arrows indicate the direction of electron flow. Enzymes are indicated by numbers as follows: (1) glyceraldehyde-3-phosphate dehydrogenase; (2) pyruvate–ferredoxin oxidoreductase; (3) NADH–ferredoxin oxidoreductase; (4) NADPH–ferredoxin oxidoreductase; (5) hydrogenase; (6) lactate dehydrogenase; (7) phosphate acetyltransferase (phosphotransacetylase); (8) acetate kinase; (9) thiolase (acetyl-CoA acetyltransferase); (10) 3-hydroxybutyryl-CoA dehydrogenase; (11) crotonase; (12) butyryl-CoA dehydrogenase; (13) phosphate butyltransferase (phosphotransbutyrylase); (14) butyrate kinase; (15) acetaldehyde dehydrogenase; (16) ethanol dehydrogenase; (17) butyraldehyde dehydrogenase; (18) butanol dehydrogenase; (19) acetoacetyl-CoA : acetate/butyrate : CoA transferase; (20) acetoacetate decarboxylase; (21) isopropanol dehydrogenase.

NADH–ferredoxin oxidoreductase and hydrogenase to produce H_2. This enables the cell to metabolize a significant proportion of the acetyl-CoA to form acetate, allowing regeneration of additional ATP. The remainder is utilized for butyryl CoA synthesis, resulting in the consumption of 2 molecules of NADH. Butyryl CoA is then metabolized to butyrate, resulting in the regeneration of another molecule of ATP (Table I). The use of this branched acid-producing pathway enables the net generation of around 3.25 moles of ATP/mole of glucose fermented (Rogers, 1984).

In common with the ethanol-producing clostridia, the butanol-producing species also possess an inducible unidirectional NADH-dependent lactate dehydrogenase enzyme (Freier and Gottschalk, 1987). In *C. acetobutylicum,* this enzyme is also activated by fructose-1,6-biphosphate and the pathway becomes operational under conditions where the cleavage of pyruvate to acetyl-CoA is inhibited. Lactate has been shown to be produced as the major fermentation product when cells are grown under limiting conditions of iron and sulfur above pH 5 (Bahl *et al.,* 1986), or when H_2 production is limited by inhibition of hydrogenase activity by CO or potassium cyanide (Simon, 1947; Kim *et al.,* 1984; Datta and Zeikus, 1985). Under these conditions pyruvate cleavage appears to be inhibited by the lack of ferredoxin to act as the electron carrier for this reaction.

At the end of the acidogenic phase, H_2 production declines sharply (Kim and Zeikus, 1984), suggesting that the cell is no longer able to dispose of its excess reducing power by this route. Under these conditions, the cell is able to continue to regenerate reduced ferredoxin by the formation of NAD(P)H using NAD(PH)H–ferredoxin oxidoreductase which can be disposed of via the formation of butanol. The shift to butanol production enables the cell to regenerate 4 moles of NAD(P) for each mole of glucose metabolized, but results in a reduction in the net amount of ATP produced to 2 moles.

The formation of butanol from butyryl-CoA involves two reduction steps which are catalyzed by butyraldehyde dehydrogenase and butanol dehydrogenase. These enzymes have proved to be difficult to detect and assay. The published data regarding the optimal conditions for activity and specificity of the coenzymes required is conflicting, and both NADH- and NADPH-dependent butanol dehydrogenase activities have been reported in cell extracts of *C. acetobutylicum* (Andersch *et al.,* 1983; George and Chen, 1983; Rogers, 1984). Recently, reliable assay techniques were described and enzymes from *C. beijerinckii* and *C. acetobutylicum* purified and characterized (Dürre *et al.,* 1987; Hiu *et al.,* 1987). Hiu *et al.* (1987) compared the alcohol dehydrogenases obtained from an acetone-producing strain of *C. beijerinckii* with that from an isopropanol-producing strain. A NADPH-linked dehydrogenase with a mol. wt. of 66,000 which was able to catalyze the final reaction for both butanol and ethanol, but not isopropanol, was obtained from the acetone-producing strain. The isopropanol-producing strain yielded an alcohol dehydrogenase of 100,000 mol. wt. which was also NADPH-dependent and was able to catalyze the final reaction for butanol, ethanol, and isopropanol. The authors concluded that these enzymes represented the major, and possibly the only, alcohol-forming enzymes in the two strains. However, Dürre *et al.* (1987) were able to characterize two alcohol dehydrogenases from *C. acetobutylicum* which could be separated by differential centrifugation. One enzyme which was partly sedimented by ultracentrifugation utilized NADH and reacted 1.7-fold faster with butyraldehyde than acetaldehyde. The other enzyme was not sedimented by ultracentrifugation, was NADPH-specific, and reacted 2.4-fold faster with butyraldehyde than acetaldehyde. Both enzymes were induced shortly before the onset of solvent production. It is not clear whether the NADH-specific enzyme is responsible specifically for ethanol formation or whether the two enzymes are involved in both butanol and ethanol production. Recently, Youngleson *et al.* (1988) reported the cloning of the NADPH-dependent alcohol dehydrogenase gene from *C. acetobutylicum* in *E. coli* This gene was not regulated from its own promoter, but expression was controlled by a downstream promoter located within the vector. Although the cloned enzyme showed a higher reaction rate with butanol than with ethanol, it resulted in enhanced ethanol formation in *E. coli*. The nucleotide sequence of the gene contained an open reading frame of 1160 nucleotides which encoded a polypeptide of 385 amino acid residues with a calculated mol. wt. of 40,000. The amino acid sequence of this alcohol dehydrogenase showed 35% homology with an NADH-specific alcohol dehydrogenase from *Z. mobilis* (Conway *et al.*, 1987) and the alcohol dehydrogenase IV from *Saccharomyces cerevisiae* (Williamson and Paquin, 1987).

The use of a variety of techniques for the isolation of allyl alcohol-resistant mutants of *C. acetobutylicum* has yielded three distinct classes of mutants which are defective in butanol production. One class of mutants produce normal levels of acetone, reduced levels of butanol, and little or no ethanol. These mutants all excreted significant amounts of butyraldehyde and appeared to be defective in alcohol dehydrogenase activity, but exhibited normal levels of aldehyde dehydrogenase activity (Rogers and Palosaari, 1987). A second class of mutants produced normal or reduced levels of ethanol and acetone, but no butanol. These mutants showed normal levels of butanol dehydrogenase activity but very reduced levels of butyraldehyde activity (Dürre *et al.*, 1986). A third class of mutants showed reduced production of all solvents (Rogers and Palosaari, 1987) and appear to be similar to the regulatory type of mutants, which are defective in both solvent production and sporulation, which were reported previously (Jones *et al.*, 1982; Long *et al.*, 1984a).

The butyraldehyde dehydrogenase from *C. acetobutylicum* has also been purified and characterized. This enzyme has a high specificity of NADH but has also been reported to react with NADPH (Durre *et al.*,

1987; Palosaari and Rogers, 1988). This enzyme also exhibits a higher specificity for butyryl-CoA than acetyl CoA. The enzyme was coinduced with butanol dehydrogenase during the switch to solvent production, but it has been reported to decay rapidly. Palosaari and Rogers (1988) postulated that this enzyme alone is responsible for both butanol and ethanol formation. However, mutants which were defective in butyraldehyde dehydrogenase activity produce normal levels of ethanol (Dürre *et al.*, 1986), and ethanol production in the absence of butanol production has been reported to occur under certain culture conditions, suggesting that at least two different enzymes are present in *C. acetobutylicum.*

In *C. acetobutylicum* and similar species, the onset of butanol production is associated with the production of acetone or isopropanol and the uptake and reutilization of acids. Recent studies have shown that the consumption of acids which occur during solventogenesis is directly coupled to the formation of acetone or isopropanol, and is dependent on the continued metabolism of sugars (Hartmanis *et al.*, 1984; Matta-El-Ammouri *et al.*, 1985). It has been suggested that this process functions as a detoxification mechanism, which brings about a decrease in the intracellular concentration of dissociated acids and allows acids to diffuse back into the cell, resulting in a reduction of acid concentration and an increase in the pH of the external medium. It has been established that *C. acetobutylicum* utilizes acetoacetyl-CoA:acetate/butyrate:CoA transferase to convert acetoacetyl-CoA to acetoacetate, which is then decarboxylated in an irreversible step to acetone by acetoacetate decarboxylase (Davies, 1943; Doelle, 1975; Andersch *et al.*, 1983). In some strains of *C. beijerinckii,* acetone is then reduced to isopropanol by the action of a NADPH-dependent isopropanol dehydrogenase (Hiu *et al.*, 1987).

The reassimilation of acids occurs only when the appropriate enzymes have been induced. The CoA transferase has a broad specificity and can use either acetate or butyrate as the CoA acceptor, resulting in the direct conversion to acetyl- or butyryl-CoA (Hartmanis *et al.*, 1984; Matta-El-Ammouri *et al.*, 1987). This reaction is energetically favorable to the cell, as the energy in the thioester bond is conserved and transferred to the acids without the requirement for ATP. The reaction results in the formation of equimolar amounts of acetone and acetyl/butyryl-CoA, and the resulting diversion of half of the acetoacetyl-CoA to acetone production requires that the cell metabolize 2 moles of glucose for each mole of acid consumed (Jones and Woods, 1986). The proportion of the acids reassimilated determines the ratio of solvent produced. Butyrate is converted directly to butanol whereas only a portion of the acetate is converted to butanol. The addition of acetate to cells grown in excess glucose results in the enhancement of acetone production (Matta-El-Ammouri *et al.*, 1985), and from these and other studies it is apparent that the production of acetone and butanol are not directly linked and the two pathways can function with some degree of independence. Mutants which are defective in CoA transferase activity no longer produce acetone but continue to synthesize normal levels of butanol and ethanol (Janati-Idrissi *et al.*, 1987). Although it is possible that acids could also be taken up by a reversal of the acid-producing pathways in which ATP is consumed, enzyme studies have shown that, with the exception of butyrate kinase, these enzymes are inactive during the solventogenic phase, and the direct uptake of acids has been confirmed by nuclear magnetic resonance studies (Hartmanis *et al.*, 1984; Ballongue *et al.*, 1986; Hartmanis, 1987). However, Meyer *et al.* (1986) reported that the uptake of butyrate can occur without acetone production, although the mechanism of uptake was not established.

Although the factors responsible for the switch from acid production to solvent production in *C. acetobutylicum* have been the subject of extensive investigation, the mechanisms responsible for triggering solvent production have not been fully elucidated. The effects of nutrient composition and concentration and the role of end product accumulation have both been identified as key factors in the onset and maintenance of solvent production.

The ability of *C. acetobutylicum* to grow on chemically defined minimal media has facilitated studies on the role of nutrients on growth and solvent production in batch and continuous culture (Gottschal and Morris, 1981; Andersch *et al.*, 1982; Monot *et al.*, 1982; Long *et al.*, 1983; Monot and Engasser, 1983b). However, these investigations have in some cases produced conflicting results. Both the initiation and maintenance of solvent production have been shown to be dependent on the availability of a carbon source and a nitrogen source (Long *et al.*, 1984b; Bahl and Gottschalk, 1985). *Clostridium acetobutylicum* can utilize a variety of organic and inorganic nitrogen sources, but the type, concentration, and ratio of nitrogen has been reported to affect the amount and ratio of the solvents (Welsh *et al.*, 1986, 1987; Masion *et al.*, 1987). In continuous culture run under nitrogen-limited conditions, little or no solvent is produced under most conditions (Gottschal and Morris, 1981; Andersch *et al.*, 1982; Monot and Engasser, 1983a). However, phosphate and sulfate limitations have proved to be a suitable growth-limiting factor for obtaining solvent production in continuous culture (Bahl *et al.*, 1982, 1986). The use of magnesium limitation has given variable results and iron limitation induces a shift to lactate production (Bahl *et al.*, 1986). Solvent production has also been obtained in continuous culture using turbidostats (Gottschal and Morris, 1982) and product-limited chemostats where all nutrients are in excess (Monot and Engasser, 1983b; Afscher *et al.*, 1986; Clarke and Hansford, 1986). These studies established that the onset and maintenance of solvent production is not dependent on the limitation of one particular nutrient and that the attainment of optimal solvent production likely depends on all nutrients being present in excess.

In the industrial ABE fermentation, end product inhibition rather than nutrient limitation appeared to be responsible for the cessation of growth and solvent production. Inhibition of solvent production at the end of the solventogenic phase occurred as a result of butanol toxicity and is discussed in Section 3.4. The inhibition of growth at the end of the acidogenic phase and the shift to solventogenesis have also been linked to end product toxicity.

Hydrogen production decreases sharply at the end of the acidogenic phase. However, it has not been established whether this decrease occurs as a result of the shift to solvent production or whether a decrease in the ability to generate hydrogen is instrumental in inducing the shift to solventogenesis (Andersch *et al.*, 1983; Kim and Zeikus, 1984). An increase in the partial pressure of hydrogen, due either to lack of mechanical agitation or to increased pressure in the head space, results in a more rapid onset of solvent production (Doremus *et al.*, 1985; Yerushalmi *et al.*, 1985; Yerushalmi and Volesky, 1985). Inhibition of hydrogenase by CO or potassium cyanide or the addition of methyl viologen to cultures also results in a shift from the production of hydrogen and acids to solvent production (Datta and Zeikus, 1985; Meyer *et al.*, 1986; Rao and Mutharasan, 1987). In all cases, the inhibition of hydrogen production would lead to the accumulation of reduced ferredoxin and NAD(P)H in the cell and appears to result in a switch to solvent production in the absence of acid end product accumulation. The use of culture fluorescence to determine NAD(P)H concentration *in situ* has indicated that intracellular NAD(P)H levels are higher during the solventogenic phase than during the acidogenic phase (Reardon *et al.*, 1987; Srinivas and Mutharasan, 1987). In a recent study, Ballongue *et al.* (1987) investigated the toxic effect of the end product gases, using vacuum or continuous bubbling with nitrogen to remove the carbon dioxide and hydrogen. The results indicated that the accumulation of these gases during the fermentation is highly toxic and they have an effect on growth and metabolism which is comparable to that of 5 g/liter of acid end products.

During the first stage of the fermentation, acetate and butyrate are produced in the ratio of about 1 : 2. These acids are able to diffuse across the membrane in their undissociated form and in the external medium, resulting in a drop in pH. In regulatory mutants of *C. acetobutylicum* (*cls* mutants), which are unable to undergo the shift to solvent production, continued acid production results in the accumulation of up to 20 g/liter in the external medium before all metabolism ceases and cell viability is lost (Clarke, 1987). Low proton permeability of the cell membrane is essential for the maintenance of the proton motive force. In their undissociated form, weak acids such as acetic and butyric are able to act as uncouplers causing an increased membrane permeability to protons, which results in the acidification of the interior of the cell and the eventual collapse of the membrane pH gradient (Kell *et al.*, 1981; Terracciano and Kashket, 1986). Monot *et al.* (1983, 1984) reported that even small increases in the acid concentration resulted in an inhibition of the specific growth rate of cultures. In a recent study, Ballongue *et al.* (1987) utilized dialysis to remove both the acidic and gaseous end products during the fermentation. Under these conditions they were able to obtain a constant specific growth rate for a 6-hr period, in contrast to a varying rate in the control culture. The maximum doubling time in the dialyzed culture was 1.9 hr compared with 3.5 hr in the control culture and the production of cell dry mass was 20 g/liter, which was 16 times greater than that in the control cultures. These studies demonstrate that both the acidic and gaseous end products produced during the initial stage of the fermentation resulted in a severe inhibition of growth when present at low concentration.

Under most conditions, the transition from the acidogenic phase to the solventogenic phase is initiated when the concentration of the acids in the culture medium reaches 3–4 g/liter (Ballongue *et al.*, 1987). The shift to solvent production enables the cells to avoid the inhibitory effect produced by high concentrations of acids, and acid reutilization during solventogenesis would further decrease the toxic effects. In addition to the inhibitory effects of the acids, their accumulation also appears to induce solvent production. The effect of acetate and butyrate addition under various culture conditions in batch, fed-batch, and continuous culture has been reported by a number of workers (Gottschal and Morris, 1981; Andersch *et al.*, 1982; Monot *et al.*, 1983; George and Chen, 1983; Yu and Saddler, 1983; Long *et al.*, 1984b; Holt *et al.*, 1984; Fond *et al.*, 1985; Matta-El-Ammouri *et al.*, 1985, 1987). Although these studies have in some cases produced conflicting results, in most circumstances the presence of additional acid has lead to the induction or enhancement of solvent production (see Fig. 3). Monot *et al.* (1984) concluded that the concentration of undissociated acids in the culture medium played a critical role in solvent induction, although results from some other studies have not supported this view (George and Chen, 1983; Holt *et al.*, 1984). Recent studies by Terracciano and Kashket (1986) indicated that a minimum threshold in the internal concentration of the undissociated form of the acids is critical for the initiation of solvent production to occur. However, Gottwald and Gottschalk (1985) argued that it is the concentration of undissociated acids within the cell which plays the key role in induction.

The concentration of both the dissociated and the undissociated acids within the cell would be markedly influenced by the external concentration of the acids and pH differential between the inside and outside of the cell. A number of studies demonstrated that during the acidogenic phase the internal pH of the cell decreases in parallel with the decrease in the external pH, while a small pH gradient is maintained across the membrane (Bowles and Ellefson, 1985; Gottwald and Gottschalk, 1985; Huang *et al.*,

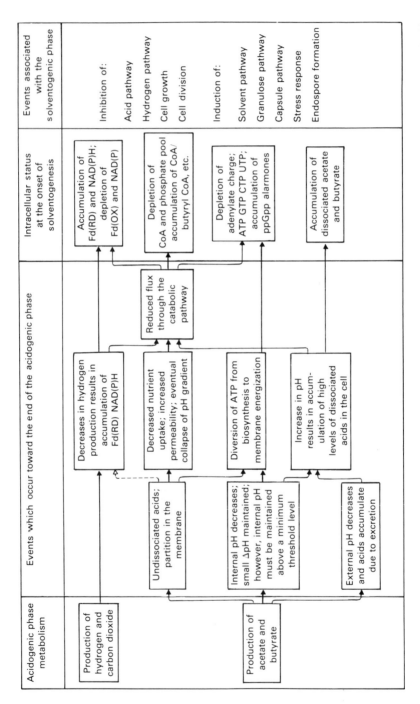

Figure 3. Model showing the possible relationship of events which occur toward the end of the acidogenic phase and events associated with the solventogenic phase.

1986; Terracciano and Kashket, 1986). However, it appears that the maintenance of the internal pH above a threshold value of at least pH 5.5 is essential to maintain cell function (Terracciano and Kashket, 1986). As the internal pH approaches the threshold level, it appears the cell would need to divert ATP from biosynthesis to membrane energization, resulting in an increase in the ΔpH across the membrane. Under these conditions, the concentration of undissociated and dissociated acids within the cell would increase rapidly and, since the catabolic pathways are reversible, would inhibit the flux through the pathways (Gottwald and Gottschalk, 1985). This would result in the depletion of the CoA and phosphate pools and lead to a further reduction in the adenylate charge in the cell, and would be accompanied by a further accumulation of the reduced ferredoxin and NAD(P)H in the cells (see Fig. 3). There is also evidence that the level of small nucleotides, which may act as alarmones, increases in these cells. It seems likely that one or more of these events are directly involved in the triggering of the various events which occur during the transition to solvent production. The solventogenic phase is characterized by the inhibition of the acid- and hydrogen-producing pathways, the induction of the pathways for solvent, granulose, and capsule production, and the production of stress-related proteins (Jones *et al.*, 1982; Andersch *et al.*, 1983; Kim and Zeikus, 1984; Long *et al.*, 1984a; Ballongue *et al.*, 1985, 1986; Reysenbach *et al.*, 1986; Hartmanis, 1987). In addition, the cell growth rate is decreased, cell division may be inhibited, and endospore formation may be induced. Little is known about the molecular mechanisms involved in the induction and repression of the various pathways either by the accumulation of acids or by changes in the concentrations of other key metabolites.

3.4. Solvent Toxicity and Tolerance

A major limitation affecting the use of clostridial species for ethanol and butanol production is the relatively low concentration of these products obtained during fermentation. The low final solvent concentration results in a high cost for recovery by distillation, the need for large-capacity fermenters, and a limitation on the amount of sugar fermented. Among wild-type alcohol-producing thermophilic clostridia, final ethanol concentrations of about 1% are obtained, although mutants with greater ethanol tolerance have been isolated (Lovitt *et al.*, 1984; Herrero *et al.*, 1985). Among the butanol-producing clostridia, butanol, which is the most hydrophobic and toxic of the neutral end products, is the only solvent produced in inhibitory concentrations (1.2–1.6%) during the fermentation (Reyden, 1958; Moreira *et al.*, 1981).

Studies on the mechanism of ethanol and butanol toxicity indicate that alcohols damage the structure of the cell membrane and thereby inhibit cellular processes (Ingram and Buttke, 1984; Ingram, 1986). The potency of these alcohols as inhibitors is directly related to their solubility in lipids, which appears to account for their action on the cell membrane. Ethanol and butanol are both amphipathic molecules, which allows them to partition in the hydrophobic region of the membrane bringing about an increase in the polarity of the hydrophobic core of the membrane. This appears to have a two-fold effect on the cell membrane. First, it results in a decrease in the efficiency of the membrane as a barrier and causes an increase in membrane leakage. Second, it results in a disruption of the organization of the phospholipids and proteins in the membrane. Although the basic modes of action of ethanol and butanol appear to show similarities, their specific modes of action appear to be somewhat different (Ingram, 1986).

The sensitivity to ethanol increases markedly with incubation temperature in both mesophilic and thermophilic species. When grown at the same temperature, *C. thermocellum* is inhibited by lower concentrations of ethanol than *C. thermohydrosulfuricum*, with a 50% reduction in growth rate by 0.5 and 1.5% (wt./vol.) ethanol, respectively, at 60°C (Herrero and Gomez, 1980; Herrero *et al.*, 1982; Lovitt *et al.*, 1984). Ethanol also caused the production of shorter chain length, monounsaturated and ante-isobranched chain fatty acids in *C. thermocellum* (Herrero *et al.*, 1985). It was suggested by Lovitt *et al.* (1984) that the higher ethanol tolerance of *C. thermohydrosulfuricum* is due to the abundance of membrane lipids containing C_{30} dicarboxylic fatty acids which are not present in *C. thermocellum*. The presence of these relatively uncommon long-chain dicarboxylic fatty acids in *C. thermohydrosulfuricum* is consistent with the hypothesis that both thermal and ethanol tolerance involves an increase in the fatty acid chain length and in the proportion of *cis*-monounsaturated fatty acids in the membrane lipids (Lovitt *et al.*, 1984).

Low ethanol tolerance in thermophilic clostridia is not only due to general solvent effects on membrane lipids but also involves ethanol-specific effects on metabolic processes. In *C. thermocellum*, ethanol has been reported to specifically inhibit glycolytic pathway enzymes, which result in an accumulation of sugar phosphates in the cell, as well as ethanol dehydrogenase and cellulase (Herrero, 1983; Herrero *et al.*, 1985). Lovitt *et al.* (1984) investigated an ethanol-tolerant mutant of *C. thermohydrosulfuricum* and concluded that direct modification of enzymes involved in controlling the electron flow are responsible for producing higher ethanol tolerance (Lovitt *et al.*, 1984).

Although the mechanisms resulting in the development of moderate ethanol tolerance in *C. thermocellum* and *C. thermohydrosulfuricum* appear to be different, both appear to involve changes which make specific enzymes less prone to end product inhibition. The development of thermophilic strains which exhibit a high degree of ethanol tolerance, however, will probably require the isolation of strains with enhanced cell membrane stability.

Among the solvent-producing clostridia, the primary action of butanol also appears to be due to its chaotrophic effects, which result in increased membrane fluidity and permeability, although other specific effects may also be involved. Vollherbst-Schneck *et al.* (1984) reported that sub-growth inhibitory levels caused a 20–30% increase in fluidity of lipid dispersion in *C. acetobutylicum*. The partitioning of butanol in the cell membrane appears to disrupt the orderly array of the fatty acid side chain of the phospholipid, which diminishes the ability of the membrane to retain or exclude electrolytes and nonelectrolytes. Butanol challenge has been reported to decrease the rate of glucose uptake by *C. acetobutylicum* (Moreira *et al.*, 1981; Bowles and Ellefson, 1985; Ounine *et al.*, 1985). However, a recent study by Hutkins and Kashket (1986) showed that butanol concentrations up to 2% did not inhibit phosphotransferase activity. The decreased rate of glucose uptake and glycolysis observed by other works appears to be due entirely to the leakage of sugar phosphates and other glycolytic intermediates from the cell.

The increased cell membrane permeability also results in the decrease of the proton gradient across the cell membrane, which dissipates the proton motive force (Bowles and Ellefson, 1985; Terracciano and Kashket, 1986). Butanol has also been reported to lower the intracellular level of ATP, but this appears to be independent of the degree of cytoplasmic acidification, which occurs as a result of increased membrane permeability (Bowles and Ellefson, 1985).

Terracciano *et al.* (1987) investigated passive H^+ conductance in deenergized clostridia by applying an acid pulse and measuring the rate of H^+ entry into the cells. In *C. acetobutylicum* and *C. thermoaceticum*, cell membrane proton permeability was increased by butanol and acetic acid, respectively. H^+ conductance was also increased in cells treated with various ionophores. *Clostridium thermoaceticum* exponential and stationary phase cells exhibited the same H^+ permeability, but in *C. acetobutylicum*, membranes from acidogenic cells were more H^+-permeable than those from solventogenic cells. Terracciano *et al.* (1987) suggest that this difference in passive proton permeability may be due to differences in phospholipid composition, which would result in the changes in membrane fluidity that have been reported for acidogenic and solventogenic phase *C. acetobutylicum* cells. There is no evidence for end product inhibition of glycolytic or solventogenic enzymes in *C. acetobutylicum* Butanol toxicity has also been linked to autolysis of cells during the solventogenic phase as a result of the induction and release of cell-free autolysin (Van der Westhuizen *et al.*, 1982).

In addition to enhanced membrane fluidity, butanol has also been reported to cause a change in the membrane phospholipid composition of *C. acetobutylicum* cells (Vollherbst-Schneck *et al.*, 1984; Baer *et al.*, 1987; Lepage *et al.*, 1987). The cell membranes of solvent-producing stationary phase cells showed an increase in the proportion of saturated fatty acids compared with actively growing acid-producing cells. In all cases, challenge with butanol and other alcohols resulted in a rapid, dose-dependent response resulting in an increase in the percentage of saturated fatty acids. A decrease in pH also resulted in an increase in saturated and cyclopropane fatty acids (Lepage *et al.*, 1987). The effect of increasing temperature was mainly to increased fatty acids chain length (Baer *et al.*, 1987; Lepage *et al.*, 1987). A butanol-tolerant mutant was found to synthesize increased amounts of saturated fatty acids, which resulted in a more stable membrane environment (Baer *et al.*, 1987).

It has been assumed that increasing the tolerance of *C. acetobutylicum* cells to butanol would result in the production of higher concentrations of solvents. Butanol-tolerant mutants have been isolated which produce more solvents than the parent strains (Van der Westhuizen *et al.*, 1982; Lin and Blaschek, 1984; Hermann *et al.*, 1985), but the solvent concentrations obtained are not markedly higher than the concentrations obtained in the industrial fermentation. Other butanol-tolerant mutants produced more butanol but less acetone than the parent strain, resulting in a decrease in the total amount of solvent produced. The butanol-tolerant *C. acetobutylicum* SA-2 mutant produced trace amounts of butanol and acetone (Baer *et al.*, 1987) and demonstrates that increased butanol tolerance is not necessarily associated with improved solvent production. A similar lack of correlation between ethanol tolerance and production capability was reported in *C. saccharolyticum* (Murray *et al.*, 1983).

4. FERMENTATION PROCESSES FOR THE PRODUCTION OF SOLVENTS

4.1. Ethanol Fermentation Processes

At present no industrial fermentation processes exist for the production of ethanol using clostridial species, although various thermophilic ethanol-producing species have been identified as having potential for use in industrial fermentation processes. The major limitations in utilizing these organisms are the low ethanol tolerance and low final solvent concentrations which are obtained. The use of mixed cultures has been investigated as a way of overcoming some of the disadvantages associated with the use of these organisms (Wiegel, 1980; Zeikus, 1980, 1985; Zeikus *et al.*, 1981). Although these studies have shown that the concept of mixed thermophilic cultures is promising and does improve substrate degradation and ethanol yields, considerably more research and development is required before such systems could be developed to the level where they are economically viable. The major process requirement still remains the development of strains which can produce high concentrations and yields of ethanol.

4.2. The Acetone–Butanol Fermentation Processes

The production of acetone and butanol (AB) by *C. acetobutylicum* flourished during the early part of this century, and details of the industrial AB fermentation process have been described by a number of authors (Beesch, 1952, 1953; McCutchan and Hickey, 1954; Ryden, 1958; Prescott and Dunn, 1959; Ross, 1961; Hastings, 1978; Spivey, 1978; Walton and Martin, 1979; Moreira, 1983; McNeil and Kristiansen, 1986). Solvents were produced in a batch process, using fermenters which lacked mechanical agitation systems with a capacity ranging from 50,000 to 200,000 gallons. The initial industrial process utilized 8–10% maize mash which was cooked for 60–90 min at 130–133°C, and normally no further nutritional additions were necessary. The use of molasses as a fermentation substrate afforded many advantages and it superseded maize mash in most industrial processes from the mid-1930s onward. Blackstrap, invert (high-test), or beet molasses were diluted to give a concentration of fermentable sugars of between 5.0 and 7.5% (wt./vol.). The molasses was cooked and sterilized at 107–120°C for 15–60 min and was usually supplemented with additional sources of organic and inorganic nitrogen, phosphorus, and a buffering agent. The use of distillation slops to replace up to 33% of the makeup water was also common practice with both molasses and maize mash.

The fermenters were filled to 90–95% of their capacity under a blanket of CO_2 and sterile CO_2 was often bubbled through, before and after inoculation, to facilitate mixing. Cultures were normally kept as spores in sterile sand or soil. Inocula were prepared by heat-activating spores at 65–100°C for 1–3 min and after two to four buildup stages, the cells were inoculated into the fermenter, either during or just after filling at a concentration of 2–4%.

Fermentations using maize mash were run at 34–39°C for 40–60 hr and produced yields of around 25–26% based on dry weight corn equivalents. The final concentration of solvents produced was generally lower than those obtained using molasses and ranged from 12 to 20 g/liter. Solvent ratios varied according to the strain and fermentation conditions, but a ratio of 6 : 3 : 1 for A : B : E was typical for the Weizmann fermentation.

Fermentations utilizing molasses as a substrate were run at a lower temperature (29–35°C) with 31–32°C being optimum for many strains. Solvent yields based on the fermentable sugars were usually around 29–33% and cell metabolism was inhibited when the concentration of solvents reached 18–22 g/liter, although in practice lower concentrations were often obtained.

In many plants, the CO_2 and H_2 produced during the fermentation were recovered, separated, and used for a variety of purposes. After the fermentation, the solvents were separated from the liquor by primary batch or continuous distillation, and the distillate obtained was then fractionally distilled to produce pure acetone, pure butanol, and a further fraction of mixed solvents. The liquid effluent after distillation had a total solids content of 4.0–4.5% (wt./vol.). The solids had a fairly high nutritional value, including about 28–30% bacterial protein and substantial quantities of B-group vitamins. The dried solids from the effluent were widely used in animal feeds (Spivey, 1978).

The cost of the fermentation substrates such as maize and molasses was a major factor leading to the abandonment of the fermentation route for solvent production. The price of these carbohydrate substrates escalated after the Second World War and the fermentation process was unable to compete with the synthetic route using petrochemical feedstocks. The cost of the substrate accounted for about 60% of the overall cost of the fermentation (Ross, 1961). Since *C. acetobutylicum* can utilize a wide range of polysaccharides including starch, glycogen, dextran, xylans, mannans, inulin, and noncrystalline cellulose, a variety of other materials have been investigated as alternative substrates (Prescott and Dunn, 1959; Jones and Woods, 1986). In France, the use of hydrolyzed inulin obtained from Jerusalem artichokes was investigated (Marchal *et al.*, 1985). Whey containing 4.5–5.0% lactose is an effluent problem in countries with large dairy industries, and it is being assessed as an alternative substrate but with limited success at present (Schoutens *et al.*, 1984; Welsh and Veliky, 1984; Ennis and Maddox, 1985; Ennis and Maddox, 1987).

The use of lignocellulose hydrolyzates or the biochemical and genetic manipulation of strains to improve lignocellulose utilization could lead to the successful utilization of biomass as a substrate (Allcock and Woods, 1981; Maddox and Murray, 1983; Volesky and Szczesny, 1983; Marchal *et al.*, 1984, 1986; Yu *et al.*, 1984; Lee *et al.*, 1985; Lemmel *et al.*, 1986; Zappe *et al.*, 1986, 1987; Lee and Forsberg, 1987). An alternative approach for the utilization of lignocellulose is the use of cocultures containing cellulolytic and solventogenic strains (Fond *et al.*, 1983; Yu *et al.*, 1985).

The AB fermentation is a complex batch process involving physiological, morphological, and developmental changes in the bacterium. The triggers controlling these changes have not been well understood and fermentation failure has occurred from time to time. In addition, there is an absolute requirement for sterile conditions, and contamination, particularly phage infections, has caused problems and reduced productivity. The fermentation process has produced large volumes of effluent which required the development of specific processes for handling, treatment, and processing. These have also contributed to the cost of running the fermentation.

Economic evaluation studies indicated that the AB fermentation process using conventional 1940s technology and agriculturally based feedstocks could not compete economically with the chemical synthesis of solvents (Lenz and Moreira, 1980; Volesky *et al.*, 1981; Gibbs, 1983; Marlatt and Datta, 1986). These conclusions were based on the high oil prices after the fuel crises of the 1970s, and the recent worldwide slump in the price of crude oil makes the fermentation route even more uncompetitive. However, in the long term, provided suitable low-cost agricultural or waste-based feedstocks are available and some of the limitations associated with the fermentation process can be overcome, the fermentation route for acetone and butanol production can be reestablished.

4.3. Continuous Solvent Production

Continuous culture techniques have been used by a number of workers to investigate fundamental aspects of the AB fermentation, including the effects of different nutrients and metabolites and to evaluate the effects of various physical factors (see Section 3.3). These studies provided useful information about the regulatory mechanisms and kinetics of solvent production. Conditions for maximizing solvent concentrations, yields, and productivity and the selection of improved strains have also been investigated.

The use of continuous fermentation processes also provides an obvious way of improving the productivity of the AB fermentation for possible commercial exploitation. A variety of types of single-stage systems have been utilized to obtain solvent production during continuous culture. Solvent production has been reported using chemostats run under carbon, nitrogen, phosphate, sulfate, and magnesium limitation (Andersch *et al.*, 1982; Monot and Engasser, 1983a; Jobses and Roels, 1983; Stephens *et al.*, 1985), as well as in product-limited chemostats and turbidostats operated under conditions of nutrient excess (Gottschal and Morris, 1982; Monot and Engasser, 1983b; Fick *et al.*, 1985; Clarke and Hansford, 1986). However, the low final solvent concentrations and low productivities obtained with most single-stage systems would make such systems unsuitable for commercial production.

The production of acids and acetone along with butanol during continuous culture indicates that the production and reutilization of acids occurs concurrently with butanol formation, demonstrating that both acid-producing and solvent-producing cells coexist in these systems. In a recent study, Clarke *et al.* (1988) showed that the two populations of cells exhibit different specific growth rates, which result in the periodic oscillation in end products and the proportion of the two cell types. This precludes the establishment of a true steady state, which limits the value of continuous culture as an experimental tool and mitigates against its use for the commercial production of solvents. A further problem been encountered has been a decrease in solvent production over prolonged periods due to culture deterioration.

To overcome some of these limitations, two-stage or multistage continuous or semicontinuous fermentation systems have been investigated. Using two-stage systems, it is possible to separate the acid-producing propagation stage from the solvent-producing stage to some extent, resulting in improved solvent production (Bahl *et al.*, 1982; Afschar *et al.*, 1985). A number of pilot scale multistage fermentation systems for the continuous or semicontinuous production of solvents were developed in the USSR after the Second World War (Dyr *et al.*, 1958; Yarovenko, 1964; Hospodka, 1966). Unfortunately, little published information is available regarding the details of these processes. Recently, Marlatt and Datta (1986) reported on a patented design for the production of solvents from maize using a multistage fermentation system (Table III).

In batch culture, solvent production normally occurs during the second phase of the fermentation when little or no increase in cell biomass occurs. This suggests that continuous flow systems designed to retain high concentrations of nongrowing, solvent-producing cells would be superior to conventional continuous culture system for economical solvent production. Two main approaches for the retention of cell biomass in flow-through systems have been utilized. These are cell immobilization and cell recycling. These systems allow the reuse of cells, so that greater cell densities are achieved resulting in greater productivity over extended periods of time. Many methods are available for immobilizing cells, but alginate entrapment has been the most commonly used method for the solvent-producing clostridia (Haggstrom and Molin, 1980; Krouwel *et al.*, 1980). The use of wild-type *C. acetobutylicum* and *C. beijerinckii* cells resulted in the production of low solvent concentrations in the effluent stream (Haggstrom and Enfors, 1982; Forberg *et al.*, 1983; Krouwel *et al.*, 1983; Schoutens *et al.*, 1985). Using sporulation mutants which allow cells to remain in the solvent-producing phase, Largier *et al.* (1985) were able to obtain much higher solvent concentrations and high levels of productivity. Cell recycling using ultrafiltration was also demonstrated as a successful method for retaining biomass and increasing productivity (Afschar *et al.*, 1985; Pierrot *et al.*, 1986; Scholote and Gottschalk, 1986) and has been used in conjunction with *in situ* recovery of solvents from the effluent stream.

4.4. Solvent Recovery

A major limitation in using the fermentation route for the production of solvents is the low final concentration of the end products produced. The standard method of recovery is distillation and although there have been significant improvements in distillation technology, the high energy cost still contributes substantially to the production cost. This has focused

Table III. Integrated Fermentation and Product Recovery Systems

Fermentation Systems

Batch—Fed-batch—Continuous fermentation—Continuous flow
using cell recycle or immobilized cells

Recovery Systems

Direct recovery—Recovery using membrane systems

Vacuum or Gaseous Extraction

Vacuum fermentation	Pervaporation
	Semipermeable membranes
	using vaccuum
Flash evaporation	Semipermeable membranes
Gas stripping	using gas stream

Liquid Extractions

Organic solvent extraction	Perstation
Corn oil	Semipermeable membranes
Dibutylphthalate	using organic solvents
Oleyl alcohol	
Fluorocarbons	Dialysis
	Semipermeable membranes
Two-phase aqueous extraction	
Dextran	Reverse osmosis
Carbopeg	
	Crossflow microfiltration
	Ultrafiltration
	Ultrafiltration and solvent
	extraction
	Ultrafiltration and reverse
	osmosis

Adsorption

Molecular sieves	Ultrafiltration and adsorp-
Ion exchange resins	tion
Zeolites	
Polymeric adsorbants	
Activated carbon	
Synthetic polymeric resins	

Chemical Extraction

Lactones	Ultrafiltration and chemical
	extraction

attention on the development of alternative methods of solvent recovery which are less energy-intensive (see review by Ennis *et al.*, 1986).

One approach is the recovery of solvent under vacuum. However, the large volumes of gases produced in the AB fermentation make this approach unattractive and flash evaporation using a second vessel employing a recycle stream appears to have greater potential (Ennis *et al.*, 1986). Solvents can also be removed directly by means of gas stripping (Moreira *et al.*, 1982; Ennis *et al.*, 1986, 1987) or by liquid-liquid extractions using a variety of organic solvents such as corn oil, dibutylphthalate, oleyl alcohol, and fluorocarbons (Griffith *et al.*, 1983; Taya *et al.*, 1985; Wayman and Parekh, 1987; Roffler *et al.*, 1987, 1988). Partial concentrations of solvents while still in the aqueous phase have been achieved by dialysis (Ballongue *et al.*, 1987), reverse osmosis (Garcia *et al.*, 1986), and the use of two-phase aqueous systems, which have utilized different polymers to retain cells in one phase only (Mattiasson *et al.*, 1982; Mattiasson, 1983). Absorbants such as activated carbon and synthetic polymeric resins as well as molecular sieves, such as zeolites (Maddox, 1983; Ennis *et al.*, 1987), and chemical extraction using lactones (Sioumis, 1987) have also been investigated as separation methods.

At present, all of these approaches suffer from some limitations. In many cases, although some separation of solvents from the fermentation broth can be achieved, feasible methods for the subsequent solvent recovery and recycling of the extractant are not available. Some of these approaches may have potential as adjunct or alternative to distillation during downstream processing of the spent fermentation broth. Most investigations, however, have concentrated on the *in situ* recovery of solvent from the fermentation broth (Ennis *et al.*, 1986). These include extractions which are designed to operate in the fermentation itself such as gap stripping, vacuum distillation, two-phase aqueous systems, or the direct addition of liquid extractants or adsorbants to the fermentation vessel. In a number of designs, recycle loops have been incorporated to allow passage through extractant columns or flash evaporation vessels.

Developments in membrane technology have provided a variety of novel approaches to the recovery of solvents. Cross-flow microfiltration and ultrafiltration have provided a feasible means of achieving cell recycle for biomass retention which allows the separation of a solvent stream (Afschar *et al.*, 1985; Pierrot *et al.*, 1986; Scholote and Gottschalk, 1986). Ultrafiltration has been used in conjunction with other extraction processes such as reverse osmosis and adsorption. The development of various types of semipermeable membranes has allowed the direct extraction of solvents *in situ*. Pervaporation, which utilizes silicone membranes to allow butanol to diffuse from the fermentation broth, which can then be removed using vacuum or a gas stream, has been one of the most promising methods investigated (Groot *et al.*, 1984a,b; Groot and Luyben, 1987; Larrayoz and Puigjaner, 1987). Perstation operates on a similar principle but utilizes organic solvents or possibly chemical extracts such as lactones for product removal. Dialysis and reverse osmosis have also been investigated as methods for concentrating solvents while still in the aqueous phase.

5. CONCLUSION

The potential for utilizing both mesophilic and thermophilic species of *Clostridium* for the microbial production of solvents has generated considerable interest in these organisms over the past decade. This research effort has lead to an increased understanding of the mechanisms involved in the regulation of solvent production and the nature of solvent toxicity and tolerances, and has suggested ways of improving fermentation technology and production strains so as to overcome some of the limitations of the fermentation route. Increasing the concentration of the solvents produced and developing less energy-intensive methods of solvent extraction are likely to be vital for potential developments. Other key areas which could play a major role in the establishment of commercial fermentation processes are the development of fermentation using low-cost substitutes and improvements in productivity and yield of solvents during the fermentation.

REFERENCES

Adler, H. I., and Crow, W., 1987, A technique for predicting the solvent-producing ability of *Clostridium acetobutylicum*. *Appl. Env. Microbiol.* **53**:2496–2499.

Afschar, A. S., Biebl, H., Schaller, K., and Schugerl, K., 1985, Production of acetone and butanol by *Clostridium acetobutylicum* in continuous culture with cell recycle, *Appl. Microbiol. Biotechnol.* **22**:394–398.

Afschar, A. S., Schaller, K., and Schurgerl, K., 1986, Continuous production of acetone and butanol with shear-activated *Clostridium acetobutylicum*, *Appl. Microbiol. Biotechnol.* **23**:315–321.

Allcock, E. R., and Woods, D. R., 1981, Carboxymethyl cellulase and cellobiase production by *Clostridium acetobutylicum* in an industrial fermentation medium, *Appl. Env. Microbiol.* **41** (2):539–541.

Andersch, W., Bahl, H., and Gottschalk, G., 1982, Acetone-butanol production by *Clostridium acetobutylicum* in an ammonium-limited chemostat at low pH values, *Biotechnol. Lett.* **4** (1):29–32.

Andersch, W., Bahl, H., and Gottschalk, G., 1983, Level of enzymes involved in acetate, butyrate, acetone and butanol formation by *Clostridium acetobutylicum*, *Eur. J. Appl. Microbiol. Biotechnol.* **18**:327–332.

Antranikian, G., Friese, C., Quentmeier, A., Hippe, H., and Gottschalk, G., 1984, Distribution of the ability for citrate utilization amongst Clostridia, *Arch. Microbiol.* **138**:179–182.

Baer, S. H., Blaschek, H. P., and Smith, T. L., 1987, Effect of butanol challenge and temperature of lipid composition and membrane fluidity of butanol-tolerant *Clostridium acetobutylicum*, *Appl. Env. Microbiol.* **53**:2854–2861.

Bahl, H., and Gottschalk, G., 1985, Parameters affecting solvent production by *Clostridium acetobutylicum* in continuous culture, *Biotechnol. Bioeng.* **514**:217–223.

Bahl, H., Andersch, W., and Gottschalk, G., 1982, Continuous production of acetone and butanol by *Clostridium acetobutylicum* in a two-stage phosphate limited chemostat, *Eur. J. Appl. Microbiol. Biotechnol.* **15**:201–205.

Bahl, H., Gottwald, M., Kuhn, A., Rale, V., Andersch, W., and Gottschalk, G., 1986, Nutritional factors affecting the ratio of solvents produced by *Clostridium acetobutylicum, Appl. Env. Microbiol.* **52**(1):169–172.

Ballongue, J., Amine, J., Masion, E., Petitdemange, H., and Gay, R., 1985, Induction of acetoacetate decarboxylase in *Clostridium acetobutylicum, FEMS Microbiol. Lett.* **29**:273–277.

Ballongue, J., Amine, J., Petitemange, H., and Gay, R., 1986, Regulation of acetate kinase and butyrate kinase by acids in *Clostridium acetobutylicum. FEMS Microbiol. Lett.***35**:295–301.

Ballongue, J., Maison, E., Amine, J., Petitdemange, H., and Gay, R., 1987, Inhibitor effects of products of metabolism on growth of *Clostridium acetobutylicum, Appl. Microbiol. Biotechnol.* **26**:568–573.

Bayer, E. A., Setter, E., and Lamed, R., 1985, Organization and distribution of the cellulosome in *Clostridium thermocellum, J. Bacteriol.* **163**(2):552–559.

Beesch, S. C., 1952, Acetone-butanol fermentation of sugars, *Eng. Proc. Dev.* **44**:1677–1682.

Beesch, S. C., 1953, Acetone-butanol fermentation of starches, *Appl. Microbiol.* **1**:85–95.

Bowles, L. K., and Ellefson, W. L., 1985, Effects of butanol on *Clostridium acetobutylicum, Appl. Env. Microbiol.* **50**(5):1165–1170.

Brener, D., and Johnson, B. F., 1984, Relationship between substrate and concentration and fermentation product ratios in *Clostridium thermocellum* cultures, *Appl. Env. Microbiol.* **47**(5):1126–1129.

Chen, J. S., and Hiu, S. F., 1986, Acetone-butanol-isopropanol production by *Clostridium beijerinckii* (synonym, *Clostridium butylicum*), *Biotechnol. Lett.* **8**(5):371–376.

Clarke, K. G., 1987, A reassessment of the production of acetone and butanol by *Clostridium acetobutylicum* in continuous culture, Ph.D. thesis, University of Cape Town, South Africa, pp. 1–195.

Clarke, K. G., and Hansford, G. S., 1986, Production of acetone and butanol by *Clostridium acetobutylicum* in a product limited chemostat, *Chem. Eng. Commun.* **45**:75–81.

Clarke, K. G., Hansford, G. S., and Jones, D. T., 1988, The nature and significance of oscillatory behavior during solvent production by *Clostridium acetobutylicum* in Continuous culture, *Biotechnol. Bioeng.* **32**:538–544.

Conway, T., Sewell, G. W., Osman, Y. A., and Ingram, L. O., 1987, Cloning and sequencing of the alcohol dehydrogenase II gene from *Zymomonas mobilis, J. Bacteriol.* **169**:2591–2597.

Corry, J. E. L., 1978, Possible sources of ethanol ante- and post-mortem: Its relationship to the biochemistry and microbiology of decomposition, *J. Appl. Bacteriol.* **44**:1–56.

Cummins, C. S., and Johnson, J. L., 1971, Taxonomy of the Clostridia: wall composition and DNA homologies in *Clostridium butyricum* and other butyric acid-producing Clostridia, *J. Gen. Microbiol.* **67**:33–46.

Datta, R., and Zeikus, J. G., 1985, Modulation of acetone-butanol-ethanol fermentation by carbon monoxide and organic acids, *Appl. Environ. Microbiol.* **49**(3):522–529.

Davies, R., 1943, Studies on the acetone-butanol fermentation. 4. Acetoacetic acid decarboxylase of *Cl. acetobutylcium* (BY), *Biochem. J.* **37**:230–238.

Doelle, H. W. (ed.), 1975, *Bacterial Metabolism*, 2nd ed., Academic Press, New York.

Doremus, M. G., Linden, J. C., and Moreira, A. R., 1985, Agitation and pressure effects on acetone-butanol fermentation, *Biotechnol. Bioeng.* **27**:852–860.

Dürre, P., Kuhn, A., and Gottschalk, G., 1986, Treatment with allyl alcohol selects specifically for mutants of *Clostridium acetobutylicum* defective in butanol synthesis, *FEMS Microbiol. Lett.* **36**:77–81.

Dürre, P., Kuhn, A., Gottwald, M., and Gottschalk, G., 1987, Enzymatic investigations on butanol dehydrogenase and butyraldehyde dehydrogenase in extracts of *Clostridium acetobutylicum, Appl. Microbiol. Biotechnol.* **26**: 268–272.

Dyr, J., Protiva, J., and Praus, R., 1958, Formation of neutral solvents in continuous fermentation by means of *Clostridium acetobutylicum*, in: *Continuous Cultivation of Microorganisms* (I. Malek, ed.), Czechoslovakian Academy of Sciences, Prague, pp. 210–226.

Ennis, B. M., and Maddox, I. S., 1985, Use of *Clostridium acetobutylicum* P262 for production of solvents from whey permeate, *Biotechnol. Lett.* **7**(8):601–606.

Ennis, B. M., and Maddox, I. S., 1987, The effect of pH and lactose concentration on solvent production from whey permeate using *Clostridium acetobutylicum, Biotechnol. Bioeng.* **29:** 329–334.

Ennis, B. M., Gutierrez, N. A., and Maddox, I. S., 1986, The acetone-butanol-ethanol fermentation: A current assessment, *Process Biochem.* **October:**131–147.

Ennis, B. M., Qureshi, N., and Maddox, I. S., 1987, In-line toxic product removal during solvent production by continuous fermentation using immobilized *Clostridium acetobutylicum, Enzyme Microb. Technol.* **9:**672–675.

Fick, M., Pierrot, P., and Engasser, J. M., 1985, Optimal conditions for long-term stability of acetone-butanol production by continuous cultures of *Clostridium acetobutylicum, Biotechnol. Lett.* **7**(7):503–508.

Finn, R. K., and Nowrey, J. E., 1958, A note on the stability of Clostridia when held in continuous culture, *Appl. Microbiol.* **7:**29–32.

Fond, O., Petitdemange, E., Petitdemange, H., and Engasser, J. M., 1983, Cellulose fermentation by a coculture of a mesophilic cellulolytic Clostridium and *Clostridium acetobutylicum, Biotechnol. Bioeng. Symp.* **13:**217–224.

Fond, O., Matta-Ammouri, G., Petitdemange, H., and Engasser, J. M., 1985, The role of acids on the production of acetone and butanol by *Clostridium acetobutylicum. Appl. Microbiol. Biotechnol.* **22**(3):195–200.

Forberg, C., Enfors, S. O., and Haggstrom, L., 1983, Control of immobilized, non-growing cells for continuous production of metabolites, *Eur. J. Appl. Microbiol. Biotechnol.* **17:**143–147.

Freier, D., and Gottschalk, G., 1987, L(+)-lactate dehydrogenase of *Clostridium acetobutylicum* is activated by fructose-1,6-bisphosphate, *FEMS Microbiol. Lett.* **43:**229–233.

Garcia, A., Iannotti, E. L., and Fischer, J. L., 1986, Butanol fermentation liquor production and separation by reverse osmosis, *Biotechnol. Bioeng.* **28:**785–791.

George, H. A., and Chen, J. S., 1983, Acidic conditions are not obligatory for onset of butanol formation by *Clostridium beijerinckii* (Synonym, *C. butylicum*), *Appl. Environ. Microbiol.* **46** (2):321–327.

George, H. A., Johnson, J. L., Moore, W. E. C., Holdeman, L. V., and Chen, J. S., 1983, Acetone, isopropanol, and butanol production by *Clostridium beijerinckii* (syn. *Clostridium butylicum*) and *Clostridium aurantibutyricum, Appl. Environ. Microbiol.* **45**(3):1160–1163.

Germain, P., Toukourou, F., and Donaduzzi, L., 1986, Ethanol production by anaerobic thermophilic bacteria: Regulation of lactate dehydrogenase activity in *Clostridium thermohydrosulfuricum, Appl. Microbiol. Biotechnol.***24:**300–305.

Gibbs, D. F., 1983, The rise and fall (. . . and rise?) of acetone/butanol fermentations, *Trends Biotechnol.* **1:**12–15.

Gottschal, J. C., and Morris, J. G., 1981, The induction of acetone and butanol production in cultures of *Clostridium acetobutylicum* by elevated concentrations of acetate and butyrate, *FEMS Microbiol. Lett.* **12:**385–389.

Gottschal, J. C., and Morris, J. G., 1982, Continuous production of acetone and butanol by *Clostridium acetobutylicum* growing in turbidostat culture, *Biotechnol. Lett.* **4**(8):477–482.

Gottschalk, G., 1979, Butyrate and butanol-acetone fermentation, in: *Bacteriol Metabolism*, (M. P. Starr, ed.), Springer-Verlag, Berlin, pp. 182–215.

Gottschalk, G., Andreesen, J. R., and Hippe, H., 1981, The genus *Clostridium* (nonmedical aspects), in: *The Prokaryotes* (M. P. Starr, H. Stolp, H. G. Truper, A. Balows, and H. G. Schlegel, eds.), Springer-Verlag, Berlin, pp. 1767–1803.

Gottwald, M., Hippe, H., and Gottschalk, G., 1984, Formation of n-butanol from D-glucose by strains of '*Clostridium tetanomorphum*' group, *Appl. Env. Microbiol.* **48**(3):573–576.

Gottwald, M., and Gottschalk, G., 1985, The internal pH of *Clostridium acetobutylicum* and its effect on the shift from acid to solvent formation, *Arch. Microbiol.* **143**:42–46.

Griffith, W. L., Compere, A. L., and Googin, J. M., 1983, Novel neutral solvents fermentations, *Dev. Ind. Microbiol.* **24**:347–352.

Groot, W. J., and Luyben, K. C. A. M., 1987, Continuous production of butanol from a glucose/xylose mixture with an immobilized cell system coupled to pervaporation, *Biotechnol. Lett.* **9**:867–870.

Groot, W. J., Schoutens, G. H., Van Beelen, P. N., Van den Oever, C. E., and Kossen, N. W. F., 1984a, Increase of substrate conversion by pervaporation in the continuous butanol fermentation, *Biotechnol. Lett.* **6**(12):789–792.

Groot, W. J., van den Oever, C. E., and Kossen, N. W. F., 1984b, Pervaporation for simultaneous product recovery in the butanol/isopropanol batch fermentation, *Biotechnol. Lett.* **6**(11):709–714.

Haggström, L., and Enfors, S. O., 1982, Continuous production of butanol with immobilized cells of *Clostridium acetobutylicum*, *Appl. Biochem. Biotechnol.* **7**:35–37.

Haggström, L., and Molin, N., 1980, Calcium alginate immobilized cells of *Clostridium acetobutylicum* for solvent production, *Biotechnol. Lett.* **2**:241–246.

Hartmanis, M. G. N., 1987, Butyrate Kinase from *Clostridium acetobutylicum*, *J. Biol. Chem.* **262**: 617–621.

Hartmanis, M. G. N., and Gatenbeck, S., 1984, Intermediary metabolism in *Clostridium acetobutylicum:* Levels of enzymes involved in the formation of acetate and butyrate, *Appl. Env. Microbiol.* **47**(6):1277–1283.

Hartmanis, M. G. N., Klason, T., and Gatenbeck, S., 1984, Uptake and activation of acetate and butyrate in *Clostridium acetobutylicum*, *Appl. Microbiol. Biotechnol.* **20**(1):66–71.

Hartmanis, M. G. N., Ahlman, H., and Gatenbeck, S., 1986, Stability of solvent formation in *Clostridium acetobutylicum* during repeated subculturing, *Appl. Microbiol. Biotechnol.* **23**: 369–371.

Hastings, J. H. J., 1978, Acetone-butyl alcohol fermentation, in: *Economic Microbiology, Primary Products of Metabolism*, Vol. 2 (A. H. Rose, ed.), Academic Press, New York, pp. 31–45.

Hermann, M., Fayolle, F., Marchal, R., Podvin, L., Sebald, M., and Vandecasteele, J. P., 1985, Isolation and characterization of butanol-resistant mutants of *Clostridium acetobutylicum*, *Appl. Env. Microbiol.* **50**(5):1238–1243.

Herrero, A. A., 1983, End-product inhibition in anaerobic fermentations, *Trends Biotechnol.* **1** (2):49–53.

Herrero, A. A., and Gomez, R. F., 1980, Development by ethanol tolerance in *clostridium thermocellum:* Effect of growth temperature, *Appl. Env. Microbiol.* **40**:571–577.

Herrero, A. A., Gomez, R. F., and Roberts, M. F., 1982, Ethanol-induced changes in the membrane lipid composition of *Clostridium thermocellum*, *Biochim. Biophys. Acta* **693**:195–204.

Herrero, A. A., Gomez, R. F., and Roberts, M. F., 1985, 31P-NMR studies of *Clostridium thermocellum:* Mechanisms of endproduct inhibition by ethanol, *J. Biol. Chem.* **260**:7442–7451.

Hiu, S. F., Zhu, C-X., Yan, R-T., and Chen, J-S., 1987, Butanol-ethanol dehydrogenase and butanol-ethanol-isopropanol dehydrogenase: Different alcohol dehydrogenases in two strains of *Clostridium beijerinckii (Clostridium butylicum)*, *Appl. Env. Microbiol.* **53**:697–703.

Holdeman, L. V., Cato, E. P., Moore, W. E. C. (eds.), 1977, *Anaerobe Laboratory Manual*, 4th Ed., Virginia Polytechnic Institute, Blacksburg, Virginia.

Holt, R. A., Stephens, G. M., and Morris, J. G., 1984, Production of solvents by *Clostridium acetobutylicum* cultures maintained at neutral pH, *Appl. Env. Microbiol.* **48**(6):1166–1170.

Hon-nami, K., Coughlan, M. P., Hon-nami, H., and Ljungdahl, L. G., 1986, Separation and characterization of the complexes constituting the cellulolytic enzyme system of *Clostridium thermocellum*, *Arch. Microbiol.* **145**:13–19.

Hospodka, J., 1966, Industrial application of continuous fermentation, in: *Theoretical and Methodological Bases of Continuous Culture of Microorganisms* (I. Malek and Z. Fencl, eds.), Academic Press, New York, pp. 611–613.

Hsu, E. J., and Ordal, Z. J., 1970, Comparative metabolism of vegetative and sporulating cultures of *Clostridium thermosaccharolyticum*, *J. Bacteriol.* **102**(2):369–376.

Huang, L., Forsberg, C. W., and Gibbins, L. N., 1986, Influence of external pH and fermentation products on *Clostridium acetobutylicum* intracellular pH and cellular distribution of fermentation products, *Appl. Env. Microbiol.* **51**(6):1230–1234.

Hugo, H. V., Schoberth, S., madan, V. K., and Gottschalk, G., 1972, Coenzyme specificity of dehydrogenases and fermentation of pyruvate by clostridia, *Arch. Mikrobiol.* **87**:189–202.

Hungate, R. E., 1944, Studies on cellulose fermentation. I. The culture and physiology of an anaerobic cellulose-digesting bacterium, *J. Bacteriol.* **48**:499–513.

Hutkins, R. W., and Kashket, E. R., 1986, Phosphotransferase activity in A. acetobutylicum from acidogenic and solventogenic phase of growth, *Appl. Env. Microbiol.* **51**:1121–1123.

Hyun, H. H., and Zeikus, J. G., 1985, Simultaneous and enhanced production of thermostable amylases and ethanol from starch by cocultures of *Clostridium thermosulfurogenes* and *Clostridium thermohydrosulfuricum*, *Appl. Env. Microbiol.* **49**(5):1174–1181.

Ingram, L. O., 1986, Microbial tolerance to alcohols: Role of the cell membrane, *TIBTECH*. Feb:40–44.

Ingram, L. O., and Buttke, T. M., 1984, Effects of alcohols on microorganisms, *Adv. Microb. Physiol.* **25**:256–300.

Janati-Idrissi, R., Junelles, A. M., El Kanouni, A., Petitdemange, H., and Gay, R., 1987, Selection de mutants de *Clostridium acetobutylicum* defectifs dans la production d'acetone, *Ann. Inst. Pasteur/Microbiol.* **138**:313–323.

Jobses, I. M. L., and Roels, J. A., 1983, Experience with solvent production by *Clostridium beijerinckii* in continuous culture, *Biotechnol. Bioeng.* **25**:1187–1194.

Jones, D. T., and Woods, D. R., 1986, The acetone butanol fermentation revisited, *Microbiol. Rev.* **50**:484–524.

Jones, D. T., van der Westhuizen, A., Long, S., Allcock, E. R., Reid, S. J., and Woods, D. R., 1982, Solvent production and morphological changes in *Clostridium acetobutylicum*, *Appl. Environ. Microbiol.* **43**(6):1434–1439.

Jungermann, K., Thauer, R. K., Leimenstoll, G., and Decker, K., 1973, Function of reduced pyridine nucleotide-ferredoxin oxidoreductases in saccharolytic Clostridia, *Biochim. Biophys. Acta.* **305**:268–280.

Kell, D. B., Peck, M. W., Rodger, G., and Morris, J. G., 1981, On the permeability to weak acids and bases of the cytoplasmic membrane of *Clostridium pasteurianum*, *Biochem. Biophys. Res. Commun.* **99**:81–88.

Kelly, W. J., Asmundson, R. V., and Hopcroft, D. H., 1987, Isolation and characterization of a strictly anaerobic, cellulolytic spore former: *Clostridium chartatabidium* sp. nov., *Arch. Microbiol.* **147**:169–173.

Kim, B. H., and Zeikus, J. G., 1985, Importance of hydrogen metabolism in regulation of solventogenesis by *Clostridium acetobutylicum*, *Dev. Ind. Microbiol.* **26**:1–14.

Kim, B. H., Bellows, P., Datta, R., and Zeikus, J. G., 1984, Control of carbon and electron flow in *Clostridium acetobutylicum* fermentations: Utilization of carbon monoxide to inhibit hydrogen production and to enhance butanol yields, *Appl. Env. Microbiol.* **48**(4):764–770.

Krouwel, P. G., Van der Laan, W. F. M., and Kossen, N. W. F., 1980, Continuous production of n-butanol and isopropanol by immobilized, growing *Clostridium butylicum* cells, *Biotechnol. Lett.* **2**:253–258.

Krouwel, P. G., Groot, W. J., Kossen, N. W. F., and van der Laan, W. F. M., 1983, Continuous isopropanol-butanol-ethanol fermentation by immobilized *Clostridium beijerinckii* cells in a packed bed fermenter, *Enzyme Microb. Technol.* **5**:46–55.

Kutzenok, A., and Aschner, M., 1952, Degenerative processes in a strain of *Clostridium butylicum, J. Bacteriol.* **64:**829–836.

Lamed, R. J., and Zeikus, J. G., 1980, Novel NADP-linked alcohol aldehyde/ketone oxidoreductase in thermophilic, ethanolgenic bacteria, *Biochem. J.* **195:**183–190.

Lamed, R., Setter, E., and Bayer, E. A., 1983, Characterization of a cellulose-binding, cellulase-containing complex in *Clostridium thermocellum, J. Bacteriol.* **156**(2):828–836.

Landuyt, S. L., Hsu, E. J., and Lu, M., 1983, Transition from acid fermentation to solvent fermentation in a continuous dilution culture of *Clostridium thermosaccharolyticum, Ann. N. Y. Acad. Sci.* **413:**473–478.

Largier, S. T., Long, S., Santangelo, J. D., Jones, D. T., and Woods, D. R., 1985, Immobilized *Clostridium acetobutylicum* P262 mutants for solvent production, *Appl. Environ. Microbiol.* **50** (2):477–481.

Larrayoz, M. A., and Puigjaner, L., 1987, Study of butanol extraction through pervaporation in acetobutylic fermentation, *Biotechnol. Bioeng.* **30:**692–696.

Lee, S. F., and Forsberg, C. W., 1987, Isolation and some properties of a B-D-xylosidase from *Clostridium acetobutylicum* ATCC 824, *Appl. Env. Microbiol.* **53:**651–654.

Lee, S. F., Forsberg, C. W., and Gibbins, L. N., 1985, Cellulolytic activity of *Clostridium acetobutylicum, Appl. Env. Microbiol.* **50**(2):220–228.

Lemmel, S. A., Datta, R., and Frankiewicz, J. R., 1986, Fermentation of xylan by *Clostridium acetobutylicum, Enzyme Microb. Technol.* **8:**217–221.

Lenz, T. G., and Moreira, A. R., 1980, Economic evaluation of the acetone-butanol fermentation, *Ind. Eug. Chem. Prod. Res. Dev.* **19:**478–483.

Lepage, C., Fayolle, F., Hermann, M., and Vandecasteele, J.-P., 1987, Changes in membrane lipid composition of *Clostridium acetobutylicum* during acetone-butanol fermentation: Effects of solvents, growth temperature and pH, *J. Gen. Microbiol.* **133:**103–110.

Lin, Y., and Blaschek, H. P., 1984, Butanol production by a butanol-tolerant strain of *Clostridium acetobutylicum* in extruded corn broth, *Appl. Env. Microbiol.* **45**(3):966–973.

Long, S., Jones, D. T., and Woods, D. R., 1983, Sporulation of *Clostridium acetobutylicum* P262 in a defined medium, *Appl. Env. Microbiol.* **45**(4):1389–1393.

Long, S., Jones, D. T., and Woods, D. R., 1984a, The relationship between sporulation and solvent production in *Clostridium acetobutylicum* P262, *Biotechnol. Lett.* **6**(8):529–534.

Long, S., Jones, D. T., and Woods, D. R., 1984b, Initiation of solvent production, clostridial stage and endospore formation in *Clostridium acetobutylicum* P262, *Appl. Microbiol. Biotechnol.* **20**(4):256–261.

Lovitt, R. W., Longin, R., and Zeikus, J. G., 1984, Ethanol production by thermophilic bacteria: Physiological comparison of solvent effects on parent and alcohol-tolerant strains of *Clostridium thermohydrosulfuricum, Appl. Environ. Microbiol.* **48**(1):171–177.

Maddox, I. S., 1983, Use of silicalite for the adsorption of n-butanol from fermentation liquors, *Biotechnol. Lett.* **5:**89–94.

Maddox, I. S., and Murray, A. E., 1983, Production of n-butanol by fermentation of wood hydrolysate, *Biotechnol. Lett.* **5**(3):175–178.

Marchal, R., Rebeller, M., and Vandecasteele, J. P., 1984, Direct bioconversion of alkali-pretreated straw using simultaneous enzymatic hydrolysis and acetone-butanol fermentation, *Biotechnol. Lett.* **6**(8):523–528.

Marchal, R., Blanchet, D., and Vandecasteele, J. P., 1985, Industrial optimization of acetone-butanol fermentation: A study of the utilization of Jerusalem artichokes, *Appl. Microbiol. Biotechnol.* **23:**92–98.

Marchal, R., Ropars, M., and Vandecasteele, J. P., 1986, Conversion into acetone and butanol of lignocellulosic substrates pretreated by steam explosion, *Biotechnol. Lett.* **8**(5):365–370.

Marlatt, J. A., and Datta, R., 1986, Acetone-butanol fermentation process development and economic evaluation, *Biotechnol. Prog.* **2:**23–28.

Masion, E., Amine, J., and Marczak, R., 1987, Influence of amino acid supplements on the metabolism of *Clostridium acetobutylicum, FEMS Microbiol. Lett.* **43:**269–274.

Matta-El-Amouri, G., Janati-Idrissi, R., Assobhei, O., Petitdemange, H., and Gay, R., 1985, Mechanism of the acetone formation by *Clostridium acetobutylicum, FEMS Microbiol. Lett.* **30:**11–16.

Matta-El-Ammouri, G., Janati-Idrissi, R., Junelles, A.-M., Petitdemange, H., and Gay, R., 1987, Effects of butyric and acetic acids on acetone-butanol formation by *Clostridium acetobutylicum, Biochimie* **69:**109–115.

Mattiasson, B., 1983, Applications of aqueous two-phase systems in biotechnology, *Trends Biotechnol.* **1**(1):16–20.

Mattiasson, B., Suominen, M., Andersson, E., Haggstrom, L., Albertsson, P. A., and Hahn-Hagerdal, B., 1982, Solvent production by *Clostridium acetobutylicum* in aqueous two-phase system, *Enzyme Eng.* **6:**153–155.

McCoy, E., and Fred, B., 1941, The stability of a culture for industrial fermentation, *J. Bacteriol.* **41:**90–91.

McCutchan, W. N., and Hickey, R. J., 1954, The butanol-acetone fermentations, *Ind. Ferment.* **1:**347–388.

McNeil, B., and Kristiansen, B., 1986, The acetone butanol fermentation, in: *Advances in Applied Microbiology,* Vol. 31 (A. Laskin, ed.), Academic Press, New York, pp. 61–92.

Meyer, C. L., Roos, J. W., and Papoutsakis, E. T., 1986, Carbon monoxide gasing leads to alcohol production and butyrate uptake without acetone formation in continuous cultures of *Clostridium acetobutylicum. Appl. Microbiol. Biotechnol.* **24:**159–167.

Monot, F., and Engasser, J. M., 1983a, Production of acetone and butanol by batch and continuous culture of *Clostridium acetobutylicum* under nitrogen limitation, *Biotechnol. Lett.* **5**(4):213–218.

Monot, F., and Engasser, J. M., 1983b, Continuous production of acetone butanol on an optimized synthetic medium, *Eur. J. Appl. Microbiol. Biotechnol.* **18:**246–248.

Monot, F., Martin, J.R., Petitdemange, H., and Gay, R., 1982, Acetone and butanol production by *Clostridium acetobutylicum* in a synthetic medium, *Appl. Env. Microbiol.* **44**(6):1318–1324.

Monot, F., Engasser, J. M., and Petitdemange, H., 1983, Regulation of acetone butanol production in batch and continuous cultures of *Clostridium acetobutylicum, Biotechnol. Bioeng. Symp.* **13:**207–216.

Monot, F., Engasser, J. M., and Petitdemange, H., 1984, Influence of pH and undissociated butyric acid on the production of acetone and butanol in batch cultures of *Clostridium acetobutylicum, Appl. Microbiol. Biotechnol.* **19**(6):422–426.

Moreira, A. R., 1983, Acetone-butanol fermentation, in: *Organic Chemicals From Biomass* (D. L. Wise, ed.), Benjamin/Cummings, Menlo Park, CA, pp. 385–406.

Moreira, A. R., Ulmer, D. C., and Linden, J. C., 1981, Butanol toxicity in the butylic fermentation, *Biotechnol. Bioeng. Symp.* **11:**567–579.

Moreira, A. R., Dale, B. E., and Doremus, M. G., 1982, Utilization of the fermentor off-gases from an acetone-butanol fermentation, *Biotechnol. Bioeng. Symp.* **12:**263–277.

Murray, W. D., and Khan, A. W., 1983a, Ethanol production by a newly isolated anaerobe, *Clostridium saccharolyticum:* Effects of culture medium and growth conditions, *Can. J. Microbiol.* **29:**342–347.

Murray, W. D., and Khan, A. W., 1983b, Growth requirements of *Clostridium saccharolyticum,* an ethanologenic anaerobe, *Can. J. Microbiol.* **29:**348–353.

Murray, W. D., Wemyss, K. B., and Khan, A. W., 1983, Increased ethanol production and tolerance by a pyruvate-negative mutant of *Clostridium saccharolyticum, Eur. J. Appl. Microbiol. Biotechnol.* **18:**71–74.

Ng, T. K., and Zeikus, J. G., 1982, Differential metabolism of cellobiose and glucose by *Clostridium thermocellum* and *Clostridium thermohydrosulfuricum, J. Bacteriol.* **150**(3):1391–1399.

Ng, T. K., Weimer, P. J., and Zeikus, J. G., 1977, Cellulolytic and physiological properties of *Clostridium thermocellum, Arch. Microbiol.* **114:**1–7.

Ng, T. K., Ben-Bassat, A., and Zeikus, J. G., 1981, Ethanol production by thermophilic bacteria fermentation of cellulose substrates by co-cultures of *Clostridium thermocellum* and *C. thermohydrosulfuricum, Appl. Env. Microbiol.* **42**:231–240.

Ounine, K., Petitdemange, H., Raval, G., and Gay, R., 1985, Regulation and butanol inhibition of D-xylose and D-glucose uptake in *Clostridium acetobutylicum, Appl. Env. Microbiol.* **49** (4):874–878.

Palosaari, N. R., and Rogers, P., 1988, Purification and properties of the inducible coenzyme A-linked butyraldehyde dehydrogenase from *Clostridium acetobutylicum, J. Bacteriol.* **170:** 2971–2976.

Parkkinen, E., 1986, Conversion of starch into ethanol by *Clostridium thermohydrosulfuricum, Appl. Microbiol. Biotechnol.* **25**:213–219.

Patni, N. J., and Alexander, J. K., 1971, Catabolism of fructose and mannitol in *Clostridium thermocellum:* Presence of phosphoenolpyruvate: Fructose phosphotransferase, fructose 1-phosphate kinase, phosphoenolpyruvate: Mannitol phosphotransferase, and mannitol 1-phosphate dehydrogenase in cell extracts, *J. Bacteriol.* **105**(1):226–231.

Petitdemange, E., Caillet, F., Giallo, J., and Gaudin, C., 1984, *Clostridium cellulolyticum* sp. nov., a cellulolytic, mesophilic species from decayed grass, *Int. J. Syst. Bacteriol.* **34**(2):155–159.

Pierrot, P., Fick, M., and Engasser, J. M., 1986, Continuous acetone-butanol fermentation with high productivity by cell ultrafiltration and recycling, *Biotechnol. Lett.* **8**(4):253–256.

Prescott, S. G., and Dunn, C. G., 1959, The acetone-butanol fermentation, in: *Industrial Microbiology,* McGraw-Hill, New York, pp. 180–214.

Qureshi, N., and Maddox, I. S., 1987, Continuous solvent production from whey permeate using cells of *Clostridium acetobutylicum* immobilized by adsorption onto bonechar, *Enzyme Microb. Technol.* **9**:668–671.

Rao, G., and Mutharasan, R., 1987, Altered electron flow in continuous cultures of *Clostridium acetobutylicum* induced by viologen dyes, *Appl. Env. Microbiol.* **53**(6):1232–1235.

Reardon, K. F., Scheper, T.-H., and Bailey, J. E., 1987, Metabolic pathway rates and culture fluorescence in batch fermentations of *Clostridium acetobutylicum, Biotechnol. Prog.* **3**:153–167.

Reysenbach, A. L., Ravenscroft, N., Long, S., Jones, D. T., and Woods, D. R., 1986, Characterization, biosynthesis, and regulation of granulose in *Clostridium acetobutylicum, Appl. Env. Microbiol.* **52**(1):185–190.

Roffler, S. R., Blanch, H. W., and Wilke, C. R., 1987, Extractive fermentation of acetone and butanol: Process design and economic evaluation, *Biotechnol. Prog.,* **3**:131–140.

Roffler, S. R., Blanch, H. W., and Wilke, C. R., 1988, *In situ* extractive fermentation of acetone and butanol, *Biotechnol. Bioeng.* **31**:135–143.

Rogers, P., 1984, Genetics and biochemistry of *Clostridium* relevant to development of fermentation processes, in: *Advances in Applied Microbiology* (A. I. Laskin, ed.), Academic Press, New York, pp. 1–89.

Rogers, P., and Palosaari, N., 1987, *Clostridium acetobutylicum* mutants that produce butyraldehyde and altered quantities of solvents, *Appl. Env. Microbiol.* **53**:2761–2766.

Ross, D., 1961, The acetone-butanol fermentation, *Prog. Ind. Microbiol.* **3**:73–85.

Rothstein, D. M., 1986, *Clostridium thermosaccharolyticum* strain deficient in acetate production, *J. Bacteriol.* **165**(1):319–320.

Ryden, R., 1958, Development of anaerobic fermentation processes: Acetone-butanol, in: *Biochemical Engineering* (R. Steel, ed.), Heywood, London, pp. 125–148.

Schink, B., and Zeikus, J. G., 1983, *Clostridium thermosulfurogenes* sp. nov., a new thermophile that produces elemental sulphur from thiosulphate, *J. Gen. Microbiol.* **129**:1149–1158.

Scholote, D., and Gottschalk, G., 1986, Effect of cell recycle on continuous butanol-acetone fermentation with *Clostridium acetobutylicum* under phosphate limitation, *Appl. Microbiol. Biotechnol.* **24**(1):1–6.

Schoutens, G. H., Nieuwenhuizen, M. C. H., and Kossen, N. W. F., 1984, Butanol from whey ultrafiltrate: Batch experiments with *Clostridium beyerinckii* LMD 27.6, *Appl. Microbiol. Biotechnol.* **19:**203–206.

Schoutens, G. H., Nieuwenhuizen, M. C. H., and Kossen, N. W. F., 1985, Continuous butanol production from whey permeate with immobilized *Clostridium beyerinckii* LMD 27.6, *Appl. Microbiol. Biotechnol.* **21:**282–286.

Simon, E., 1947, The formation of lactic acid by *Clostridium acetobutylicum* (Weizman), *Arch. Biochem.* **13:**237–243.

Sioumis, A. A., 1987, Recovery of alcohols: A chemical approach utilizing lactones, *TIBTECH* **5:**215–217.

Spivey, M. J., 1978, The acetone/butanol/ethanol fermentation, *Process Biochem.* **13**(11):2–5.

Srinivas, S. P., and Mutharasan, R., 1987, Culture fluorescence characteristics and its metabolic significance in batch cultures of *Clostridium acetobutylicum, Biotechnol. Lett.* **9:**139–142.

Stephens, G. M., Holt, R. A., Gottschal, J. C., and Morris, J. G., 1985, Studies on the stability of solvent production by *Clostridium acetobutylicum* in continuous culture, *J. Appl. Bacteriol.* **59:**597–605.

Taya, M., Ishii, S., and Kobayashi, T., 1985, Monitoring and control for extractive fermentation of *Clostridium acetobutylicum, J. Ferment. Technol.* **63**(2):181–187.

Terracciano, J. S., and Kashket, E. R., 1986, Intracellular conditions required for initiation of solvent production by *Clostridium acetobutylicum, Appl. Env. Microbiol.* **52**(1):86–91.

Terracciano, J. S., Schreurs, W. J. A., and Kasket, E. R., 1987, Membrane H $^+$ conductance of *Clostridium thermoaceticum* and *Clostridium acetobutylicum:* Evidence for electrogenic Na $^+$/H $^+$ antiport in *Clostridium thermoaceticum, Appl. Env. Microbiol.* **53:**782–786.

Turunen, M., Parkkinen, E., Londesborough, J., and Korhola, M., 1987, Distinct forms of lactate dehydrogenase purified from ethanol- and lactate-producing cells of *Clostridium thermohydrosulfuricum, J. Gen. Microbiol.* **133:**2865–2873.

Van der Westhuizen, A., Jones, D. T., and Woods, D. R., 1982, Autolytic activity and butanol tolerance of *Clostridium acetobutylicum, Appl. Env. Microbiol.* **44**(6):1277–1282.

Volesky, B., and Szczesny, T., 1983, Bacterial conversion of pentose sugars to acetone and butanol, *Adv. Biochem. Eng./Biotechnol.* **27:**101–117.

Volesky, B., Mulchandani, A., and Williams, J., 1981, Biochemical production of industrial solvents (acetone-butanol-ethanol) from renewable resources, *Ann. N.Y. Acad. Sci.* **369:**205–218.

Vollherbst-Schneck, K., Sands, J. A., and Montenecourt, B. S., 1984, Effect of butanol on lipid composition and fluiditty of *Clostridium acetobutylicum* ATCC 824, *Appl. Env. Microbiol.* **47:**193–194.

Walther, R., Hippe, H., and Gottschalk, G., 1977, Citrate, a specific substrate for the isolation of *Clostridium sphenoides, Appl. Env. Microbiol.* **33:**955–962.

Walton, M. T., and Martin, J. L., 1979, Production of butanol-acetone by fermentation, in: *Microbial Technology*, Vol. 1 (H. J. Peppler and D. Perlman, eds.), Academic Press, New York, pp. 187–209.

Wayman, M., and Parekh, R., 1987, Production of acetone-butanol by extractive fermentation using dibutylphthalate as extractant, *J. Ferment. Technol.* **65:**295–300.

Welsh, F. W., and Veliky, I. A., 1984, Production of acetone-butanol from acid whey, *Biotechnol. Lett.* **6**(1):61–64.

Welsh, F. W., Williams, R. E., and Veliky, I. A., 1986, A note on the effect of nitrogen source on growth of and solvent production by *Clostridium acetobutylicum, J. Appl. Bacteriol.* **61:**413–419.

Welsh, F. W., Williams, R. E., and Veliky, I. A., 1987, Organic and inorganic nitrogen source effects on the metabolism of *Clostridium acetobutylicum, Appl. Microbiol. Biotechnol.* **26:**369–372.

Wiegel, J., 1980, Formation of ethanol by bacteria. A pledge for the use of extreme thermophilic anaerobic bacteria in industrial ethanol fermentation process, *Experientia* **36:** 1434–1446.

Wiegel, J., Ljungdahl, L. G., and Rawson, J. R., 1979, Isolation from soil and properties of the extreme thermophile *Clostridium thermohydrosulfuricum, J. Bacteriol.* **139**(3):800–801.

Williamson, V. M., and Paquin, C. E., 1987, Homology of *Saccharomyces cerevisiae* ADH4 to an iron-activated alcohol dehydrogenase from *Zymomonas mobilis, Mol. Gen. Genet.* **209:**374–381.

Yarovenko, V. L., 1964, Principles of the continuous alcohol and butanol-acetone fermentation processes, in: *Continuous Cultivation of Microorganisms* (I. Malek, ed.), Czechoslovakian Academy of Sciences, Prague, pp. 205–217.

Yerushalmi, L., and Volesky, B., 1985, Importance of agitation in acetone-butanol fermentation, *Biotechnol. Bioeng.* **28:**1297–1305.

Yerushalmi, L., Volesky, B., and Szczesny, T., 1985, Effect of increased hydrogen partial pressure on the acetone-butanol fermentation by *Clostridium acetobutylicum, Appl. Microbiol. Biotechnol.* **22:**103–107.

Youngleson, J. S., Santangelo, J. D., Jones, D. T., and Woods, D. R., 1988, Cloning and expression of a *Clostridium acetobutylicum* alcohol dehydrogenase gene in *Escherichia coli, Appl. Env. Microbiol.* **54:**676–682.

Yu, E. K. C., and Saddler, J. N., 1983, Enhanced acetone-butanol fermentation by *Clostridium acetobutylicum* grown on D-xylose in the presence of acetic or butyric acid, *FEMS Microbiol. Lett.* **18:**103–107.

Yu, E. K. C., Deschatelets, L., and Saddler, J. N., 1984, The bioconversion of wood hydrolyzates to butanol and butanediol, *Biotechnol. Lett.* **6**(5):327–332.

Yu, E. K. C., Chan, M. K. H., and Saddler, J. N., 1985, Butanol production from cellulosic substrates by sequential coculture of *Clostridium thermocellum* and *C. acetobutylicum, Biotechnol. Lett.* **7**(7):509–514.

Zappe, H., Jones, D. T., and Woods, D. R., 1986, Cloning and expression of *Clostridium acetobutylicum* endoglucanase, cellobiase and amino acid biosynthesis genes in *Escherichia coli, J. Gen. Microbiol.* **132:**1367–1372.

Zappe, H., Jones, D. T., and Woods, D. R., 1987, Cloning and expression of a xylanase gene from *Clostridium acetobutylicum* P262 in *Escherichia coli, J. Microbiol. Biotechnol.* **27:**57–63.

Zeikus, J. G., 1979, Thermophilic bacteria: Ecology, physiology and technology, *Enzyme Microb. Technol.* **1:**243–252.

Zeikus, J. G., 1980, Chemical and fuel production by anaerobic bacteria, *Annu. Rev. Microbiol.* **34:**423–464.

Zeikus, J. G., 1985, Biology of spore-forming anaerobes, in: *Biology of Industrial Microorganisms* (A. L. Demain and N. A. Solomon, eds.), Benjamin Cummings, Menlo Park, CA, pp. 79–114.

Zeikus, J. G., Ben-Bassat, A., and Hegge, P.W., 1980, Microbiology of methanogenesis in thermal, volcanic environments, *J. Bacteriol.* **143**(1):432–440.

Zeikus, J. G., Ben-Bassat, A., Ng, T. K., and Lamed, R. J., 1981, Thermophilic ethanol fermentations, in: *Trends in the Biology of Fermentations* (A. Hollander, ed.), Plenum Press, New York, pp. 441–461.

Acetogenic and Acid-Producing Clostridia

5

LARS G. LJUNGDAHL, JEROEN HUGENHOLTZ, and JUERGEN WIEGEL

1. INTRODUCTION

Clostridia degrade a wide variety of organic compounds to mixtures of products that include acids, alcohols, CO_2, and H_2. The most commonly formed acids are acetate, propionate, and butyrate, but formate, lactate, succinate, and caproate are also observed as products. In addition, clostridia that ferment amino acids produce valerate, isovalerate, isobutyrate, and other similar acids. Here we review clostridia that form acids as the only or most prominent products. We have excluded clostridia that form acids in addition to solvents such as ethanol, acetone, and butanol (see Chapter 4). However, it should be noted that some clostridia can be manipulated physiologically to form mainly acids or solvents. Typical examples of such bacteria are *C. thermosaccharolyticum, C. thermocellum, C. thermohydrosulfuricum, C. saccharolyticum,* and *C. acetobutylicum.*

Acid-generating clostridia generally produce a mixture of acids, mostly acetate and butyrate. In industrial fermentations, the formation of a single product is advantageous because its recovery would be simplified. In this respect the homoacetogenic clostridia are unique. They form essentially only acetate in fermentations of hexoses and pentoses. They have been considered heterotrophs but it is now established that they also grow autotrophically on a mixture of molecular hydrogen and CO_2 as the only energy and carbon source. Energy for cell growth is obtained by the reduction of CO_2 to acetate via a newly established autotrophic pathway. The versatility of the acetogenic clostridia is even greater; they synthesize acetate

---

LARS G. LJUNGDAHL and JEROEN HUGENHOLTZ ● Center for Biological Resource Recovery, and Department of Biochemistry, University of Georgia, Athens, Georgia 30602, U.S.A. JUERGEN WIEGEL ● Center for Biological Resource Recovery, and Department of Microbiology, University of Georgia, Athens, Georgia 30602, U.S.A.

from several one-carbon compounds, including the methyl groups of methoxyphenols. The homoacetogens, forming a single product from several substrates, appear to be ideal for the microbial production of acetate. Consequently, a large portion of this chapter concerns the acetogens. We will, however, briefly consider the production of propionate, butyrate, caproate, formate, succinate, and products of degradations of amino acids by clostridia.

2. ACETOGENIC AND ACIDOGENIC CLOSTRIDIA

The genus *Clostridium* contains a large number of spore-forming, rod-shaped, anaerobic bacteria. Cato *et al.* (1986), in the last edition of *Bergey's Manual of Systematic Bacteriology*, list 90 species. All of them produce acids and most yield solvents. We prepared Table I based on the description by Cato *et al.* (1986), the summary by Gottschalk *et al.* (1981), and a few more recent publications. We only list clostridial species that are considered non-pathogenic and produce acids as their main product. The list is not complete but it illustrates the diversity of substrates used and products formed by acid-producing clostridia. It should be noted that the bacterial names to be used throughout this chapter are those now in common use. However, Cato *et al.* (1986) assigned new names (correct Latin spellings) to several clostridia. Examples are (the new names are within parentheses) *C. acidiurici (acidurici), C. formicoaceticum (formicaceticum), C. purinolyticum (purinilyticum), C. thermoaceticum (thermaceticum),* and *C. thermoautotrophicum (thermautotrophicum).*

With the exception of *C. kluyveri*, the bacteria listed in Table I produce acetic acid. The term acetogen is commonly applied to a bacterium that forms acetate whether the acetate is produced by a catabolic process such as fermentation or by an autotrophic-type synthesis. However, as before (Ljungdahl, 1986), we will classify bacteria producing acetate by both fermentative and synthetic processes as acetogenic, whereas those bacteria forming acetate only by fermentation will be referred to as acidogenic. Invariably, the acidogenic bacteria produce other acids in addition to acetate. Therefore, it is appropriate to subgroup the acetogens and the acidogens.

2.1. The Acetogens

Acetogens are either saccharolytic or purinolytic. The saccharolytic acetogens, often called homoacetate-fermenting bacteria, include the mesophiles *C. aceticum* (Wieringa, 1940; Adamse, 1980; Braun *et al.*, 1981), *C. formicoaceticum* (Andreesen *et al.*, 1970), and *C. magnum* (Schink, 1984), as well as the thermophiles *C. thermoaceticum* (Fontaine *et al.*, 1942) and *C. thermoautotrophicum* (Wiegel *et al.*, 1981). Purinolytic acetogens are *C. acid-*

Table I. *Clostridium* Species Producing Acids as Major Products

Clostridial species	Substrates	Products Ac[a]	Pr	But	Cap	Suc	For	Lac	Other acids[b]	CO$_2$	H$_2$	NH$_3$	Ref.
aceticum	Fructose, ribose, C-1 compounds	+											Wieringa, 1940; Braun et al., 1981
acidiurici	Purines	+								+		+	Barker and Beck, 1942, Vogels and Van der Drift, 1976
aminobutyricum	γ-Aminobutyrate	+		+								+	Hardman and Stadtman, 1960a, 1963
aminovalericum	δ-Aminovalerate	+	+						+			+	Hardman and Stadtman, 1960b
arcticum	Hexoses, xylose	+	+										Jordan and McNicol, 1979
baratii[c]	Hexoses, disaccharides	+		+				+			+		Nakamura et al., 1973; Cato et al., 1982
barkeri	Hexoses, dissaccharides, nicotinic acid	+		+				+		+	+	+	Stadtman et al., 1972, Imhoff and Andreesen, 1979
bifermentans	Glucose, proteins, amino acids	+					+		+	+	+	+	Mead, 1971; Holdeman et al., 1977
butyricum	Starch, disaccharides, sugars	+		+		+	+			+	+		Cummins and Johnson, 1971; Jungerman et al., 1973
cellulovorans	Cellobiose	+		+			+	+		+	+		Sleat et al., 1984
clostridioforme	Hexoses, xylose	+		+			+	+		+	+		Kaneuchi et al., 1976a; Cato and Salmon, 1976
coccoides	Disaccharides, sugar	+				+							Kaneuchi et al., 1976b
cochlearium	Glutamate, histidine	+		+								+	Barker, 1939; Laanbroek et al., 1979
cocleatum	Hexoses, some disaccharides	+					+	+			+		Kaneuchi et al., 1979
cyclindrosporum	Purines	+					+		+	+		+	Barker and Beck, 1942; Andreesen et al., 1985

(continued)

Table I. (*Continued*)

Clostridial species	Substrates	Ac[a]	Pr	But	Cap	Suc	For	Lac	Other acids[b]	CO$_2$	H$_2$	NH$_3$	Ref.
formicoaceticum	Fructose, gluconate, fumarate, C-1 compounds	+				+				+			Andreesen et al., 1970; O'Brien and Ljungdahl, 1972
kluyveri	Ethanol, acetate (+ CO$_2$), propanol, succinate		+	+	+						+		Barker and Taha, 1942; Schobert and Gottschalk, 1969
lortetii	Amino acids, glucose, fructose, maltose, starch	+	+	+					+		+	+	Oren, 1983
magnum	Sugars, butanediol, citrate, malate	+											Schink, 1984
malenominatum	Peptone, yeast extract, uric acid, lactate	+	+	+			+	+			+	+	Holdeman et al., 1977; Buckel, 1980
mangenotii	Amino acids, proteins, pyruvate	+	+	+			+		+		+	+	Prévot and Zimmès–Chaverou, 1947; Elsden and Hilton, 1978
oceanicum	Peptone, yeast extract, glucose	+	+					+	+		+		Smith, 1970
paraputrificum	Hexose, disaccharides, starch, steroids	+	+					+			+		Snyder, 1936; MacDonald et al., 1983
pasteurianum	Hexoses, disaccharides	+		+						+	+	+	Mortenson, 1966; Malette et al., 1974

Species	Substrates	Ac	Pr	But	Cap	Suc	For	Lac	Reference
pfennigii	Pyruvate, CO, methylated phenols	+		+				+	Krumholz and Bryant, 1985
polysaccharolyticum	Cellulose, starch	+	+	+				+	van Gylswyk et al., 1980
populeti	Cellulose, xylan, pectin, sugars	+	+	+		+	+	+	Sleat and Mah, 1985
propionicum	Amino acids, lactate	+	+	+			+	+	Cardon and Barker, 1947; Johns, 1952
purinolyticum	Purines	+					+	+	Dürre et al., 1981; Dürre and Andreesen, 1982a
putrefaciens	Glucose, proteins, amino acids	+		+	+	+	+	+	Sturgess and Drake, 1927; Roberts and Hobbs, 1968
sticklandii	Amino acids	+	+	+	+		+	+	Stadtman and McClung, 1957; Barker, 1981
subterminale	Amino acids, proteins	+		+			+	+	Holdeman et al., 1977; Elsden and Hilton, 1978
symbiosum	Glucose, fructose, glutamate	+		+		+	+	+	Kaneuchi et al., 1976a; Buckel, 1980
thermoaceticum	Glucose, fructose, xylose, C-1 compounds	+							Fontaine et al., 1942; Andreesen et al., 1973
thermoautotrophicum	Glucose, fructose, xylose, C-1 compounds	+							Wiegel et al., 1981
thermobutyricum	Carbohydrates, hexoses, pentoses	+		+			+	+	Wiegel et al., 1988
thermolacticum	Xylan, starch, cellobiose	+					+	+	Le Ruyet et al., 1985
tyrobutyricum	Hexoses, lactate, acetate	+	+	+			+	+	Roux and Bergere, 1977; Bryant and Barkey, 1956

[a] Ac = acetate, Pr = propionate, But = butyrate, Cap = caproate, Suc = succinate, For = formate, Lac = lactate.
[b] Valerate, glycine, isovalerate, isobutyrate.
[c] *C. baratii* (syn. *perenne* and *paraperfringens*).

iurici, C. cylindrosporum (Barker and Beck, 1942; Andreesen *et al.*, 1985), and *C. purinolyticum* (Dürre *et al.*, 1981). They are all mesophilic.

A third group of acetogenic clostridia, that was recently discovered, differs from the homoacetogens and purinolytic species. This group is exemplified by *C. pfennigii* (Krumholz and Bryant, 1985). Like the acetogens, it produces acetate from CO, but butyrate is also formed. It also metabolizes the methoxy group of a number of methoxylated compounds to butyrate in the presence of CO_2. However, in contrast to the homoacetogens, it is not saccharolytic and does not grow on methanol or H_2 plus CO_2.

2.1.1. The Homoacetogenic Clostridia

The homoacetogenic bacteria grow both heterotrophically, on several organic substrate, and autotrophically. They have in common the conversion of fructose to acetate (Reaction 1).

$$C_6H_{12}O_6 \rightarrow 3CH_3COOH \tag{1}$$

Glucose is similarly utilized, but only by *C. thermoaceticum*, *C. thermoautotrophicum*, and *C. magnum*. Galactose is used as substrate by *C. thermoautotrophicum*. Xylose is fermented by *C. magnum*, *C. thermoaceticum*, and *C. thermoautotrophicum*, and ribose by *C. formicoaceticum* and *C. aceticum* according to Reaction 2.

$$2C_5H_{10}O_5 \rightarrow 5CH_3COOH \tag{2}$$

Other substrates include oxidized sugars, such as glucuronate and gluconate, mannitol, glycerol, some organic acids and amino acids. Carbon sources metabolized by acetogenic clostridia are detailed by Ljungdahl (1983) and Fuchs (1986).

The most intriguing property of the homoacetogens is their ability to grow autotrophically on a gas mixture of H_2 and CO_2 as the only energy and carbon source. In addition, they grow on other one-carbon compounds, including CO, formate, and methanol with acetate as the product (Reactions 3–6).

$$2CO_2 + 4H_2 \rightarrow CH_3COOH + 2H_2O \tag{3}$$

$$4CO + 2H_2O \rightarrow CH_3COOH + 2CO_2 \tag{4}$$

$$4HCOOH \rightarrow CH_3COOH + 2CO_2 + 2H_2O \tag{5}$$

$$4CH_3OH + 2CO_2 \rightarrow 3CH_3COOH + 2H_2O \tag{6}$$

The homoacetogens use the newly discovered acetyl-CoA pathway for the synthesis of acetate from one-carbon precursors. Acetyl-CoA, which is the first two-carbon product of the autotrophic fixation of CO_2, is either used for the synthesis of cell carbon or converted to acetate. The acetyl-CoA pathway established in investigations carried out with *C. thermoacet-icum* growing heterotrophically on glucose was recently reviewed (Wood *et al.*, 1986a,b; Ljungdahl, 1986; Fuchs, 1986, see also Section 3.1.1).

Autotrophy was only recently considered a common property of acetogens, even though in 1940 Wieringa demonstrated that *C. aceticum* could grow on H_2 and CO_2 and, subsequently, total synthesis of acetate from CO_2 was demonstrated in *C. thermoaceticum* (Fontaine *et al.*, 1942; Wood, 1952a,b). The former bacterium was considered lost until re-discovered by Braun *et al.* (1981) and *C. thermoaceticum* was known to grow only heterotrophically as shown in Reactions 7 and 8. Reaction 8 represents the total synthesis of acetate from CO_2 (Wood, 1952a) and demonstrates that CO_2 is used as an acceptor of electrons generated in the fermentation of glucose (Reaction 7).

$$C_6H_{12}O_6 + 2H_2O \rightarrow 2CH_3COOH + 2CO_2 + 8H^+ + 8e^- \quad (7)$$

$$2CO_2 + 8H^+ + 8e^- \rightarrow CH_3COOH + 2H_2O \quad (8)$$

$$\text{Sum: } C_6H_{12}O_6 \rightarrow 3CH_3COOH \quad (1)$$

However, hydrogenase was demonstrated in *C. thermoaceticum* (Drake, 1982; Martin *et al.*, 1983). Subsequently, it was established that H_2 serves as a source of electrons, and that *C. thermoaceticum* grows on H_2 plus CO_2 as well as on CO (Kerby and Zeikus, 1983) and methanol (Wiegel and Garrison, 1985). Autotrophic growth by *C. formicoaceticum* has now also been demonstrated (Bryson and Drake, 1988) although it was previously not considered to have this capacity (Braun *et al.*, 1981). It should be noted that growth of *C. magnum* on H_2 plus CO_2 has not been shown (Schink, 1984).

A further important metabolic potential of homoacetogens is their ability to use the *O*-methyl group of methoxylated aromatic acids, as demonstrated by the formation of acetate and gallic acid from syringic acid and CO_2 (Reaction 9). The capacity to use the *O*-methyl group of a large variety of phenylmethylethers was first observed with *Acetobacterium woodii*, a non-spore-forming homoacetogen (Bache and Pfennig, 1981). This capacity has now also been observed with *C. pfennigii* (Krumholz and Bryant, 1985)

$$2 \text{ (syringic acid)} + 2CO_2 + 2H_2O \longrightarrow 2 \text{ (gallic acid)} + 3CH_3COOH \quad (9)$$

Syringic acid   Gallic acid

and other acetogens (Cornish-Frazer and Young, 1985) including *C. formicoaceticum* (Bryson and Drake, 1988) and *C. thermoaceticum* (Daniel and Drake, 1988). In Reaction 9, it is indicated that CO_2 is required for the formation of acetate from *O*-methyl groups. Recently, Wu *et al.* (1988) discovered that in *C. thermoaceticum*, CO rather than CO_2 is involved. The reaction may therefore occur according to Reaction 10.

$$\underset{\text{Syringic acid}}{\text{H}_3\text{CO}\underset{\text{OH}}{\overset{\text{COOH}}{\diagdown}}\text{OCH}_3} + 2\text{CO} + 2\text{H}_2\text{O} \longrightarrow \underset{\text{Gallic acid}}{\text{HO}\underset{\text{OH}}{\overset{\text{COOH}}{\diagdown}}\text{OH}} + 2\text{CH}_3\text{COOH} \quad (10)$$

It was realized only recently that acetogenic bacteria are ubiquitous in anaerobic ecological systems (Braun *et al.*, 1979; Wiegel *et al.*, 1981). Their importance in these systems has yet to be assessed. It has been pointed out that acetate plays a key role in the global cycle of methane (Ohwaki and Hungate, 1977; Vogels, 1979; Mah, 1981) and that about 75% of methane obtained by anaerobic degradation of organic material in nature is formed with acetate as an intermediate. Clearly, the homoacetogens are important in this process, as has been suggested for the anaerobic degradation of cellulose to methane (Ljungdahl and Eriksson, 1985).

Recent findings that the homoacetogens are able to use *O*-methyl groups of phenylmethylethers indicate a role for these bacteria in the degradation of lignin (Young and Frazer, 1987). This degradation is considered to be exclusively an aerobic process involving H_2O_2-requiring lignin peroxidases (Kirk and Farrell, 1987). However, anaerobic degradation of lignin has also been observed (Benner *et al.*, 1984). The oxidative degradation of lignin produces several metabolic products including cinnamic, caffeic, benzoic, vanillic, and ferulic acids. These compounds are then degraded in the anaerobic environment by bacterial consortia that include acetogenic bacteria (Colberg and Young, 1985a,b). The complete degradation of syringic acid to methane and CO_2 has been accomplished with a defined coculture consisting of *Acetobacterium woodii*, *Pelobacter acidigallici*, and *Methanosarcina barkeri* (Schink and Pfennig, 1982). The syringic acid is first demethylated by the acetogen to yield acetate and gallic acid, the latter being completely fermented to acetate by *P. acidigallici*. Acetate formed by the action of the two bacteria is then converted to methane and CO_2 by the methanogen. It should be possible to replace *A. woodii* in the defined coculture with either *C. formicoaceticum*.

It should be emphasized that many homoacetogens have now been described that are not spore formers (see review by Ljungdahl, 1986). Consequently, they have not been placed in the genus *Clostridium*. They are species of the genera *Acetoanaerobium* (Sleat *et al.*, 1985), *Acetobacterium*

(Balch *et al.*, 1977), and *Sporomusa* (Möller *et al.*, 1984). These as well as clostridial acetogens have been found to be of importance in the intestinal tracts of man, animals, and insects (Prins and Lankhorst, 1977; Lajoie *et al.*, 1988; Breznak *et al.*, 1988). Their roles in these small ecosystems are apparently to serve as acceptors in interspecies transfer of H_2—a process important for syntrophic associations of bacteria in anaerobic environments (Iannotti *et al.*, 1973).

2.1.2. The Purinolytic Acetogenic Clostridia

The fermentation of uric acid and other purines to acetate, CO_2, and ammonia is the common property of the purinolytic clostridia. The fermentation is accompanied by the reduction of CO_2 to acetate (Karlsson and Barker, 1949; Schulman *et al.*, 1972). This is shown in Reactions 11–13 which summarize the fermentation of hypoxanthine.

$$4C_5H_4N_4O + 28H_2O \rightarrow 16NH_3 + 12CO_2 + 4CH_3COOH + 8H^+ + 8e^- \quad (11)$$

$$2CO_2 + 8H^+ + 8e^- \rightarrow CH_3COOH + 2H_2O \quad (12)$$

$$\text{Sum: } 4C_5H_4N_4O + 26H_2O \rightarrow 16NH_3 + 10CO_2 + 5CH_3COOH \quad (13)$$

Reaction 12 is identical with Reaction 8 and involves a total synthesis of acetate from CO_2. However, it is pointed out later in this chapter that the pathway for acetate synthesis in the purinolytic bacteria is partly different from that in the homoacetogenic bacteria (Ljungdahl, 1984).

Table I lists three species of purinolytic acetogenic clostridia. There has been some doubt as to whether these are strains of the same species or of different species (Cato *et al.*, 1986). Phenotypically they are very similar; however, 16S rRNA cataloging (Tanner *et al.*, 1982) and DNA-DNA hybridization data (Schiefer-Ullrich *et al.*, 1984; Andreesen *et al.*, 1985) demonstrate clearly that *C. acidiurici*, *C. cylindrosporum*, and *C. purinolyticum* are different species.

The purinolytic clostridia appear to be very specialized. They utilize uric acid and a large variety of other purines as substrates. They are also able to grow on glycine, which is a degradation product of the purines, and on derivatives of glycine (Schiefer-Ullrich *et al.*, 1984). Since uric acid is a main component of droppings from birds, reptiles, and insects, it is not surprising that the purinolytic clostridia are widely distributed and easily isolated from a large variety of soils (Barker and Beck, 1942). They are also present in fecal material of birds and may be responsible for the degradation of uric acid in the intestinal tract of man and animals (Sørensen, 1978).

It is noteworthy that the metabolism of purinolytic bacteria is dependent on selenium (Dürre and Andreesen, 1982b) and that a lack of selenium in the diet may affect the metabolism of purines in the intestinal tract.

2.2. The Acidogens

The acidogenic clostridia can be considered as two groups: the saccharolytic and the amino acid-fermenting (including the proteolytic). This division is very artificial since several species will fall into both groups. The first group contains the clostridia that ferment carbohydrates consisting of simple sugars, disaccharides, oligosaccharides, and polymers such as cellulose (*C. populeti*), starch (*C. polysaccharolyticum*), or xylan (*C. thermolacticum*). Also included in this group are *Clostridia* that grow on such compounds as ethanol, acetate, and lactate (*C. kluyveri* and *C. tyrobutyricum*). The second group contains proteolytic and/or amino acid-degrading clostridia, some of which can also ferment sugars.

2.2.1. The Saccharolytic Clostridia

The saccharolytic clostridia grow heterotrophically on a variety of carbohydrates, producing a mixture of acids. Most species convert glucose to acetate according to Reaction 13. They differ from the acetogens in that they do not synthesize acetate from CO_2 or other C_1 compounds by the Wood autotrophic acetyl-CoA pathway.

$$C_6H_{12}O_6 + 2H_2O \rightarrow 2CH_3COOH + 2CO_2 + 4H_2 \qquad (13)$$

Reaction (13) involves the Embden–Meyerhof glycolytic pathway, a pyruvate:ferredoxin oxidoreductase (Von Hugo *et al.*, 1972) and a ferredoxin-linked hydrogenase (Mortenson and Chen, 1974). Acetate and hydrogen are indeed found as major fermentation products with saccharolytic clostridia but not in the amounts predicted by Reaction 13. Usually H_2 (or reducing equivalents originating from ferredoxin) is used for the reduction of acetate in the synthesis of organic acids such as butyrate, caproate, and succinate.

Many of the species can also utilize other hexoses such as fructose, galactose, and mannose; disaccharides including cellobiose, lactose, and maltose; and pentoses represented by xylose. Other substrates that are fermented by some saccharolytic clostridia include glycerol, mannitol, sorbitol, inositol, and polymeric carbohydrates.

Several species are specialists, e.g., *C. cellulovorans*, which degrades cellobiose (Sleat *et al.*, 1984), and *C. polysaccharolyticum*, which uses only the polymers cellulose and starch (van Gylswyk *et al.*, 1980). A special case is *C. kluyveri*, which can metabolize ethanol (or propanol), in the absence or presence of acetate (or succinate), to butyrate and caproate (Kenealy and Waselefsky, 1985).

The fermentation pathway performed by saccharolytic clostridia is exemplified by the clostridial-type strain *C. butyricum*. That produces butyric acid as the major fermentation product together with CO_2, acetate, and H_2 (Reaction 13). The pathway of this fermentation, which is found in approximately 50% of all clostridial species that have been isolated to date, will be discussed later in this chapter (Section 3.2.1). Other fermentations found in saccharolytic clostridia are those leading to the production of propionate by *C. arcticum* (Jordan and McNicoll, 1979), succinate by *C. coccoides* (Kaneuchi *et al.*, 1976a), and lactate by *C. barkeri* (Stadtman *et al.*, 1972).

Saccharolytic clostridia are ubiquitous in nature (being spore formers), and have been isolated from a large variety of sources. Saccharolytic clostridia seem to play an important role in the breakdown of plant material under anaerobic conditions (Ljundahl and Eriksson, 1985). Many species, such as *C. butyricum* (Schink *et al.*, 1981), are able to degrade pectin allowing access to the inside of the plant. Furthermore, several clostridia contain cellulolytic enzymes, allowing the breakdown of plant cell walls and ultimately the whole plant. clostridia can thus be found in large numbers in silages (Gibson *et al.*, 1958) and they have been isolated from numerous plant and vegetable sources (Veldkamp, 1965). They can cause major problems during the conservation of fruits and vegetables by their production of butyric acid, whose smell and taste is most undesirable in foods (Gottschalk *et al.*, 1981). They are also a nuisance in the dairy industry where *C. butyricum* is often found in butter. *Clostridium tyrobutyricum* is the major cause of fouling or late blowing of cheese (Goudkov and Sharpe, 1966) because it is able to convert lactate to butyrate in the presence of acetate (Roux and Bergere, 1977). Saccharolytic clostridia have the ability to grow in most environments where the oxygen tension is sufficiently low. They are usually not fastidious and require no or very few growth factors (often only biotin, *p*-aminobenzoate). Some species, e.g., *C. arcticum*, *C. butyricum*, and *C. pasteurianum*, fix molecular nitrogen and consequently do not need another nitrogen source. Most known acidogenic saccharolytic clostridia are mesophilic. Only two thermophilic acidogen are well recognized (Table I). *Clostridium thermolacticum* produces lactate as the major product (Le Ruyet *et al.*, 1985), while *C. thermobutyricum* produces predominantly butyrate (Wiegel *et al.*, 1988). However, as was already discussed, several saccharolytic acetogens are thermophilic. One acidogenic species—*C. arcticum* —has psychrophilic properties. Its optimum growth temperature is 22–25°C, but it is able to grow at temperatures as low as 5°C (Jordan and McNicoll, 1979).

2.2.2. The Amino Acid-Fermenting and Proteolytic Clostridia

The proteolytic clostridia are able to hydrolyze proteins and ferment amino acids. The presence of these clostridia in cultures can be recognized by the appearance of branched chain fatty acids and other products of deaminated amino acids in the culture fluid. Most proteolytic clostridia can degrade a large number of substrates. Besides utilizing proteins and most amino acids, they often ferment carbohydrates. However, several protein and amino acid specialists do exist, which vary widely in their substrate specificities. For instance, *C. sticklandii* degrades or converts 14 different amino acids and four purines (Stadtman and McClung, 1957). These act either as electron donors or as electron acceptors in a coupled redox reaction called the Stickland reaction (see Section 3.2.5). *Clostridium cochlearium*, on the other hand, only ferments glutamate, glutamine, and histidine (Barker, 1939) and *C. malenominatum* grows only on threonine, glutamate, and tyrosine (Holdeman *et al.*, 1977). *Clostridium propionicum* has an unusual fermentation strategy for the degradation of alanine, threonine, pyruvate, and lactate. It uses the so-called acrylate pathway (Cardon and Barker, 1947) to produce propionate as the main fermentation product (see Section 3.2.2). In addition, it is able to ferment the amino acids valine, leucine, and isoleucine.

The known proteolytic clostridia are mesophiles, except *C. putrefaciens*, which is the only *Clostridium* species described as having true psychrophilic properties. Its optimum growth temperature is 15–22°C and growth is not observed at 37°C (Roberts and Hobbs, 1968). It should be noted that there are no known examples of thermophilic proteolytic clostridia. Although many clostridia have been isolated from marine environments, growing cells are generally not found in the sea because of the salt sensitivity of most clostridia. There are, however, two exceptions. *Clostridium oceanicum* was isolated from marine sediment and is able to grow in media containing up to 4% NaCl (Smith, 1970). It is not an obligate halophile, however, as it also grows in freshwater. Recently, a true obligate halophile, *C. lortetii*, was isolated from Dead Sea sediments (Oren, 1983). This species requires NaCl at concentrations of 1–2 M (5.8–11.6%). It is the only known obligate halophile within the genus.

3. METABOLIC PATHWAYS OF ACID-PRODUCING CLOSTRIDIA

The classical concept of clostridia is that, under anaerobic conditions, they carry out fermentations of organic compounds to form cell material and end products that include alcohols and acids, producing ATP by substrate level phosphorylation. The energy gain is comparatively small, with comparatively large amounts of fermentation products being formed per unit of bacterial cell yield. In contrast, the aerobes generally oxidize organic compounds completely to carbon dioxide and water, in a process involving a respiratory chain, producing ATP by electron transport-linked phosphorylation. It follows that aerobic bacteria have high growth yields and form very little, if any, fermentation products besides CO_2 and H_2O. If the goal is to produce useful products from complex organic matter (biomass), the use of anaerobic bacteria, including clostridia, is the obvious choice. Here we will discuss the fermentative pathways used by clostridia for the production of acids.

It should be realized, however, that the classical concepts of fermentation and respiration as outlined above are no longer applicable to at least some clostridia. As already noted, the acetogenic clostridia and other anaerobic bacteria are able to grow autotrophically by the reduction of carbon dioxide to acetyl-CoA (Wood *et al.*, 1986a,b; Ljungdahl, 1986; Fuchs, 1986). The acetyl-CoA is used for cell material and for the production of acetate and, in some bacteria, butyrate and caproate. Furthermore, the reduction of CO_2 to acetyl-CoA must be coupled to electron transport and ATP synthesis. Thus anaerobic bacteria are able to form products by fermentative processes but, in addition, some have the capacity to use CO_2 as an electron acceptor and to convert it to acetate and other compounds. A number of different electron donors can be utilized, particularly hydrogen, carbon monoxide, methanol, and a number of organic compounds including sugars. The pathway of autotrophic synthesis of acetyl-CoA in acetogenic clostridia will now be considered in more detail.

3.1. Metabolism in Acetogenic Bacteria

3.1.1. The Acetyl-CoA Autotrophic Pathway

The only homoacetogenic bacterium available for investigations until 1970 was *C. thermoaceticum*. Consequently, the acetyl-CoA pathway was largely elucidated using this bacterium. Fontaine *et al.* (1942), who first described *C. thermoaceticum*, found that it carried out the complete conversion of glucose, fructose, and xylose to acetate (Reactions 1 and 2). They postulated that the conversion proceeded according to Reactions 7 and 8 with Reaction 7 representing the fermentation of the hexoses via the glycolytic pathway and Reaction 8 the total synthesis of acetic acid from CO_2 (Fig. 1). Their postulate was confirmed by the experiments conducted by Barker and Kamen (1945) and Wood (1952a,b).

The pathway of acetate synthesis from CO_2 proved difficult to elucidate. Formate, the C_1 derivative of tetrahydrofolate, and the methylcorrinoid 5-methoxybenzimidazolyl-Co-methylcobamide (Co-methyl factor III_m) were shown to be intermediates in the reduction of CO_2 to the methyl group of acetate (Poston *et al.*, 1966; Ljungdahl *et al.*, 1966). A pathway was

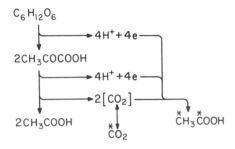

Figure 1. Fermentation of glucose to 3 moles of acetate showing CO_2 as an electron acceptor and as a source of one-third of the acetate. (From Ljungdahl, 1986, with permission of publisher.)

postulated for the formation of the methyl group, which essentially is the one accepted today (Ljungdahl and Wood, 1969). However, the final step of acetate synthesis, involving the condensation of the methyl group with "carbon monoxide" and coenzyme A to form acetyl-CoA, is a recent finding. It was based on the discovery of carbon monoxide dehydrogenase in *C. thermoaceticum* (Diekert and Thauer, 1978) and the realization that this enzyme catalyzes the condensation reaction (Hu *et al.*, 1982; Ragsdale and Wood, 1985).

The present concept of the fermentation of glucose and the autotrophic pathway of acetyl-CoA synthesis is outlined with heavy arrows in Fig. 2. The reduction of CO_2 to the methyl group involves the enzymes formate dehydrogenase (Reaction 6), formyl–H_4 folate synthetase (Reaction 7), methenyl–H_4 folate cyclohydrolase (Reaction 8), methylene–H_4 folate dehydrogenase (Reaction 9), and methylene–H_4 folate reductase (Reaction 10). The enzymes have been purified from *C. thermoaceticum* and other acetogenic clostridia. The formate dehydrogenase is a tungsten-selenium-iron protein (Yamamoto *et al.*, 1983). The cyclohydrolase and dehydrogenase reactions are catalyzed by a single bifunctional enzyme (Ljungdahl *et al.*, 1980) and the reductase is a nonheme iron-flavin protein (Clark and Ljungdahl, 1984; Han, 1988).

The incorporation of the methyl group of methyl–H_4 folate into acetate occurs via a corrinoid enzyme (CoE in Fig. 2; Hu *et al.*, 1984; Ragsdale *et al.*, 1987), and involves methyltransferase (Reaction 11) (Drake *et al.*, 1981) and carbon monoxide dehydrogenase (CO-Ni-E in Fig. 2). The latter enzyme catalyzes the reduction of CO_2 to CO and the condensation reaction of the methyl group on the corrinoid enzyme, CO, and coenzyme A to form acetyl-CoA (Reactions marked 4 in Fig. 2) (Ragsdale and Wood, 1985). Acetyl-CoA may then be used for cell material (Ljungdahl, 1986) or, through the actions of phosphotransacetylase and acetate kinase, be converted to acetate with generation of ATP (Schaupp and Ljungdahl, 1974; Drake *et al.*, 1981).

The corrinoid enzyme CoE and the methyltransferase that catalyzes Reaction 11 in Fig. 2 are two separate proteins. The latter is a dimer of mol.

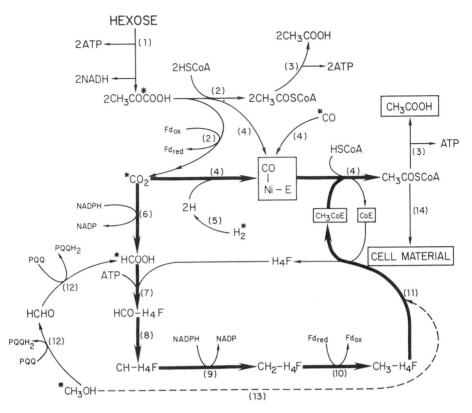

Figure 2. The autotrophic acetyl-CoA pathway (heavy arrows) and connected metabolism of hexoses, methanol, and CO: H_4F, tetrahydrofolate; CoE, corrinoid enzyme; CO-Ni-E, carbon monoxide dehydrogenase with CO moiety bound to nickel; PQQ, pyrroloquinoline quinone; Fd, ferredoxin. Enzymes or reaction sequences are as follows: 1, glycolysis; 2, pyruvate–ferredoxin oxidoreductase; 3, phosphotransacetylase and acetate kinase; 4, carbon monoxide dehydrogenase; 5, hydrogenase; 6, formate dehydrogenase; 7, formyl–H_4 folate synthetase; 8, methenyl–H_4 folate cyclohydrolase; 9, methylene–H_4 folate dehydrogenase; 10, methylene–H_4 folate reductase; 11, transmethylase; 12, methanol dehydrogenase; 13, methanol–cobamide methyltransferase; 14, anabolism.

wt. 58,800 (Drake *et al.*, 1981). It has not been investigated in detail. Ragsdale *et al.* (1987) reported that the corrinoid enzyme is also a dimer containing different subunits with mol. wt. of 33,000 and 55,000. It contains 5-methoxybenzimidazolylcobamide and an $[Fe_4S_4]$ cluster. It can be assumed that the nonheme iron cluster is involved in the reduction of the cobamide, in which the cobalt atoms must be reduced to a Co^+ state before methylation from methyl–H_4 folate can occur.

Carbon monoxide dehydrogenase has been purified from two acetogens: *C. thermoaceticum* (Ragsdale *et al.*, 1983a) and *Acetobacterium woodii*

(Ragsdale *et al.*, 1983b). In these bacteria, the enzyme has the composition $\alpha_3\beta_3$, the subunits having molecular weights of about 80,000 and 70,000, respectively. The hexameric enzyme contains six nickel atoms, three zinc or magnesium atoms, at least six [Fe_4S_4] clusters, and additional iron atoms. In addition to catalyzing the CO dehydrogenase and the condensation reactions, the enzyme catalyzes an exchange reaction as shown in Reaction 14 (Ragsdale and Wood, 1985; Pezacka and Wood, 1986).

$$^{14}CO + CH_3{}^{12}CO\text{-}SCoA \rightarrow CH_3{}^{14}CO\text{-}SCoA + {}^{12}CO \qquad (14)$$

The occurrence of this exchange reaction demonstrates that the methyl, carboxy, and CoA moieties must separate, and that each has a separate binding site on the enzyme. This has been confirmed experimentally. Electron proton resonance studies show that Ni and CO form a complex (Ragsdale *et al.*, 1983c, 1985). More recently, Shanmugasundaram *et al.* (1988) presented evidence that a tryptophan residue is involved in the binding site of CoA, and Pezacka and Wood (1988) identified the SH group of a cysteine residue to be the methyl binding site. It should be noted that CO dehydrogenase is activated by a novel enzyme termed CO dehydrogenase disulfide reductase, isolated from *C. thermoaceticum* (Pezacka and Wood, 1986), and that it is associated with the membrane. The latter has also been demonstrated for CO dehydrogenase in *C. thermoautotrophicum* by Hugenholtz *et al.* (1987). The importance of this finding for energy generation in acetogenic autotrophic bacteria will be discussed below.

By definition, autotrophic organisms require only inorganic compounds for growth, notably CO_2 as the only source of carbon. A source of electrons is also required and in acetogens it may be molecular hydrogen. It is oxidized by hydrogenase to yield protons and electrons that are accepted by electron carriers such as ferredoxin. Reduced ferredoxin may then be used by the CO dehydrogenase to reduce CO_2 to the nickel-bound "CO" moiety (Reaction 4, Fig. 2). Reduced ferredoxin is also used directly for the reduction of methylene–H_4 folate to methyl–H_4 folate (Reaction 10, Fig. 2). A NADP–ferredoxin oxidoreductase is present in *C. thermoaceticum* and it can generate NADPH from reduced ferredoxin. The NADPH is required in the formate dehydrogenase (Reaction 6, Fig. 2) and the methylene–H_4 folate dehydrogenase-catalyzed reactions (Reaction 9, Fig. 2). Hydrogenase (Reaction 5, Fig. 2) has been demonstrated in several acetogenic clostridia (Drake, 1982; Clark *et al.*, 1982), but the enzyme has not been well characterized from any of them. Two ferredoxins have been purified from *C. thermoaceticum*, one containing a single [Fe_4S_4] cluster (Elliott *et al.*, 1982) and a second containing two such clusters (Elliott and Ljungdahl, 1982).

Growth on CO [Reaction 4, Section 2.1.1], which serves both as a source of carbon and electrons, must be considered to be autotrophy. Here

CO is directly taken up by the CO dehydrogenase. Part of the CO is oxidized to CO_2 to obtain electrons for the reduction of CO_2 to the methyl group, via the H_4 folate pathway, and acetyl-CoA is formed in the condensation reaction (Reaction 4, Fig. 2) catalyzed by the CO dehydrogenase.

During heterotrophic growth on hexoses, two-thirds of the acetate produced is formed directly from the sugars, whereas one-third is synthesized via the acetyl-CoA pathway (Reactions 7, 8, and Fig. 1, and Section 2.1.1). The fermentation in *C. thermoaceticum* (Wood, 1952b) and *C. formicoaceticum* (O'Brien and Ljungdahl, 1972) follows the Embden–Meyerhof pathway to yield pyruvate. Pyruvate is further metabolized in the pyruvate–ferredoxin oxidoreductase reaction to yield acetyl-CoA (Reaction 2, Fig. 2). In clostridia, the carboxyl group of the pyruvate is generally considered to be oxidized to CO_2. This is also true for the acetogens. In addition, evidence has been presented that the carboxyl group may directly react with CO dehydrogenase, without going through CO_2 as shown in Fig. 2, to yield [CO-Ni-E] (Schulman *et al.*, 1972; Pezacka and Wood, 1984). The latter reaction requires pyruvate–ferredoxin oxidoreductase, thiamin pyrophosphate, ferredoxin, and CO dehydrogenase. It is apparent that the CO group of [CO-Ni-E] has three sources: CO, CO_2, and the carboxyl group of pyruvate.

The finding that clostridial acetogens grow on methanol is relatively recent (Braun *et al.*, 1981; Wiegel *et al.*, 1981; Wiegel and Garrison, 1985). The fermentations may be summarized as follows (Reactions 6 and 15–17):

$$CH_3OH + H_2O \rightarrow CO_2 + 6H^+ + 6e \qquad (15)$$
$$3CO_2 + 6H^+ + 6e \rightarrow 3CO + 3H_2O \qquad (16)$$

$$3CH_3OH + 3CO \rightarrow 3CH_3COOH \qquad (17)$$

$$\text{Sum: } 4CH_3OH + 2CO_2 \rightarrow 3CH_3COOH + 2H_2O \qquad (6)$$

The oxidation of methanol is pyrroloquinoline quinone-dependent (Duine *et al.*, 1984). The enzyme has been purified by Winters-Ivey (1987). It is an oxygen-sensitive protein of mol. wt. 110,000 composed of two apparently identical subunits. It catalyzes the oxidation of methanol and formaldehyde to formate (Reactions 12, Fig. 2). Other primary alcohols and aldehydes, including ethanol and acetaldehyde, also serve as substrates. Formate is then further oxidized to CO_2 (Reaction 6, Fig. 2) before being reduced to CO of [CO-Ni-E] and serving as the precursor of the carboxy group of acetyl-CoA (Reaction 4, Fig. 2). Methanol is a direct precursor of the methyl group of acetate. The incorporation may proceed either via methyl–H_4 folate or, more likely, by direct methylation of the corrinoid enzyme (Reaction 13, Fig. 2). The latter alternative is suggested by the finding of a methanol: 5-hydroxybenzimidazolylcobamide methyltransferase in methanogenic bacteria and the possible presence of a similar enzyme in the acetogen *Eubacterium limosum* (van der Meijden *et al.*, 1984). The oxidation products of methanol, formaldehyde, and formate, may also be precursors of the methyl group of acetyl-CoA. Incorporation of formate occurs via formyl–H_4 folate synthetase (Reaction 7, Fig. 2). Formaldehyde reacts chemically with H_4 folate to yield methylene–H_4 folate, which, of course, is an excellent precursor of methyl–H_4 folate (Reaction 10, Fig. 2).

3.1.2. Energetics of Homoacetogenic Clostridia

In the metabolism of hexoses, homoacetogenic bacteria generate 4 moles of ATP per mole of hexose by substrate level phosphorylation; two ATP molecules are formed during glycolysis to pyruvate and one for each pyruvate converted to acetate (Fig. 2). Growth studies of several acetogens, however, suggest the generation of additional energy in these organisms. The acetogens have unusually high growth yields when grown on sugars (Andreesen *et al.*, 1973; Tschech and Pfennig, 1984) and most acetogens are able to grow under autotrophic conditions with only H_2/CO_2, CO, HCOOH, or methanol/CO_2 as sources of carbon and energy (Fuchs, 1986; Ljungdahl, 1986). This strongly suggests that generation of energy is coupled to acetate synthesis via the autotrophic pathway as described in Fig. 2. No overall substrate level phosphorylation results from this pathway, so energy (ATP) must be generated by chemiosmotic mechanisms. Two separate mechanisms are required for this so-called oxidative phosphorylation: (1) proton pumping, presumably by membrane-associated electron transport, and (2) a membrane-associated H^+-ATPase capable of catalyzing ATP synthesis.

An H^+-ATPase was recently purified from *C. thermoaceticum* and *C. thermoautotrophicum* (Ivey and Ljungdahl, 1986, 1989; Mayer *et al.*, 1986). It is a typical F_1F_0 ATPase with the F_1 part composed of four different subunits in the stoichiometric $\alpha_3\beta_3\gamma\delta$. It lacks the supposedly regulatory ϵ subunit, present in F_1-ATPase of *Escherichia coli* and several other aerobic bacteria (Munoz, 1982). The F_0 part is different from that of other bacteria. It is composed of two different subunits (*a* 21,000 and *b* 12,300 mol. wt.) with the smaller *b* subunit presumably being the H^+ channel. It contains the DCCD (dicyclohexylcarbodiimide) binding site. The purified F_1F_0 complex has been reconstituted into proteoliposomes.

The other required mechanism for oxidative phosphorylation is the presence of electron transport in the cytoplasmic membrane, with simultaneous proton pumping to the outside. This requires membrane-associated enzymes that catalyze electron-donating and electron-accepting reactions, plus membrane-associated electron carriers. Those carriers identified in homoacetogenic clostridia include cytochromes, menaquinones (Gottwald *et al.*, 1975), and ferredoxin (Elliot and Ljungdahl, 1982). *Clostridium thermoaceticum* and *C. thermoautotrophicum* each contain two b-type cytochromes (Ivey 1987). Cytochrome b_{560} from *C. thermoautotrophicum* is a dimer of mol. wt. 70,000 with a midpoint redox potential of -200 mV, and cytochrome b_{556}, from the same bacterium, is a 160,000 mol. wt. protein with a redox potential of -48 mV. Almost identical cytochromes are present in *C. thermoaceticum*.

Several enzymes that catalyze redox reactions have been demonstrated either in association with, or integral components of, membranes prepared from *C. thermoautotrophicum* or *C. thermoaceticum*. Ivey (1987) obtained from *C. thermoautotrophicum* a flavoprotein of mol. wt. 15,000 and E_m -220 mV that catalyzes the oxidation of NADH in the reaction sequence outlined in Reaction 18, in which MQ is menaquinone.

$$NADH \rightarrow flavoprotein \rightarrow cyt\ b_{560} \rightarrow MQ \rightarrow Cyt\ b_{556} \qquad (18)$$

The reaction sequence involves carriers of protons and electrons alternating with carriers of only electrons, and seems to provide a mechanism for proton translocation as described by Mitchell (1966). Hydrogenase has been demonstrated in membrane preparations of *C. thermoautotrophicum* and has been shown to catalyze the reduction of the cytochromes using molecular hydrogen (Hugenholtz and Ljungdahl, 1988; and unpublished results). The reduction of membrane-bound cytochromes by hydrogenase has also been observed with *C. thermoaceticum* (Drake, 1982).

Recently, it was established that CO dehydrogenase and methylene–H_4 folate reductase are membrane-associated enzymes in *C. thermoautotrophicum* (Hugenholtz *et al.*, 1987). This is of particular significance because it has been postulated that these enzymes may be involved in energy generation in acetogenic bacteria (Thauer *et al.*, 1977; Fuchs, 1986). When membrane preparations containing the enzymes were exposed to CO, both the two b-type cytochromes and a flavoprotein were reduced. Further observations allowed the postulate (Hugenholtz *et al.*, 1987) that electrons generated in the CO dehydrogenase reaction reduce cytochrome b_{560} (E_m -200 mV) and, in turn, are ultimately used for the reduction of methylene–H_4 folate to methyl–H_4 folate (E_m -120 mV). The reduction of the high potential cytochrome b_{556} (E_m -48 mV) was always observed (Hugenholtz *et al.*, 1987). Presumably the electrons for this reduction originate from cytochrome b_{560} and reach the cytochrome b_{556} via menaquinone (E_m -74 mV). The role of this high potential cytochrome is not known. It is possible that it serves as an electron donor for the reduction of rubredoxin (E_m 0 mV), which is also an electron carrier with unknown function found in many anaerobes, including acetogens (Yang *et al.*, 1980).

All of these electron transport steps are summarized in Figure 3. Figure 3a focuses on the relative location in the membrane of the different components of the electron transport chain, while Figure 3b illustrates the (downhill) flow of electrons in thermodynamic terms. A hypothetical flow of electrons to NADPH (via ferredoxin) is also noted in the figure. There is currently no evidence to support this reaction. However, since *C. thermoaceticum* and *C. thermoautotrophicum* use NADPH in two reduction steps during acetate synthesis (see Fig. 2) and only NADH is generated in glycolysis, some mechanism must occur that transfers reducing equivalents from NADH to NADPH. Attempts to demonstrate this reaction in cell-free extracts of *C. thermoaceticum* have been unsuccessful, possibly because the enzymes responsible were membrane-associated and had been removed during the preparation of extracts for the assay.

We recently demonstrated that proton pumping occurs simultaneously with electron transport (Hugenholtz and Ljungdahl, 1988). When the artificial electron acceptor $K_3[Fe(CN)_6]$ is added to membranes of *C. thermoautotrophicum* together with the electron donor carbon monoxide, a proton motive force (Δp) is generated of approx. 150 mV, consisting mostly of a membrane potential ($\Delta \Psi$) and a small pH gradient (10–40 mV). The Δp can drive not only ATP synthesis (Ivey and Ljungdahl, 1986), but also the uptake and accumulation (up to 400-fold) of several amino acids (alanine, glycine, serine), and probably many other compounds. Results presented by Diekert *et al.* (1986) using *Acetobacterium woodii* whole-cell suspensions also demonstrate that energy is generated by the oxidation of CO.

3.1.3. The Purinolytic Fermentation

The purinolytic clostridia have in common with the homoacetogens the ability to synthesize acetate from CO_2 and a dependence on H_4 folate for this synthesis. They also share a dependence on trace elements including iron, tungsten, molybdenum, and selenium that are constituents of enzymes. However, there is no evidence for autotrophy and a part of the H_4 folate pathway may actually be used in reverse to oxidize methenyl–H_4 folate to CO_2 to gain energy. Be that as it may, the main products of fermentations of purines by the purinolytic bacteria are acetate, ammonia, and CO_2 (Reaction 13, Section 2.1.2), and a part of the acetate is formed entirely from CO_2, as shown by using $^{13}CO_2$ and mass analysis of the produced acetate (Schulmann *et al.*, 1972).

The pathway of the purine fermentation outlined in Figure 4 is covered in reviews by Barker (1961), Vogels and van der Drift (1976), and Ljungdahl (1984), and the reactions leading to acetate synthesis were considered by Ljungdahl and Wood (1982). Different purines are converted to xanthine (Dürre and Andreesen, 1983). In the case of uric acid, this is done by xanthine dehydrogenase, of which selenium is a required component (Wagner and Andreesen, 1979). The pyrimidine ring of xanthine is then

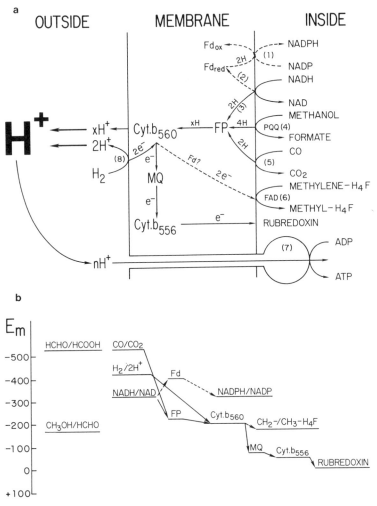

Figure 3. Summary of electron transport and energy generation in the homoacetogen *C. thermoautotrophicum.* (a) Membrane associated enzyme reactions, electron transport, proton translocation and ATP synthesis. Enzyme and abbreviations are 1, NADPH–ferredoxin oxidoreductase; 2, NADH–ferredoxin oxidoreductase; 3, NADH dehydrogenase; 4, methanol dehydrogenase; 5, carbon monoxide dehydrogenase; 6, methylene–H_4 folate reductase; 7, H^+-ATPase; 8, hydrogenase; Cyt, cytochrome; MQ, menaquinone; Fd, ferredoxin; FP, flavoprotein; PQQ, pyrroloquinoline quinone. (b) Electron acceptors with redox potentials and suggested flow of electrons.

opened with loss of NH_3 and CO_2 and formation of 4-aminoimidazole. The imidazole ring is now opened to yield formiminoglycine (Rabinowitz, 1963), which in turn is converted to glycine, NH_3, and methenyl–H_4 folate (Reaction 9, Fig. 4). The latter reaction is catalyzed by glycine formimino-transferase and formimino–H_4 folate cyclodeaminase (Uyeda and

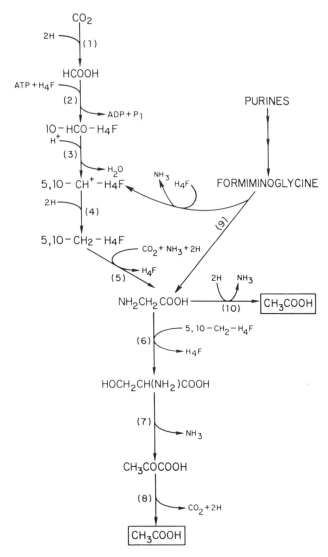

Figure 4. Pathway of the clostridial purinolytic fermentation. Enzymes are 1, formate dehydrogenase; 2, formyl–H_4 folate synthetase; 3, methenyl–H_4 folate cyclohydrolase; 4, methylene–H_4 folate dehydrogenase; 5, glycine decarboxylase; 6, serine hydroxymethyltransferase; 7, serine dehydratase; 8, pyruvate–ferredoxin oxidoreductase; 9, glycine formiminotransferase and formimino–H_4 folate cyclodeaminase; 10, glycine reductase.

Rabinowitz, 1965, 1967). Methenyl–H_4 folate may then be oxidized to CO_2 by reversal of Reactions 3, 2, and 1 of Fig. 4, or it may be reduced to methylene–H_4 folate (Reaction 4, Fig. 4). The first alternative allows the generation of ATP in the formyl–H_4 folate synthetase reaction (Reaction 2, Fig. 4) (Himes and Harmony, 1973).

Glycine is a key intermediate in the metabolism of purinolytic clostridia. As mentioned, it is formed from formiminoglycine. It can also be formed from methylene–H_4 folate, CO_2, NH_3, and reducing equivalents (NADH), as indicated in Reaction 5 (Fig. 4). This reaction, which is reversible, is catalyzed by the enzyme complex glycine decarboxylase consisting of four proteins (Champion and Rabinowitz, 1977; Waber and Wood, 1979). The formation of acetate from glycine may occur directly via Reaction 10 (Fig. 4), which is catalyzed by glycine reductase (Dürre and Andreesen, 1982a–c). Alternatively, acetate may be produced via Reactions 6–8 (Fig. 4), catalyzed by serine hydroxymethyltransferase, serine dehydratase, and pyruvate–ferredoxin oxidoreductase (Raeburn and Rabinowitz, 1971; Uyeda and Rabinowitz, 1971). The direct conversion of glycine to acetate appears to be the main pathway (Dürre and Andreesen, 1982c, 1983). Glycine reductase, which has been extensively studied (Stadtman, 1978; Tanaka and Stadtman, 1979), is a selenium-dependent enzyme and is ATP forming (Stadtman *et al.*, 1958).

Of the enzyme reactions involved in the purine fermentation, three are selenium-dependent, namely, formate dehydrogenase (Reaction 1, Fig. 4), glycine reductase (Reaction 10, Fig. 4), and xanthine dehydrogenase. Clearly the availability of selenium will very much influence the metabolism by purinolytic clostridia (Dürre and Andreesen, 1982b).

3.2. Metabolism in Acidophilic Bacteria

In this section we will discuss fermentation patterns used by acidophilic clostridia for the production of butyrate, caproate, propionate, and acrylate. Furthermore several other fermentation strategies will be considered which lead to the formation of succinate, formate, and some other organic acids. A substantial amount of knowledge regarding metabolism in acidogenic clostridia accumulated following the application of radioactive isotopes and the development of methods for protein purification and enzymology. This information is available in many microbiology and biochemistry textbooks (e.g., Gottschalk, 1985; Doelle, 1975) and a recent excellent review of clostridial fermentations by Rogers (1986).

3.2.1. Butyrate and Caproate

Clostridium butyricum is the type strain of the genus *Clostridium*. It is saccharolytic and metabolizes glucose producing butyrate, acetate, CO_2, and molecular hydrogen as fermentation products. This type of fermentation is typical for many *Clostridium* species (Table I).

The fermentation pathway is presented in Figure 5. Glucose is converted to pyruvate by the Embden–Meyerhof–Parnas glycolytic pathway (Reaction 1, Fig. 5). Pyruvate, which is the intermediate of hexose metabolism in all clostridia, is then simultaneously decarboxylated and oxidized by

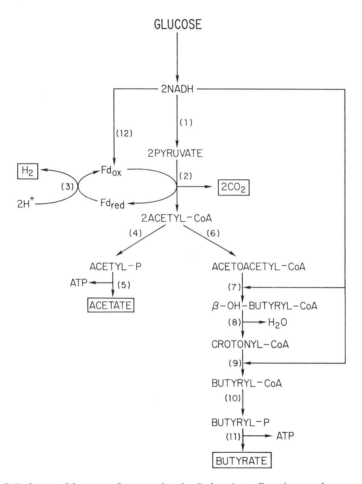

Figure 5. Pathway of butyrate fermentation in *C. butyricum.* Reactions and enzyme are 1, glycolysis; 2, pyruvate–ferredoxin oxidoreductase; 3, hydrogenase; 4, phosphotransacetylase; 5, acetate kinase; 6, acetyl-CoA acetyltransferase; 7, β-hydroxybutyryl-CoA dehydrogenase; 8, enoyl-CoA hydratase and 3-hydroxyacyl-CoA hydrolyase (crotonase); 9, butyryl-CoA dehydrogenase; 10, phosphotransbutyrylase; 11, butyrate kinase; 12, NADH–ferredoxin oxidoreductase.

the enzyme complex pyruvate–ferredoxin oxidoreductase to yield acetyl-CoA, CO_2, and reduced ferredoxin (Reaction 2, Fig. 5) (Uyeda and Rabinowitz, 1971). This is a rather typical clostridial reaction that is also present in the acetogenic *Clostridia.* The reduced ferredoxin is reoxidized in several reactions of which the most important involves H_2 evolution catalyzed by hydrogenase (Reaction 3, Fig. 5). It may also be used for NADPH production, i.e., catalyzed by NADPH–ferredoxin oxidoreductase (Jungermann *et al.,* 1973; Thauer *et al.,* 1977).

Acetyl-CoA is the central intermediate in product formation. In the butyrate fermentation, two acetyl-CoA molecules are combined to form acetoacetyl-CoA (Reaction 6, Fig. 5). It is subsequently reduced in three steps (Reactions 7–9, Fig. 5) to butyryl-CoA, with β-hydroxybutyryl-CoA and crotonyl-CoA as intermediates. In this reduction, the two equivalents of NADH generated in the glycolysis are used as electron donors. Finally, butyryl-CoA is converted to butyrate. The enzymes responsible for this are phosphotransbutyrylase (Twarog and Wolfe, 1962) and butyrate kinase (Valentine and Wolfe, 1960) (Reactions 10 and 11, Fig. 5). An alternative mechanism for the generation of butyrate from butyryl-CoA suggested in some publications would involve CoA transferase (Reaction 19).

$$CH_3CH_2CH_2CO\text{-}CoA + CH_3COOH \rightarrow$$
$$CH_3CH_2CH_2COOH + CH_3CO\text{-}CoA \qquad (19)$$

Although this mechanism is used by some clostridia, notably *C. kluyveri* (see below), it appears not to be the prevalent pathway in *C. butyricum* (Hartmanis, 1987).

It can be noted that the formation of butyrate and acetate can be considered to occur according to Reactions 20 and 21, respectively.

$$C_6H_{12}O_6 \rightarrow CH_3CH_2CH_2COOH + 2CO_2 + 2H_2 \qquad (20)$$

$$C_6H_{12}O_6 + 2H_2O \rightarrow 2CH_3COOH + 2CO_2 + 4H_2 \qquad (21)$$

In the butyrate fermentation 3 ATP equivalents are generated, whereas in the acetate fermentation 4 ATP are produced. Although acetate production seems to be energetically more favorable than butyrate production, *in vivo* most of the glucose is converted to butyrate with the stoichiometry shown in Reaction 22, yielding 3.3 ATP molecules (Wood, 1961).

$$4C_6H_{12}O_6 + 2H_2O \rightarrow$$
$$3CH_3CH_2CH_2COOH + 2CH_3COOH + 8CO_2 + 10H_2 \qquad (22)$$

It is not completely understood why clostridia choose the less energy-efficient butyrate fermentation. One explanation is that the formation of one equivalent of butyrate leads to less acidification of the organisms' environment than the formation of two equivalents of acetate. The formation of molecular hydrogen from NADH can be energetically unfavorable, affecting the production of acetate. Similarly, the production of higher proportions of butyrate may serve to consume excess reducing equivalents. The varying amounts of butyrate, acetate, and H_2 produced indicate that this fermentation is fairly flexible and can be influenced by changing the environmental conditions. Indeed, it has been observed in *C. butyricum* that butyrate production was stimulated by high external H_2 pressure (Junger-mann *et al.*, 1973, Gottschalk *et al.*, 1981). In principle, if the external H_2 concentration could be kept low, acetate production, would increase.

Clostridium kluyveri also produces butyric acid as a major fermentation product. However, it is not saccharolytic and it produces butyrate and caproate from ethanol and acetate as substrates as shown in Figure 6 (Barker and Taha, 1942). Ethanol is oxidized in two steps (Reactions 1 and 2, Fig. 6) to acetyl-CoA with NAD and NADP, respectively, as electron acceptors. The conversion of two acetyl-CoA molecules into butyryl-CoA is basically the same as found in *C. butyricum* (Fig. 5), except that NADPH instead of NADH is used as electron donor in the reduction of acetoacetyl-CoA to β-hydroxybutyryl-CoA. The production of butyrate from butyryl-CoA in *C. kluyveri* occurs via a cyclic mechanism catalyzed by CoA transferase (Stadtman, 1953). This enzyme transfers the CoA group from butyryl-CoA to acetate, regenerating acetyl-CoA (Reaction 7, Fig. 6).

The formation of butyrate from ethanol and acetate by *C. kluyveri* can be expressed by Reaction 23:

$$C_2H_5OH + CH_3COOH \rightarrow CH_3CH_2CH_2COOH + H_2O \qquad (23)$$

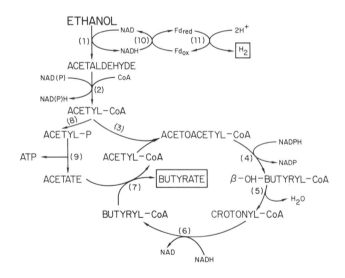

Figure 6. Pathway of butyrate formation from ethanol and acetate in *C. kluyveri*. Enzymes are 1, alcohol dehydrogenase; 2, acetaldehyde dehydrogenase; 3, acetyl-CoA acetyltransferase; 4, β-hydroxybutyryl-CoA dehydrogenase; 5, crotonase; 6, butyryl-CoA dehydrogenase; 7, CoA transferase; 8, phosphotransacetylase; 9, acetate kinase; 10, NADH-ferredoxin oxidoreductase; 11, hydrogenase.

When ethanol is present in the medium in excess of acetate, caproate can be formed instead of butyrate. Butyrate takes the place of acetate in the CoA transferase reaction, leading to the regeneration of butyryl-CoA instead of acetyl-CoA which, together with acetyl-CoA, is converted into β-oxocaproyl-CoA. This is subsequently reduced in two steps, via β-hydroxycaproyl-CoA to caproyl-CoA, resulting in the production of caproate. The overall equation for caproate formation is given in Reaction 24:

$$2C_2H_5OH + CH_3COOH \rightarrow CH_3(CH_2)_4COOH + 2H_2O \qquad (24)$$

The fermentation of ethanol and acetate to butyrate and caproate does not yield any ATP by substrate level phosphorylation or chemiosmotic mechanisms (Thauer *et al.*, 1977). However, in *C. kluyveri* cultures production of molecular hydrogen has been observed (Schoberth and Gottschalk, 1969). Thus, some acetyl-CoA is hydrolyzed to produce acetate and ATP, and the excess reducing equivalents lead to H_2 production.

Clostridium kluyveri is able to use other alcohols instead of ethanol in its fermentation. When propanol is used as a substrate, propionate and valerate are produced instead of butyrate and caproate (Kenealy and Waselefsky, 1985). The fermentation of *C. tyrobutyricum* is similar to that of *C. kluyveri*. It also uses acetate as electron acceptor, resulting in the production of butyrate, but uses lactate instead of ethanol as electron donor (Roux and Bergere, 1977). This organism is the cause of many problems in the dairy industry because of its ability to convert lactate to butyrate (Goudkov and Sharpe, 1966).

3.2.2. Propionate and Acrylate

Clostridium arcticum (Jordan and McNicoll, 1979), *C. novyi* (Holdemann *et al.*, 1977), and *C. propionicum* (Cardon and Barker, 1947) produce propionate as a major fermentation product. The pathway of the formation of propionate has been most extensively studied in *C. propionicum*. This organism is not saccharolytic but uses mostly C_3 compounds such as lactate and alanine as substrates, converting them to propionate and acetate. It also grows on threonine. The pathway of this conversion is completely different from that of the very thoroughly investigated pathway of the propionate fermentation, carried out by propionic acid bacteria, which involves several unique enzymes, including the biotin-dependent transcarboxylase (Wood and Kumar, 1985) and a B_{12}-dependent methylmalonyl-CoA mutase (Kellermeyer *et al.*, 1964). The clostridial pathway shown in Figure 7 is the so-called acrylate pathway (Cardon and Barker, 1947). The name is based on the fact that acrylate can be used as a substrate and that it was considered to be an intermediate in the formation of propionate from C_3 compounds. When lactate is metabolized, it serves as both electron donor and elec-

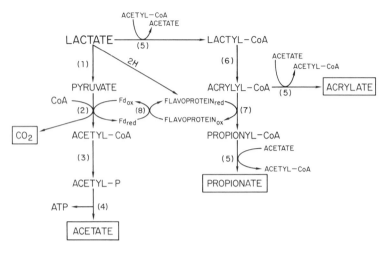

Figure 7. The acrylate pathway of propionate formation in *C. propionicum.* Enzymes are 1, lactate dehydrogenase; 2, pyruvate : ferredoxin oxidoreductase; 3, phosphotransacetylase; 4, acetate kinase; 5, CoA transferase; 6, lactyl-CoA dehydratase; 7, acrylyl-CoA reductase; 8, ferredoxin: flavoprotein oxidoreductase.

tron acceptor. Part of it is oxidized to pyruvate and undergoes the usual oxidation/decarboxylation reactions to acetyl-CoA, yielding acetate and CO_2, with the generation of ATP (Reactions 1–4, Fig. 7). The other part is converted to propionate. It involves the enzyme CoA transferase (Reaction 5, Fig. 7) to form lactyl-CoA, which is dehydrated and reduced to yield propionyl-CoA (Reactions 6 and 7, Fig. 7). Acrylyl-CoA has not been identified as a free intermediate in the oxidation of lactyl-CoA to propionyl-CoA. Recently, however, an enzyme was purified from *C. propionicum* which converts acrylyl-CoA to lactyl-CoA, but not the reverse reaction (Kuchta and Abeles, 1985). It was concluded that acrylyl-CoA was formed not as a free but as an enzyme-bound intermediate. The reduction of acrylyl-CoA to propionyl-CoA involves a flavoprotein that can be reduced either by reduced ferredoxin or directly by the oxidation of lactate catalyzed by lactate dehydrogenase. The final step in the formation of propionate involves CoA transferase. The fermentation of lactate to propionate and acetate can be summarized according to Reaction 25:

$$3CH_3CHOHCOOH \rightarrow$$
$$2CH_3CH_2COOH + CH_3COOH + CO_2 + H_2O \quad (25)$$

The bioenergetics of propionate fermentation seems to be clear-cut. An ATP equivalent is produced with every acetate formed. Possibly, additional energy could be generated in the coupling of lactate or pyruvate oxidation to acrylyl-CoA reduction by membrane-associated electron transport (Thauer *et al.*, 1977; Rogers, 1986).

Under certain conditions, *C. propionicum* can produce some acrylate during fermentation of lactate or alanine. When acrylyl-CoA reductase is inhibited by a specific inhibitor such as 3-butynyl-CoA or by a high redox potential (in the presence of O_2 or methylene blue), acrylate accumulation is observed in cultures and extracts of *C. propionicum* (Sinskey *et al.*, 1981; Akedo *et al.*, 1983). The fermentation of β-alanine may involve acrylyl-CoA aminase, which has been purified from *C. propionicum* and shown to catalyze Reaction (26) (Vagelos *et al.*, 1959):

$$CH_2{=}CHCO\text{-}CoA + NH_3 \rightarrow CH_2NH_2CH_2CO\text{-}CoA \qquad (26)$$

Threonine, also fermented by *C. propionicum*, is converted to butyrate and propionate as shown in Reaction 27 (Cardon and Barker, 1947; Barker, 1961):

$$CH_4CHOHCHNH_2COOH + H_2O \rightarrow CH_3CH_2CH_2COOH +$$
$$2CH_3CH_2COOH + 2CO_2 + 3NH_3 \quad (27)$$

The fermentation is initiated by the enzyme threonine dehydratase.

3.2.3. Formate

Many clostridia produce formate during fermentation. Although it is not a major product, metabolism of formate in clostridia is of major importance. We have already discussed its central roles for the production of acetyl-CoA via the autotrophic pathway in homoacetate-fermenting bacteria (Fig. 2, Section 3.1.1) and for the purinolytic fermentation (Fig. 4(1), Section 3.1.3). In these fermentations, it is not unusual to observe formate as a minor product (Barker, 1961; Andreesen *et al.*, 1970). Formate is also a precursor of C_1 units of the H_4 folate derivatives needed for the biosynthesis of purines, pyrimidines, and several amino$^-$acids. Thus it plays a major role in building precursors for the synthesis of DNA, RNA, and proteins (Thauer *et al.*, 1972). Furthermore, recent studies (Thiele and Zeikus, 1988; Zindel *et al.*, 1988) indicate that formate is used under anaerobic conditions in bacterial interspecies energy transfer (syntrophy), similar to that of hydrogen (Ianotti *et al.*, 1973).

In clostridia formate is produced by four mechanisms, of which the following three are of importance for the acetogens: the reduction of CO_2 by formate dehydrogenase (Thauer, 1972; Wagner and Andreesen, 1977; Yamamoto *et al.*, 1983), the conversion of formyl–H_4 folate to formate with the generation of ATP as catalyzed by formyl–H_4 folate synthetase (Himes and Harmony, 1973; Whitehead and Rabinowitz, 1986; Lovell *et al.*, 1988), and the oxidation of methanol by methanol dehydrogenase (Winters-Ivey, 1987). It is of interest that the clostridial formate dehydrogenases depend on selenium, iron, and tungsten, or molybdenum for activity (see Adams and Mortenson, 1985).

The fourth way of formate production is by the pyruvate-formate-lyase reaction (Stadtman *et al.*, 1951; Thauer *et al.*, 1972). In this reaction, pyruvate is not oxidized but cleaved to form acetyl-CoA and formate (Reaction 28).

$$CH_3COCOOH + HSCoA \rightarrow CH_3CO\text{-}SCoA + HCOOH \qquad (28)$$

The reaction is important for the anaerobic catabolism of glucose in facultative *Enterobacteriaceae* and the enzyme of *E. coli* has been studied extensively (Pecher *et al.*, 1982). The enzyme has been demonstrated in *C. butyricum*, *C. kluyveri*, and *C. acetobutylicum*, where it is involved in C_1 metabolism (Thauer *et al.*, 1972).

3.2.4. Succinate

Succinate is produced in small quantities by many clostridia. It is a major fermentation product in the saccharolytic *C. coccoides* (Kaneuchi *et al.*, 1976b) and the proteolytic *C. putrefaciens* (Roberts and Hobbs, 1968). The fermentation pathways in these organisms have not been studied.

There are three pathways that lead to succinate production in anaerobes. One of these involves the transcarboxylase reaction which has been studied extensively in *Propionibacterium shermanii* (Wood, 1982; Wood and Kumar, 1985; see also Section 3.2.2). The second pathway is exemplified by *Bacteroides fragilis* (Macy *et al.*, 1978). A third pathway has been suggested to occur in *C. kluyveri* (Doelle, 1975). It would involve the synthesis of propionyl-CoA from acetate via acrylyl-CoA (Fig. 7), and the subsequent conversion of propionyl-CoA by carboxylation to form methylmalonyl-CoA and its isomerization to succinyl-CoA by a B_{12}-dependent methylmalonyl-CoA mutase. However, experimental evidence for this pathway is lacking.

The first two pathways are very similar in that pyruvate is carboxylated to oxalacetate, which is subsequently reduced via malate and fumarate to succinate, involving the enzymes malate dehydrogenase, fumarase, and fumarate reductase of the tricarboxylic acid (TCA) cycle. The two pathways differ in the mechanism of formation of oxaloacetate. In *P. shermanii*, it is formed either from phosphoenolpyruvate (PEP) by PEP carboxytransphosphorylase (Wood *et al.*, 1977), or in a transcarboxylation reaction involving pyruvate by transfer of the carboxyl group of methylmalonyl-CoA (Reaction 29) (Wood and Kumar, 1985).

$$CH_3COCOOH + CH_3CHCOOHCO\text{-}CoA \rightarrow$$
$$HOOCCH_2COCOOH + CH_3CH_2CO\text{-}CoA \qquad (29)$$

In *B. fragilis*, oxaloacetate is formed by carboxylation of PEP using the enzyme PEP carboxykinase (Macy *et al.*, 1978). *Clostridium formicoaceticum* uses enzymes of the TCA cycle and is able to metabolize malate and fumarate to yield succinate and acetate (Dorn *et al.*, 1978a), and it has also been shown to contain a fumarate reductase (Dorn *et al.*, 1978b).

3.2.5. Fermentation of Amino Acids

Several clostridia are proteolytic and many ferment amino acids (Table I). It is impossible to discuss here the fermentation pathways of individual amino acids. Instead we refer to the book by Gottschalk (1985), some general references (Gottschalk *et al.*, 1981; Cato *et al.*, 1986), and reviews by Barker (1961, 1981). A good account for the essence of amino acid fermentations, pathways, and methods of study is the paper by Zindel *et al.* (1988), who describe a new anaerobe, *Eubacterium acidaminophilum*, that metabolizes alanine, serine, glycine, several derivatives of glycine, aspartate, valine, and leucine. The pattern of metabolism in this bacterium seems to be very close to that of many clostridia, including the purinolytic acetogens.

One of the major discoveries concerning amino acid degradations in *Clostridia* and other anaerobes is the Stickland reaction (Stickland, 1934). In this reaction, pairs of amino acids are fermented with one serving as electron donor and the other as acceptor. The general principle is demonstrated by Reaction 30 for the donor and by Reaction 31 for the acceptor.

$$R\text{-}CHNH_2COOH \rightarrow R\text{-}COCOOH + NH_3 + 2H \qquad (30)$$
$$\quad \rightarrow R\text{-}COOH + CO_2 + 2H$$
$$2R\text{-}CHNH_2COOH + 4H \rightarrow 2R\text{-}CH_2COOH + 2NH_3 \qquad (31)$$

It can be generalized that alanine, leucine, isoleucine, valine, histidine, tyrosine, phenylalanine, and tryptophan serve as electron donors, whereas glycine, proline, hydroxyproline, ornithine, and arginine are electron acceptors (Stadtman *et al.*, 1958; Mead, 1971; Elsden and Hilton, 1979; Barker, 1981). During oxidation of the amino acids serving as electron donors, they are deaminated, either by oxidation or by transamination. This is generally followed by oxidation of the corresponding α-keto acids. Examples of products formed are isobutyrate from valine, 3-methylbutyrate from leucine, 2-methylbutyrate from isoleucine, and phenylacetate, phenylpropionate, and phenyllactate and corresponding hydroxyphenylates from phenylalanine and tyrosine, respectively. Tryptophan yields indoleacetate and indolepropionate, and alanine is oxidized to acetate.

Of the amino acids that are reduced, ornithine is converted to proline by ornithine cyclodeaminase (Muth and Costilov, 1974). Proline is reduced to δ-aminovalerate by proline reductase (Seto, 1980), which is fermented to valerate, propionate, acetate, and ammonia. Proline reductase has been shown to be associated with the membrane in *C. sporogenes* and has now been found to take part in vectorial proton translocation (Lovitt *et al.*, 1986). Ornithine may also serve as a reductant. When *C. sticklandii* is grown on a mixture of proline and ornithine, proline is reduced as mentioned above, whereas ornithine is oxidized to acetate, alanine, ammonia, and CO_2. The oxidation occurs after ornithine has been converted to 2,4-diaminovalerate by a coenzyme B_{12}-dependent ornithine mutase. It is then deaminated by an NAD-dependent dehydrogenase to 2-amino-4-ketopentanoate (Somack and Costilow, 1973), which is cleaved to yield the final products. Lysine is fermented by a pathway similar to that of ornithine involving conversion to 3,5-diaminohexanoate by a coenzyme B_{12}-dependent lysine mutase and oxidation to 3-keto-5-aminohexanoate. The products are acetate, butyrate, and NH_3. A review of these fermentations with special consideration of the amino mutases was published by Baker and Stadtman (1981).

The importance of glycine in the metabolism of purinolytic bacteria is discussed in Section 3.1.3 (see also Fig. 4). In these bacteria the main route of glycine oxidation is the glycine reductase reaction (Tanaka and Stadtman, 1979). This reaction is also used when glycine is an oxidant in the Stickland reaction by several clostridia and other anaerobic bacteria that ferment glycine (Costilow, 1977; Stadtman, 1978; Zindel *et al.*, 1988).

Studies of the Stickland reaction have been performed with cultures containing a single bacterial species. It is fair to say that the Stickland reaction is considered an intraspecies redox reaction. However, results presented by Zindel *et al.* (1988) indicate that, in a broader sense, it can be carried out syntrophically between two different bacterial species. This was demonstrated by coculturing *E. acidaminophilum* with a H_2-utilizing bacterium (*Desulfovibrio* species, *Methanosprillum hungatei*, or *Acetobacterium woodii*). H_2 produced in the glycine fermentation was used by the other bacteria. It allowed a vigorous growth of *E. acidaminophilum*, which otherwise was inhibited by the H_2 it produced (see Iannotti *et al.*, 1973, for interspecies H_2 transfer and syntrophy). Although the above-cited experiments were with bacterial species other than clostridia, it indicates the possibility of clostridia also using the Stickland reaction in syntrophic partnerships.

Some clostridia ferment single amino acids. *Clostridium propionicum*, for instance, utilizes threonine or alanine as a single growth substrate. Alanine is shuttled into the acrylate pathway (Fig. 6) at the level of acrylyl-CoA via malonate semialdehyde, β-hydroxypropionate, and β-hydroxypropionyl-CoA (Sinskey *et al.*, 1981). Other typical fermenters of single amino acids are *C. tetanomorphum* (aspartate, glutamate, lysine, arginine), *C. cochlearium* (glutamate, aspartate, histidine, serine), *C. sticklandii* (lysine, arginine, ornithine), and *C. sporogenes* (tryptophan) (Wachsman and Barker, 1955; Mead, 1971; Holdeman *et al.*, 1977; Elsden and Hilton, 1978, 1979).

The fermentation of glutamate is of special interest. In many clostridia it is fermented by a pathway involving methylaspartate as an intermediate (Barker, 1981). The conversion of glutamate to β-methylaspartate involves a novel B_{12}-dependent enzyme glutamate mutase (Switzer, 1982). The enzyme was discovered in *C. tetanomorphum* and studies of it led to the discovery of coenzyme B_{12} (Barker, 1976). The products of the glutamate fermentation via the methylaspartate pathway are acetate, butyrate, CO_2, and H_2.

4. INDUSTRIAL ORGANIC ACID PRODUCTION

Presently, organic acids are produced from petrochemicals, natural gas, or methanol. However, prices for these feedstocks have increased, the supply may be erratic (as demonstrated by the 1972–1973 energy crisis), and it is realized that the sources of these feedstocks are finite. Therefore, an interest has evolved in using clostridia for the production of organic acids by fermentations of renewable resources such as starch, lignocellulose, or other plant biomass materials (Zeikus, 1980; Wise, 1983; Busche, 1985a; Rogers, 1986; Wiegel and Ljungdahl, 1986).

Acids considered for production using clostridia include acetic, acrylic, butyric, fumaric, propionic, and succinic acid. Of these, most attention has been given to acetic acid because it is produced in the largest amount (1.3 × 10^9 kg in the United States per year; U.S. Int. Trade Comm., 1986). Furthermore, the U.S. Federal Highway Administration has found that calcium-magnesium acetate (CMA) is a good, and apparently environmentally safe, deicer (Chollar, 1984). If CMA is going to be used as a deicer, it has been predicted that the consumption of acetate should at least double. A second reason for the focus on acetic acid production by fermentation is that the homoacetogenic bacteria are almost ideal catalysts for the complete conversion of glucose and xylose to acetate as the only product (Section 2.1.1) (Ljungdahl, 1983).

In developing industrial methods for acid production using clostridia, one has to consider: (1) that they are anaerobes; (2) that we have limited knowledge of their growth requirements, substrate spectra, physiologies, and genetics; (3) that the products produced are at low concentrations in the beers; and (4) that they generally produce mixtures of products.

The use of anaerobic conditions should not be an obstacle. Anaerobic digestors have been used for decades to treat sewage and industrial waste.

Discussions of designs of fermenters for different types of fermentations are found in a series of books edited by Wise (1983, 1984). Methods for working in the laboratory with anaerobic bacteria are found in the review by Ljungdahl and Wiegel (1986), including a description of a setup for continuous and cell-recycling studies under anaerobic conditions that was used to investigate acetate production by *C. thermoaceticum* and *C. thermoautotrophicum* (Ljungdahl *et al.*, 1986).

Acetate production using *C. thermoaceticum* was also studied by Reed (1985), Reed and Bogdan (1985), and Wang and Wang (1984). Reed and Bogdan (1985) used cell recycling together with bacterial cell adsorption on corn cob granules and activated carbon. They reported production rates of 14.3 g of acetic acid per liter/hr and a concentration of 7.1 g/liter. At the slower production rate of 3.9 g/liter/hr, a concentration of 13.4 g/liter was obtained. Wang and Wang (1984) used a batch approach, adding the glucose in increments to overcome substrate inhibition observed in fermentations using *C. thermoaceticum*. They obtained a production rate of 0.8 g/liter/hr and reached 45 g of acetic acid/liter. Wiegel and Garrison (quoted by Ljungdahl *et al.*, 1986) constructed a rotary fermenter based on a fermenter designed by Clyde (1983) that had been used for ethanol fermentation. The fermentation vessel consisted of a vertical tube with inserted pads or disks of a white felt material (35% wool, 65% rayon) intermingled with disks of a stiffer Dupont Reemay 2033 material. The disks were fastened to a central rod and rotated during the fermentation. The bacterial cells adhered to the disks. The design allowed a large area for cell attachment and cell-to-substrate interaction. A production rate of 11 g of acetic acid/liter/hr was reached in a medium of pH 6.6.

Although fermentation rates are of economical importance in the industrial production of acids, the major cost is that of product recovery, separation, and concentration. In the acetic acid fermentation using the homoacetogenic bacteria, we are fortunate in that acetic acid is the only product and that a separation step is not required. The newly isolated *C. thermobutyricum* is approaching this situation whereby 90% of the acid produced is butyrate, and no alcohol is formed (Wiegel *et al.*, 1988). *Clostridium thermolacticum* is another species that yields only one major product: lactate (Le Ruyet *et al.*, 1985). Several processes have been developed for the recovery of acetic acid from fermentation liquids (Busche, 1985a). The simplest way seems to be distillation after acidification of the fermentation broth. However, the energy requirement is so high that the cost becomes prohibitive (Busche, 1985b). More economical methods involve acidification using CO_2 under pressure, followed by extraction with an organic solvent and separation of the solvent from the acetic acid (Shimshick, 1981; Yates, 1981). Other methods for the recovery of organic acids from salt solutions (reviewed by Busche, 1985a,b) involve the use of membranes or of a membrane process combined with electrodialysis. An electrodialysis method was developed to recover acetic acid from fermentations of ethanol by *Acetobacter aceti* (Nomura *et al.*, 1988). This method may be applicable to clostridial fermentations.

Since the recovery of acetic acid, rather than acetate, from fermentation beers has been considered the method of choice, attempts have been made to develop strains of *C. thermoaceticum* which can grow and produce acetic acid at a low pH. The pK_a of acetic acid is 4.76. The pH optimum for *C. thermoaceticum* is about 6.8 (Ljungdahl *et al.*, 1985); thus, in fermentations under optimum conditions, acetate rather than acetic acid is the product. Results obtained by Baronowsky *et al.* (1984) indicate that the undissociated acetic acid inhibits *C. thermoaceticum*. The free acid diffuses through the membrane into the bacterial cell where it dissociates. This decreases the pH inside the cell, which inhibits energy production, and metabolism in the cell, and results in a breakdown of the membrane potential. It follows that at a low pH only very limited acetic acid production can occur before the cell ceases to function. However, a mutant strain of *C. thermoaceticum* has been obtained that grows slowly at pH 4.5, with a doubling time of 35 hr (Schwartz and Keller, 1982a,b; Keller *et al.*, 1985). The strain produces 75 mM acetic acid at pH 4.5. In fermentations using the rotary fermenter described above, we used a medium with very low buffering capacity (5 mM phosphate, pH 6.8). The effluent from the fermenter had a low pH and we obtained (at pH 4.8 and 4.5) rates of 3.9 and 1.2 g of acetic acid/liter/hr, at concentrations of 135 and 141 mM (8.1 and 8.6 g/liter) acetic acid, respectively.

A rich medium is normally used for growing clostridia. The medium used in most of the experiments discussed above was based on that described by Ljungdahl (1983). It is a very rich medium containing yeast extract and tryptone, and is therefore expensive. For industrial use, a more cost-effective medium is desirable. It has now been shown by Lundie and Drake (1984) that *C. thermoaceticum* can be adapted to grow on glucose, ammonium sulfate, nicotinic acid as the only required vitamin, a reductant, and mineral salts in a phosphate-bicarbonate buffer under CO_2 atmosphere. A similar medium was shown to support growth of *C. thermoautotrophicum*, using glucose or methanol as carbon source (Savage and Drake, 1986). With this medium, the concentration of acetate reached was higher per unit of biomass than in the undefined richer medium.

5. CONCLUSION

Clostridia are perhaps unique in their ability to convert a wide variety of complex organic material (biomass) to rather simple alcohols and acids. They interact with each other and with other types of organisms in the earth's carbon cycle. They help us in cleaning the environment. Given their versatility, it is not surprising that they are considered for industrial usage. In this review, we have pointed out their abilities to produce a variety of acids. However, the most in-depth discussion concerns the acetogens. An obvious reason for this is that we have been studying them for many years and know them better than the other clostridia; we have a bias. A second, perhaps more compelling reason is our belief that the homoacetogenic clostridia will be the first to be used in an industrial process. It may involve the production of calcium-magnesium acetate (CMA) from hydrolyzed cornstarch. Calculations made by us indicate that CMA can be produced for around 0.20 U.S. $/lb (Ljungdahl *et al.*, 1986)—a figure that is close to the present cost. With more knowledge of the homoacetogens, the fermentation process certainly can be improved and made competitive with other methods of acetate production. Basic research is also the key to applying other clostridia for industrial usage and for our understanding of their role in the environment.

ACKNOWLEDGMENT. We express our thanks to several investigators including H. G. Wood, H. Drake, A. Cornish Frazer, J. A. Breznak, and M. J. Wolin and their co-workers for sending preprints. We also express our appreciation for financial support for studies of acetogens from the U.S. National Institutes of Health, Public Health Service grant AM-27323, the U.S. Department of Energy projects DE-FG09-86ER13614 and DE-FG09-84ER13248, the U.S. Department of Transportation contract RFD DTFH-61-83-R-00124, and the Georgia Power Company.

REFERENCES

Adams, M. W. W., and Mortenson, L. E., 1985, Mo reductases: Nitrate reductase and formate dehydrogenase, in: *Molybdenum Enzymes* (T. G. Spiro, ed.), John Wiley and Sons, New York, pp. 519–593.

Adamse, A. D., 1980, New isolation of *Clostridium aceticum* (Wieringa), *Antonie van Leeuwenhoek J. Microbiol. Serol.* **46**:523–531.

Akedo, M., Cooney, C. L., Sinskey, A. J., 1983, Direct evidence for lactate-acrylate interconversion in *Clostridium propionicum*, *Abstr. Ann. Meet. Am. Soc. Microbiol.*, p. 240.

Andreesen, J. R., Gottschalk, G., and Schlegel, H. G., 1970, *Clostridium formicoaceticum* nov. spec. isolation, description, and distinction from *C. aceticum* and *C. thermoaceticum*, *Arch. Mikrobiol.* **72**:154–174.

Andreesen, J. R., Schaupp, A., Neurauter, C., Brown, A., and Ljungdahl, L. G., 1973, Fermentation of glucose, fructose, and xylose by *Clostridium thermoaceticum*: Effect of metals on growth yield, enzymes, and the synthesis of acetate from CO_2. *J. Bacteriol.* **114**:743–751.

Andreesen, J. R., Zindel, U., and Dürre, P., 1985, *Clostridium cylindrosporum* (ex Barker and Beck 1942) nom. rev, *Int. J. Syst. Bacteriol.* **35**:206–208.

Bache, R., and Pfennig, N., 1981, Selective isolation of *Acetobacterium woodii* on methoxylated aromatic acids and determination of growth yields, *Arch. Microbiol.* **130**:255–261.

Baker, J. J., and Stadtman, T. C., 1981, Amino mutases, in: B_{12}, Vol. 2 (D. Dolphin, ed.), John Wiley and Sons, New York, pp. 203–232.

Balch, W. E., Schoberth, S., Tanner, R. S., and Wolfe, R. S. 1977, *Acetobacterium* a new genus of hydrogen-oxidizing, carbon dioxide-reducing, anaerobic bacteria, *Int. J. Syst. Bacteriol.* **27:**355–361.

Barker, H. A., 1939, The use of glutamic acid for the isolation and identification of *Clostridium cochlearium* and *Cl. tetanomorphum, Arch. Mikrobiol.* **10:**376–384.

Barker, H. A., 1961, Fermentations of nitrogenous organic compounds, in: *The Bacteria,* Vol. 2, *Metabolism* (I. C. Gunsalus and R. Y. Stainer, eds.), Academic Press, New York, pp. 151–207.

Barker, H. A., 1976, Glutamate fermentation and the discovery of B_{12} coenzymes, in: *Reflections on Biochemistry* (A. Kornberg, B. L. Horecker, L. Cornudella, and J. Oro., eds.), Pergamon Press, New York, p. 75.

Barker, H. A., 1981, Amino acid degradation of anaerobic bacteria, *Ann. Rev. Biochem.* **50:**23–40.

Barker, H. A., and Beck, J. V., 1942, *Clostridium acidiurici* and *Clostridium cylindrosporum,* organisms fermenting uric acid and some other purines, *J. Bacteriol.* **43:**291–304.

Barker, H. A., and Kamen, M. D., 1945, Carbon dioxide utilization in the synthesis of acetic acid by *Clostridium thermoaceticum, Proc. Natl. Acad. Sci. USA* **31:**219–225.

Barker, H. A., and Taha, S. M., 1942, *Clostridium kluyveri,* an organism concerned in the formation of caproic acid from ethyl alcohol, *J. Bacteriol.* **43:**347–363.

Baronowsky, J. J., Schreurs, W. J. A., Kaskhet, E. R., 1984, Uncoupling by acetic acid limits growth and acetogenesis by *Clostridium thermoaceticum, Appl. Env. Microbiol.* **48:**1134–1139.

Benner, R., Maccubin, A. E., and Hodson, R. E., 1984, Anaerobic biodegradation of the lignin and polysaccharide components of lignocellulose and synthetic lignin by sediment microflora, *Appl. Env. Microbiol.* **47:**998–1004.

Braun, M., Schoberth, S., and Gottschalk, G., 1979, Enumeration of bacteria forming acetate from H_2 and CO_2 in anaerobic habitats, *Arch. Microbiol.* **120:**201–204.

Braun, M., Mayer, F., and Gottschalk, G., 1981, *Clostridium aceticum* (Wieringa), a microorganism producing acetic acid from molecular hydrogen and carbon dioxide, *Arch. Microbiol.* **128:**288–293.

Breznak, J. A., Switzer, J. M., and Seitz, H. J., 1988, *Sporomusa termitida* sp. nov., and H_2/CO_2-utilizing acetogen isolated from termites, *Arch. Microbiol.* **150:**282–288.

Bryant, M. P., and Burkey, L. A., 1956, The characteristics of lactate-fermenting sporeforming anaerobes from silage, *J. Bacteriol.* **71:**43–46.

Bryson, M. F., and Drake, H. L., 1988, A reevaluation of the metabolic potential of *Clostridium formicoaceticum, Abstr. Annu. Meet. Am. Soc. Microbiol.* **I-107,** p. 198.

Buckel, W., 1980, Analysis of the fermentation pathways of clostridia using double labeled glutamate, *Arch. Microbiol.* **127:**167–169.

Busche, R. M., 1985a, Acetic acid manufacture-fermentation alternatives, in: *Biotechnology Applications and Research* (P. N. Cheremisinoff and R. P. Oulette, P., eds.), Technicon, Lancaster, PA, pp. 88–102.

Busche, R. M., 1985b, The business of biomass, *Biotechnol. Progr.* **1:**165–180.

Cardon, B. P., and Barker, H. A., 1947, Amino acid fermentations by *Clostridium propionicum* and *Diplococcus glycinophilus, Arch. Biochem.* **12:**165–180.

Cato, E. P., and Salmon, C. W., 1976, Transfer of *Bacteroides clostridiiformis* subsp. *clostridiiformis* (Burri and Ankersmit) Holdeman and Moore and *Bacteroides clostridiiformis* subsp. *girans* (Prevot) Holdeman and Moore to the genus *Clostridium* as *Clostridium clostridiiforme* (Burri and Ankersmit) comb. nov. Emendation of description and designation of neotype strain, *Int. J. Syst. Bacteriol.* **26:**205–211.

Cato, E. P., Holdeman, L. V., and Moore, W. E. C., 1982, *Clostridium perenne* and *Clostridium paraperfringens* later subjective synonyms of *Clostridium barati, Int. J. Syst. Bacteriol.* **32:**77–81.

Cato, E. P., George, W. L., and Finegold, S. M., 1986, Genus *Clostridium* Prazmowski, 1980, in: *Bergey's Manual of Systematic Bacteriology,* Vol. 2 (P. H. A. Sneath, N. S. Mair, M. E. Sharpe, and J. G. Holt, eds.), Williams and Wilkins, Baltimore, pp. 1141–1200.

Champion, A. B., and Rabinowitz, J. C., 1977, Ferredoxin and formyltetrahydrofolate synthetase comparative studies with *Clostridium acidiurici* and *Clostridium cylindrosporum*, and a newly isolated anaerobic uric acid-fermenting strain, *J. Bacteriol.* **132**:1003–1020.

Chollar, B. H., 1984, Federal Highway Administration research on calcium magnesium acetate: An alternative deicer, *Public Roads* **47**:113–118.

Clark, J. E., and Ljungdahl, L. G., 1984, Purification and properties of 5,10-methylenetetrahydrofolate reductase, an iron-sulfur flavoprotein from *Clostridium formicoaceticum*, *J.Biol. Chem.* **259**:10845–10849.

Clark, J. E., Ragsdale, S. W., Ljungdahl, L. G., and Wiegel, J., 1982, Levels of enzymes involved in the synthesis of acetate from CO_2 in *Clostridium thermoautotrophicum*, *J. Bacteriol.* **151**:507–509.

Clyde, R. A., 1983, Fiber Fermenter, U.S. Patent 4,407,954.

Colberg, P. J., and Young, L. Y., 1985a, Anaerobic degradation of soluble fractions of [^{14}C-lignin] lignocellulose, *Appl. Env. Microbiol.* **49**:345–349.

Colberg, P. J., and Young, L. Y., 1985b, Aromatic and volatile acid intermediates observed during anaerobic metabolism of lignin-derived oligomers, *Appl. Env. Microbiol.* **49**:350–358.

Cornish-Frazer, A., and Young, L. Y., 1985, A gram-negative anaerobic bacterium that utilizes O-methyl substituents of aromatic acids. *Appl. Env. Microbiol.* **49**:1345–1347.

Costilow, R. N., 1977, Selenium requirement for the growth of *Clostridium sporogenes* with glycine as the oxidant in Stickland reaction, *J. Bacteriol.* **131**:366–368.

Cummins, C. S., and Johnson, J. L., 1971, Taxonomy of the clostridia: Cell-wall composition and DNA homologies in *Clostridium butyricum* and other butyric acid-producing clostridia, *J. Gen. Microbiol.* **67**:33–46.

Daniel, S. L., and Drake, H. L., 1988, Acetogenesis from methoxylated aromatic acids by *Clostridium thermoaceticum*, *Abstr. Annu. Meet. Am. Soc. Microbiol.* **I-105**, p. 198.

Diekert, G. B., and Thauer, R. K., 1978, Carbon monoxide oxidation by *Clostridium thermoaceticum* and *Clostridium formicoaceticum*, *J. Bacteriol.* **136**:597–606.

Diekert, G. B., Schräder, E., and Harder, W., 1986, Energetics of CO formation and CO oxidation in cell suspensions of *Acetobacterium woodii*, *Arch. Microbiol.* **144**:386–392.

Doelle, H. W., 1975, *Bacterial Metabolism*, 2nd ed., Academic Press, New York.

Dorn, M., Andreesen, J. R., and Gottschalk, G., 1978a, Fermentation of fumarate and L-malate by *Clostridium formicoaceticum*, *J. Bacteriol.* **133**:26–32.

Dorn, M., Andreesen, J. R., and Gottschalk, G., 1978b, Fumarate reductase of *Clostridium formicoaceticum*, a peripheral membrane protein, *Arch. Microbiol.* **119**:7–11.

Drake, H. L., 1982, Demonstration of hydrogenase in extracts of the homoacetate-fermenting bacterium *Clostridium thermoaceticum*, *J. Bacteriol.* **150**:702–709.

Drake, H. L., Hu, S.-I., Wood, H. G., 1981, Purification of five components from *Clostridium thermoaceticum* which catalyze synthesis of acetate from pyruvate and methyltetrahydrofolate. Properties of phosphotransacetylase, *J. Biol. Chem.* **256**:11137–11144.

Duine, J. A., Frank Izn, J., Jongejan, J. A., and Dijkstra, M., 1984, Enzymology of the bacterial methanol step, in: *Microbial Growth on C₁ Compounds* (R. L. Crawford, and R. S. Hanson, eds.) Am. Soc. Microbiol., Washington, D.C., pp. 91–96.

Dürre, P., and Andreesen, J. R., 1982a, Pathway of carbon dioxide reduction to acetate without a net energy requirement in *Clostridium purinolyticum*,. *FEMS Microbiol. Lett.* **15**:51–56.

Dürre, P., and Andreesen, J. R., 1982b, Anaerobic degradation of uric acid via pyrimidine derivatives by selenium-starved cells of *Clostridium purinolyticum*,. *Arch. Microbiol.* **131**:255–260.

Dürre, P., and Andreesen, J. R., 1982c, Selenium dependent growth and glycine fermentation by *Clostridium purinolyticum*, *J. Gen. Microbiol.* **128**:1457–1466.

Dürre, P., and Andreesen, J. R., 1983, Purine and glycine metabolism by purinolytic clostridia, *J. Bacteriol.* **154**:192–199.

Dürre, P., Andersch, W., and Andreesen, J. R., 1981, Isolation and characterization of an adenine-utilizing anaerobic sporeformer, *Clostridium purinolyticum* sp. nov., *Int. J. Syst. Bacteriol.* **31**:184–194.

Elliott, J. I., and Ljungdahl, L. G., 1982, Isolation and characterization of an Fe_8-S_8 ferredoxin (ferredoxin II) from *Clostridium thermoaceticum, J. Bacteriol.* **151**:328–333.

Elliott, J. I., Yang, S.-S., Ljungdahl, L. G., Travis, J., and Reilly, C. F., 1982, Complete amino acid sequence of the 4Fe-4S thermostable ferredoxin from *Clostridium thermoaceticum, Biochemistry* **21**:3294–3298.

Elsden, S. R., and Hilton, M. G., 1978, Volatile acid production from threonine, valine, leucine, and isoleucine by clostridia, *Arch. Microbiol.* **117**:165–172.

Elsden, S. R. and Hilton, M. G., 1979, Amino acid utilization patterns in clostridial taxonomy, *Arch. Microbiol.* **123**:137–141.

Fontaine, F. E., Peterson, W. H., McCoy, E., and Johnson, M. J., 1942, A new type of glucose fermentation by *Clostridium thermoaceticum* n. sp., *J. Bacteriol.* **43**:701–715.

Fuchs, G., 1986, CO_2 fixation in acetogenic bacteria: Variations on a theme, *FEMS Microbiol. Rev.* **39**:181–213.

Gibson, T., Stirling, A. C., Keddi, R. M., Rosenberger, R. F., 1958, Bacteriological changes in silage made at controlled temperatures, *J. Gen. Microbiol.* **19**:112–119.

Gottschalk, G., 1985, *Bacterial Metabolism*, 2nd. ed., Springer-Verlag, New York.

Gottschalk, G., Andreesen, J. R., and Hippe, H., 1981, The genus *Clostridium* (nonmedical aspects). in: *The Prokaryotes: A Handbook on Habitats, Isolation and Identification of Bacteria* (M. P. Starr, H. Stolp, H. G. Trüper, A. Balows, and H. G. Schlegel, eds.), Springer-Verlag, Berlin, pp. 1767–1803.

Gottwald, M., Andreesen, J. R., LeGall, J., and Ljungdahl, L. G., 1975, Presence of cytochromes and menaquinone in *Clostridium formicoaceticum* and *Clostridium thermoaceticum, J. Bacteriol.* **122**:325–328.

Goudkov, A. V. and Sharp, M. E., 1966, A preliminary investigation of the importance of clostridia in the production of rancid flavor in cheddar cheese, *J. Dairy Res.* **33**:139–149.

Han, E. Y., 1988, Purification and Characterization of methylenetetrahydrofolate reductase from *Clostridium thermoaceticum*, Thesis, University of Georgia, Athens.

Hardman, J. K., and Stadtman, T. C., 1960a, Metabolism of ω-amino acids. I. Fermentation of γ-aminobutyric acid by *Clostridium aminobutyricum* n. sp., *J. Bacteriol.* **79**:544–548.

Hardman, J. K., and Stadtman, T. C., 1960b, Metabolism of ω-amino acids. II. Fermentation of δ-aminovaleric acid by *Clostridium aminovalericum, J. Bacteriol.* **79**:549–552.

Hardman, J. K., and Stadtman, T. C., 1963, Metabolism of ω-amino acids. IV. γ-Aminobutyrate fermentation by cell-free extracts of *Clostridium aminobutyricum, J. Biol.Chem.* **238**:2088–2093.

Hartmanis, M. G. N., 1987, Butyrate kinase from *Clostridium acetobutylicum, J. Biol. Chem.* **262**:617–621.

Himes, R. H., and Harmony, J. A. K., 1973, Formyltetrahydrofolate synthetase, *CRC Crit. Rev. Biochem.* **1**:501–535.

Holdeman, L. V., Cato, E. P., and Moore, W. E. C., 1977, Anaerobe Laboratory Manual, 4th. ed., Anaerobe Laboratory, Virginia Polytechnic Institute and State University, Blacksburg, pp. 1–156.

Hu, S.-I., Drake, H. L., and Wood, H. G., 1982, Synthesis of acetyl coenzyme A from carbon monoxide, methyltetrahydrofolate and coenzyme A by enzymes from *Clostridium thermoaceticum, J. Bacteriol.* **149**:440–448.

Hu, S.-I., Pezacka, E. and Wood, H. G., 1984, Acetate synthesis from carbon monoxide by *Clostridium thermoaceticum*, Purification of the corrinoid protein, *J. Biol. Chem.* **259**:8892–8897.

Hugenholtz, J., Ivey, D. M., and Ljungdahl, L. G., 1987, Carbon monoxide-driven electron transport in *Clostridium thermoautotrophicum* membranes, *J. Bacteriol.* **169**:5845–5847.

Hugenholtz, J., and Ljungdahl, L. G., 1988, The bioenergetics of *Clostridium thermoautotrophicum*, *Abstr. Ann. Meet. Am. Soc. Microbiol.* **K153**, p. 232.

Iannotti, E. L., Kafkewitz, D., Wolin, M. J., and Bryant, M. P., 1973, Glucose fermentation products of *Ruminococcus albus* grown in continuous culture with *Vibrio succinogenes:* Changes caused by interspecies transfer of H₂, *J. Bacteriol.* **114**:1231–1249.

Imhoff, D., and Andreesen, J. R., 1979, Nicotinic acid hydroxylase from *Clostridium barkeri:* Selenium-dependent formation of active enzyme, *FEMS Microbiol. Lett.* **5**:155–158.

Ivey, D. M., 1987, Generation of energy during CO_2 fixation in acetogenic bacteria, Dissertation, University of Georgia, Athens.

Ivey, D. M., and Ljungdahl, L. G., 1986, Purification and characterization of the F_1-ATPase from *Clostridium thermoaceticum*, *J. Bacteriol.* **165**:252–257.

Ivey, D. M., and Ljungdahl, L. G., 1989, Purification and reconstitution into proteoliposomes of the F_1F_0-ATPase from *Clostridium thermoautotrophicum* (submitted for publication).

Johns, A. T., 1952, The mechanism of propionic acid formation by *Clostridium propionicum*, *J. Gen. Microbiol.* **6**:123–127.

Jordan, D. C., and McNicoll, P. J., 1979, A new nitrogen-fixing *Clostridium* species from a high arctic ecosystem, *Can. J. Microbiol.* **25**:947–948.

Jungermann, K., Thauer, R. K., Leimenstoll, G., and Decker, K., 1973, Function of reduced pyridine nucleotide-ferredoxin oxidoreductases in saccharolytic clostridia, *Biochim. Biophys. Acta* **305**:268–280.

Kaneuchi, C., Benno, Y., and Mitsuoka, T., 1976a, *Clostridium coccoides*, a new species from the feces of mice, *Int. J. Syst. Bacteriol.* **26**:482–486.

Kaneuchi, C., Watanabe, K., Terada, A., Benno, Y., and Mitsuoka, T., 1976b, Taxonomic study of *Bacteroides clostridiiformis* subsp. *Clostridiiformis* (Burri and Ankersmit) Holdeman and Moore and of related organisms: Proposal of *Clostridium clostridiiformis* (Burri and Ankersmit) comb. nov. and *Clostridium symbiosum* (Stevens) comb. nov., *Int. J. Syst. Bacteriol.* **26**:195–204.

Kaneuchi, C., Miyazato, T., Shinjo, T., and Mitsuoka, T., 1979, Taxonomic study of helically coiled, sporeforming anaerobes isolated from the intestines of humans and other animals: *Clostridium cocleatum* sp. nov. and *Clostridium spiroforme* sp. nov., *Int. J. Syst. Bacteriol.* **29**:1–12.

Karlsson, J. L., and Barker, H. A., 1949, Tracer experiments on the mechanism of uric acid decomposition and acetic acid synthesis by *Clostridium acidi-urici*, *J. Biol. Chem.* **178**:891–902.

Keller, F. A., Ganoung, J. S., Luenser, S. J., 1985, Mutant strain of *Clostridium thermoaceticum* useful for the preparation of acetic acid, U.S. Patent 4,513,084.

Kellermeyer, R. W., Allen, S. H. G., Stjernholm, R., and Wood, H. G., 1964, Methylmalonyl isomerase. IV. Purification and properties of the enzyme from propionibacteria, *J. Biol. Chem.* **239**:2562–2569.

Kenealy, W. R. and Waselefsky, D. M., 1985, Studies on the substrate range of *Clostridium kluyveri;* the use of propanol and succinate, *Arch. Microbiol.* **141**:187–194.

Kerby, R., and Zeikus, J. G., 1983, Growth of *Clostridium thermoaceticum* on H_2/CO_2 or CO as energy source, *Curr. Microbiol.* **8**:27–30.

Kirk, T. K., and Farrell, R. L., 1987, Enzymatic "Combustion": The microbial degradation of lignin, *Ann. Rev. Microbiol.* **41**:465–505.

Krumholz, L. R., and Bryant, M. P., 1985, *Clostridium pfennigii* sp. nov. uses methoxyl groups of monobenzenoids and produces butyrate, *Int. J. Syst. Bacteriol.* **35**:454–456.

Kuchta, R. D., and Abeles, R. H., 1985, Lactate reduction in *Clostridium propionicum*, Purification and properties of lactyl-CoA dehydratase, *J. Biol. Chem.* **260**:13181–13189.

Laanbroek, H. J., Smit, A. J., Klein Nulend, G., and Veldkamp, H., 1979, Competition for L-glutamate between specialized and versatile *Clostridium* species, *Arch. Microbiol.* **120**:61–66.

Lajoie, S. F., Bank, S., Miller, T. L., Wolin, M. J., 1988, Acetate production from hydrogen and [^{13}C] carbon dioxide by the microflora of human feces, *Appl. Env. Microbiol.* **54:** 2723–2727.

Le Ruyet, P., Dubourguier, C., Albagnac, G., and Prensier, G., 1985, Characterization of *Clostridium thermolacticum* sp. nov., a hydrolytic thermophilic anaerobe producing high amounts of lactate, *Appl. Microbiol.* **6,** 196–202.

Ljungdahl, L. G., 1983, Formation of acetate using homoacetate fermenting anaerobic bacteria, in: *Organic Chemicals from Biomass* (D. L. Wise, ed.), Benjamin Cummings, Menlo Park, CA, pp. 219–248.

Ljungdahl, L. G., 1984, Other functions of folates, in: *Folates and Pterins*, Vol. 1, *Chemistry and Biochemistry of Folates* (R. L. Blakley and S. J. Benkovic, eds.), John Wiley and Sons, New York, pp. 555–579.

Ljungdahl, L. G., 1986, The autotrophic pathway of acetate synthesis in acetogenic bacteria, *Ann. Rev. Microbiol.* **40:**415–450.

Ljungdahl, L. G., and Eriksson, K. E., 1985, Ecology of microbial cellulose degradation, *Adv. Microbial Ecol.* **8:**237–299.

Ljungdahl, L. G., and Wiegel, J., 1986, Working with anaerobic bacteria, in: *Manual of Industrial Microbiology and Biotechnology* (A. L. Demain, and N. A. Solomon, eds.), Am. Soc. Microbiol., Washington, D.C., pp. 84–96.

Ljungdahl, L. G., and Wood, H. G., 1969, Total synthesis of acetate from CO$_2$ by heterotrophic bacteria, *Ann. Rev. Microbiol.* **23:**515–538.

Ljungdahl, L. G., and Wood, H. G., 1982, Acetate biosynthesis, in: *B$_{12}$*, Vol. 2 (D. Dolphin, ed), John Wiley and Sons, New York, pp. 165–202.

Ljungdahl, L. G., Irion, E., and Wood, H. G., 1966, Role of corrinoids in the total synthesis of acetate from CO$_2$ by *Clostridium thermoaceticum*, *Fed. Proc.* **25:**1642–1648.

Ljungdahl, L. G., O'Brien, W. E., Moore, M. R., and Liu, M.-T., 1980, Methylenetetrahydrofolate dehydrogenase from *Clostridium formicoaceticum* and methylenetetrahydrofolate dehydrogenase, methenyltetrahydrofolate cyclohydrolase (combined) from *C. thermoaceticum*, *Methods Enzymol.* **66:**599–609.

Ljungdahl, L. G., Carreira, L. H., Garrison, R. J., Rabek, N. E., and Wiegel, J., 1985, Comparison of three thermophilic acetogenic bacteria for production of calcium magnesium acetate, *Biotechnol. Bioeng. Symp.* **15:**207–223.

Ljungdahl, L. G., Carreira, L. H., Garrison, R. J., Rabek, N. E., Gunter, L. F., and Wiegel, J., 1986, CMA manufacture [II]. Improved bacterial strain for acetate production, U.S. Dept. of Transportation, Federal Highway Administration Report No. FHWA/RD-86/117, available from the National Technical Information Service, Springfield, VA 22161.

Lovell, C. R., Przybyla, A., and Ljungdahl, L. G., 1988, Cloning and expression in *Escherichia coli* of the *Clostridium thermoaceticum* gene encoding thermostable formyltetrahydrofolate synthetase, *Arch. Microbiol.* **149:**280–285.

Lovitt, R. W., Kell, D. B., and Morris, J. G., 1986, Proline reduction by *Clostridium sporogenes* is coupled to vectorial proton ejection, *FEMS Microbiol. Lett.* **36:**269–273.

Lundie, L. L., Jr., and Drake, H. L., 1984, Development of a minimally defined medium for the acetogen *Clostridium thermoaceticum*, *J. Bacteriol.* **159:**700–703.

MacDonald, I. A., Bokkenheuser, V. D., Winter, J., McLernon, A. M., and Mosbach, E. H., 1983, Degradation of steroids in the human gut, *J. Lipid Res.* **24:**675–700.

Macy, J. M., Ljungdahl, L. G., and Gottschalk, G., 1978, Pathway of succinate and propionate formation in *Bacteroides fragilis*, *J. Bacteriol.*, **134:**84–91.

Mah, R. H., 1981, The methanogenic bacteria, their ecology and physiology, in: *Trends in the Biology of Fermentation for Fuels and Chemicals* (A. Hollaender, ed.), Plenum Press, New York, pp. 357–374.

Martin, D. R., Lundie, L. L., Kellum, R., and Drake, H. L., 1983, Carbon monoxide-dependent evolution of hydrogen by the homoacetate-fermenting bacterium *Clostridium thermoaceticum*, *Curr. Microbiol.* **8:**337–340.

Mallette, M. F., Reece, P., and Dawes, E. A., 1974, Culture of *Clostridium pasteurianum* in defined medium and growth as a function of sulfate concentration, *Appl. Microbiol.* **28:** 999–1003.

Mayer, F., Ivey, D. M., and Ljungdahl, L. G., 1986, Macromolecular organization of F_1-ATPase isolated from *Clostridium thermoaceticum* as revealed by electron microscopy, *J. Bacteriol.* **166:**1128–1130.

Mead, G. C., 1971, The amino acid-fermenting clostridia, *J. Gen. Microbiol.* **67:**47–56.

Mitchell, P., 1966, Chemiosmotic coupling in oxidative and photosynthetic phosphorylation, *Biol. Rev.* **41:**445–502.

Möller, B., Ossmer, R., Howard, B. H., Gottschalk, G., and Hippe, H., 1984, *Sporomusa*, a new genus of Gram negative anaerobic bacteria including *Sporomusa spheroides* spec. nov. and *Sporomusa ovata*, spec. nov., *Arch. Microbiol.* **139:**388–396.

Mortenson, L. E., 1966, Components of cell-free extracts of *Clostridium pasteurianum* required for ATP-dependent H_2 evolution from dithionite and for N_2 fixation, *Biochim. Biophys. Acta* **127:**18–25.

Mortenson, L. E., and Chen, J. S., 1974, Hydrogenase, in: *Microbial Iron Metabolism: A Comprehensive Treatise* (J. B. Neilands, ed.), Academic Press, New York, pp. 231–282.

Munoz, E., 1982, Polymorphism and conformational dynamics of F_2 ATPases from bacterial membranes, a model for the regulation of these enzymes on the basis of molecular plasticity, *Biochim. Biophys. Acta* **650:**233–265.

Muth, W. L., and Costilow, R. N., 1974, Ornithine cyclase (deaminating), II. Properties of the homogeneous enzyme, *J. Biol. Chem.* **249:**7457–7462.

Nakamura, S., Shimamura, T., Hayase, M., and Nishida, S., 1973, Numerical taxonomy of saccharolytic clostridia, particularly *Clostridium perfringens*-like strains: Description of *Clostridium absonum* sp. n. and *Clostridium paraperfringens*, *Int. J. Syst. Bacteriol.* **23:**419–429.

Nomura, Y., Iwahara, M., and Hongo, M., 1988, Acetic acid production by an electrodialysis fermentation method with a computerized control system, *Appl. Env. Microbiol.* **54:**137–142.

O'Brien, W. E., and Ljungdahl, L. G., 1972, Fermentation of fructose and synthesis of acetate from carbon dioxide by *Clostridium formicoaceticum*, *J. Bacteriol.* **109:**626–632.

Ohwaki, K., and Hungate, R. E., 1977, Hydrogen utilization by clostridia in sewage sludge, *Appl. Env. Microbiol.* **33:**1270–1274.

Oren, A., 1983, *Clostridium lortetii* sp. nov., a halophilic obligatory anaerobic bacterium producing endospores with attached gas vacuoles, *Arch. Microbiol.* **136:**42–48.

Pecher, A., Blaschkowski, H. P., Knappe, K., and Bock, A., 1982, Expression of pyruvate formate-lyase of *Escherichia coli* from the cloned structural gene, *Arch. Microbiol.* **132:**365–371.

Pezacka, E., and Wood, H. G., 1984, Role of carbon monoxide dehydrogenase in the autotrophic pathway used by acetogenic bacteria, *Proc. Natl. Acad. Sci. USA* **81:**6261–6265.

Pezacka, E., and Wood, H. G., 1986, The autotrophic pathway of acetogenic bacteria: Role of CO dehydrogenase disulfide reductase, *J. Biol. Chem.* **261:**1609–1615.

Pezacka, E., and Wood, H. G., 1988, Acetyl-CoA pathway of autotrophic growth: Identification of the methyl-binding site of the CO-dehydrogenase, *J. Biol. Chem.* **263:**16000–16006.

Poston, J. M., Kuratomi, K., and Stadtman, E. R., 1966, The conversion of carbon dioxide to acetate. 1. The use of cobalt-methylcobalamin as a source of methyl groups for the synthesis of acetate by cell-free extracts of *Clostridium thermoaceticum*, *J. Biol. Chem.* **241:** 4209–4216.

Prévot, A. R. and Zimmès-Chaverou, J., 1947, Etude d'une nouvelle espece anaerobic de Cote d'Ivoire: *Inflabilis mangenotii*, *Ann. Inst. Pasteur* (Paris) **73:**602–604.

Prins, R. A. and Lankhorst, A., 1977, Synthesis of acetate from CO_2 in the cecum of some rodents, *FEMS Microbiol. Lett.* **1:**255–258.

Rabinowitz, J. C., 1963, Intermediates in purine breakdown, *Methods Enzymol.* **6:**703–713.

Raeburn, S., and Rabinowitz, J. C., 1971, Pyruvate:ferredoxin oxidoreductase. II. Characteristics of the forward and reverse reactions and properties of the enzyme, *Arch. Biochem. Biophys.* **146**:21–33.

Ragsdale, S. W., and Wood, H. G., 1985, Acetate biosynthesis by acetogenic bacteria. Evidence that carbon monoxide dehydrogenase is the condensing enzyme that catalyzes the final steps in the synthesis, *J. Biol.Chem.* **260**:3970–3977.

Ragsdale, S. W., Clark, J. E., Ljungdahl, L. G., Lundie, L. L., and Drake, H. L., 1983a, Properties of purified carbon monoxide dehydrogenase from *Clostridium thermoaceticum*, a nickel, iron-sulfur protein, *J. Biol. Chem.* **258**:2364–2369.

Ragsdale, S. W., Ljungdahl, L. G., and DerVartanian, D. V., 1983b, Isolation of carbon monoxide dehydrogenase from *Acetobacterium woodii* and comparison of its properties with those of the *Clostridium thermoaceticum* enzyme, *J. Bacteriol.* **255**:1224–1237.

Ragsdale, S. W., Ljungdahl, L. G., and DerVartanian, D. V., 1983c, ^{13}C and [61]Ni isotope substitutions confirm the presence of a nickel (II)-carbon species in acetogenic CO dehydrogenase, *Biochem. Biophys. Res. Commun.* **115**:658–665.

Ragsdale, S. W., Wood, H. G., and Antholine, W. E., 1985, Evidence that an iron-nickel-carbon complex is formed by reaction of CO with the CO dehydrogenase from *Clostridium thermoaceticum*, *Proc. Natl. Acad. Sci. USA* **82**:6811–6814.

Ragsdale, S. W., Lindahl, P. A., and Münck, E., 1987, Mössbauer, EPR, and optical studies of the corrinoid/iron sulfur protein involved in the synthesis of acetyl coenzyme A by *Clostridium thermoaceticum*, *J. Biol. Chem.* **262**:14289–14297.

Reed, W. M., 1985, Production of organic acids by a continuous fermentation process, U.S. Patent 4,506,012.

Reed, W. M., and Bogdan, M. E., 1985, Application of cell recycling to continuous fermentative acetic acid production, *Biotech. Bioeng. Symp.* **15**:641–647.

Roberts, T. A., and Hobbs, G., 1968, Low temperature growth characteristics of clostridia, *J. Appl. Bacteriol.* **31**:75–88.

Rogers, P., 1986, Genetics and biochemistry of *Clostridium* relevant to development of fermentation process, *Adv. Appl. Microbiol* **31**:1–60.

Roux, C., and Bergère, J.-L., 1977, Charactères taxonomiques de *Clostridium tyrobutyricum*, *Ann. Microbiol.* (Inst. Pasteur) **128A**:267–276.

Savage, M. D., and Drake, H. L., 1986, Adaptation of the acetogen *Clostridium thermoautotrophicum* to minimal medium, *J. Bacteriol.* **165**:315–318.

Schaupp, A., and Ljungdahl, L. G., 1974, Purification and properties of acetate kinase from *Clostridium thermoaceticum*, *Arch. Microbiol.* **100**:121–129.

Schiefer-Ullrich, H., Wagner, R., Dürre, P., and Andreesen, J. R., 1984, Comparative studies on physiology and taxonomy of obligately purinolytic clostridia, *Arch. Microbiol.* **138**:345–353.

Schink, B., 1984, *Clostridium magnum* sp. nov., a non-autotrophic homoacetogenic bacterium, *Arch. Microbiol.* **137**:250–255.

Schink, B., and Pfennig, N., 1982, Fermentation of trihydroxybenzenes by *Pelobacter acidigallici* gen. nov. sp. nov., a new strictly anaerobic, non-sporeforming bacterium, *Arch. Microbiol.* **133**:195–201.

Schink, B., Ward, J. C., and Zeikus, J. G., 1981, Microbiology of wetwood: Importance of pectin degradation and *Clostridium* species in living trees, *Appl. Env. Microbiol.* **42**:526–532.

Schobert, S., and Gottschalk, G., 1969, Considerations on the energy metabolism of *Clostridium kluyveri*, *Arch. Mikrobiol.* **65**:318–328.

Schulman, M., Parker, D., Ljungdahl, L. G., and Wood, H. G., 1972, Total synthesis of acetate from CO_2. V. Determination by mass analysis of the different types of acetate formed from [13]CO_2 by heterotrophic bacteria, *J. Bacteriol.* **109**:633–644.

Schwartz, R. D. and Keller, F. A., Jr., 1982a, Isolation of a strain of *Clostridium thermoaceticum* capable of growth and acetic acid production at pH 4.5, *Appl. Env. Microbiol.* **43**:117–123.

Schwartz, R. D., and Keller, F. A., Jr., 1982b, Acetic acid production by *Clostridium ther-*

moaceticum in pH-controlled batch fermentation at acidic pH, *Appl. Env. Microbiol.* **43:** 1385–1392.

Seto, B., 1980, Chemical characterization of an alkali-labile bond in the polypeptide of proline reductase from *Clostridium sticklandii, J. Biol. Chem.* **255:**5004–5006.

Shanmugasundaram, T., Kumar, G. K., and Wood, H. G., 1988, Involvement of tryptophan residues of the coenzyme A binding site of carbon monoxide dehydrogenase from *clostridium thermoaceticum, Biochemistry* **27:**6499–6503.

Shimshick, E. J., 1981, Removal of organic acids from aqueous solutions of salts of organic acids by super critical fluids, U.S. Patent 4,250,331.

Sinskey, A. J., Adedo, M., and Cooney, C. L., 1981, Acrylate Fermentations, in: *Trends in the Biology of Fermentations for Fuels and Chemicals*, (A. Hollaender, Rabson, R., Rogers, P., A. San Pietro, R. Valentine, and R. Wolfe, eds.), Plenum Press, New York, pp. 473–492.

Sleat, R., and Mah, R. A., 1985, *Clostridium populeti* sp. nov., a new cellulolytic species from a woody-biomass digestor, *Int. J. Syst. Bacteriol.* **35:**160–163.

Sleat, R., Mah, R. A., and Robinson, R., 1984, Isolation and characterization of an anaerobic, cellulolytic bacterium, *Clostridium cellulovorans* sp. nov., *Appl. Env. Microbiol.* **48:**88–93.

Sleat, R., Mah, R. A., and Robinson, R., 1985, *Acetoanaerobium noterae* gen. nov., sp. nov; an anaerobic bacterium that forms acetate from H_2 and CO_2, *Int. J. Syst. Bacteriol.* **35:**10–15.

Smith, L. D. S., 1970, *Clostridium oceanicum*, sp. n., a spore-forming anaerobe isolated from marine sediments, *J. Bacteriol.* **103:**811–813.

Snyder, M. L., 1936, The serological agglutination of the obligate anaerobes *Clostridium paraputrificum* (Bienstock) and *Clostridium capitovalis* (Snyder and Hall), *J. Bacteriol.* **32:**401–410.

Somack, R., and Costilow, R. N., 1973, 2,4-Diaminopentanoic acid C_4 dehydrogenase, *J. Biol. Chem.* **248:**385–388.

Sørensen, L. B., 1978, Extrarenal disposal of uric acid, in: *Uric Acids (Handbook of Experimental Pharmacology)*, Vol. 51 (W. N. Kelly and I. M. Weiner, eds.), Springer-Verlap, Berlin, pp. 325–336.

Stadtman, E. R., 1953, The coenzyme A transphorase system in *Clostridium kluyveri, J. Biol. Chem.* **203:**501–512.

Stadtman, E. R,. Novelli, G. D., and Lipmann, F., 1951, Coenzyme A function in an acetyl transfer by phosphotransacetylase system, *J. Biol. Chem.* **191:**365–376.

Stadtman, E. R., Stadtman, T. C., Pastan, I., and Smith, L. D. S., 1972, *Clostridium barkeri* sp. n., *J. Bacteriol.* **110:**758–760.

Stadtman, T. C., 1978, Selenium-dependent clostridial glycine reductase, *Methods Enzymol.* **53:** 372–382.

Stadtman, T. C., and McClung, L. S., 1957, *Clostridium sticklandii* nov. spec., *J. Bacteriol.* **73:** 218–219.

Stadtman, T. C., Elliott, P., and Tiemann, L., 1958, Studies on the enzymic reduction of amino acids. III. Phosphate esterification coupled with glycine reduction, *J. Biol. Chem.* **231:**961–973.

Stickland, L. H., 1934, Studies in the metabolism of the strict anaerobes (genus *Clostridium*). I. The chemical reactions by which *Cl. sporogenes* obtains its energy, *Biochem. J.* **28:**1746–1759.

Sturges, W. S., and Drake, E. T., 1927, A complete description of *Clostridium putrefaciens* (McBryde), *J. Bacteriol.* **14:**175–179.

Switzer, R. L., 1982, Glutamate mutase, in: B_{12}, Vol. 2 (D. Dolphin, ed.), John Wiley and Sons, New York, pp. 289–305.

Tanaka, H., and Stadtman, T. C., 1979, Selenium dependent clostridial glycine reductase. Purification and characterization of the two membrane-associated protein components, *J. Biol. Chem.* **254:**447–452.

Tanner, R. S., Stackebrandt, E., Fox, G. E., Gupta, R., Magrum, L. J., and Woese, C. R., 1982,

A phylogenetic analysis of anaerobic eubacteria capable of synthesizing acetate from carbon dioxide, *Curr. Microbiol.* **7:**127–132.

Thauer, R. K., 1972, CO_2-reduction to formate by NADPH. The initial step in the total synthesis of acetate from CO_2 in *Clostridium thermoaceticum, FEBS Lett.* **27:**111–115.

Thauer, R. F., Kirchniawy, F. H., and Jungermann, K. A., 1972, Properties and function of the pyruvate-formate-lyase reaction in clostridia, *Eur. J. Biochem.* **27:**282–290.

Thauer, R. K., Jungerman, K., and Decker, K., 1977, Energy conservation in chemotrophic anaerobic bacteria, *Bacteriol. Rev.* **41:**100–180.

Thiele, J. H., and Zeikus, J. G., 1988, Control of interspecies electron flow during anaerobic digestion:significance of formate transfer versus hydrogen transfer during syntrophic methanogenesis in flocs, *Appl. Env. Microbiol.* **54:**20–29.

Tschech, A., and Pfennig, N., 1984, Growth yield increase linked to caffeate reduction in *Acetobacterium woodii, Arch. Microbiol.* **137:**163–167.

Twarog, R., and Wolfe, R. S., 1962, Enzymatic phosphorylation of butyrate, *J. Biol. Chem.* **237:** 2474–2477.

U.S. International Trade Commission, 1986, Synthetic organic chemicals. United States Production and Sales, USITC Publ. 2009, pp. 209–213.

Uyeda, K., and Rabinowitz, J. C., 1965, Metabolism of formiminoglycine. Glycine formiminotransferase, *J. Biol. Chem.* **240:**1701–1710.

Uyeda, K., and Rabinowitz, J. C., 1967, Metabolism of formiminoglycine. Formiminotetrahydrofolate cyclodeaminase, *J. Biol. Chem.* **242:**24–31.

Uyeda, K., and Rabinowitz, J. C., 1971, Pyruvate-ferredoxin oxidoreductase. III. Purification and properties of the enzyme, *J. Biol. Chem.* **245:**3111–3119.

Vagelos, P. R., Earl, J. M., and Stadtman, E. R., 1959, Propionic acid metabolism. I. The purification and properties of a acrylyl coenzyme A aminase, *J. Biol. Chem.* **234:**490–497.

Valentine, R. C., and Wolfe, R. S., 1960, Purification and role of phosphotransbutyrylase, *J. Biol. Chem.* **255:**1948–1952.

Van der Meijden, van der Drift, C., and Vogels, G. D., 1984, Methanol conversion in *Eubacterium limosum, Arch. Microbiol.* **138:**360–364.

Van Gylswyk, N. O., Morris, E. J., and Els, H. J., 1980, Sporulation and cell wall structure of *Clostridium polysaccharolyticum* comb. nov. (Formerly *Fusobacterium polysaccharolyticum*), *J. Gen. Microbiol.* **121:**491–493.

Veldkamp, H., 1965, Enrichment cultures of procaryotic organisms, in: *Methods in Microbiology* (J. R. Norris and D. W. Ribbons, eds.), Vol. 3A, Academic Press, London, pp. 305–361.

Vogels, G. D., 1979, The global cycle of methane, *Antonie van Leeuwenhoek, J. Microbiol. Serol.* **45:**347–352.

Vogels, G. D., and van der Drift, C., 1976, Degradations of purines and pyrimidines by microorganisms, *Bacteriol. Rev.* **40:**403–468.

Von Hugo, H., Schoberth, S., Madan, V. K., Gottschalk, G., 1972, Coenzyme specificity by dehydrogenases and fermentation of pyruvate by clostridia, *Arch. Microbiol.* **87:**189–202.

Waber, J. L., and Wood, H. G., 1979, Mechanism of acetate synthesis from CO_2 by *Clostridium acidiurici, J. Bacteriol.* **140:**468–478.

Wachsman, J. T., and Barker, H. A., 1955, Tracer experiments on glutamate fermentation by *Clostridium tetanomorphum, J. Biol. Chem.* **217:**695–702.

Wagner, R., and Andreesen, J. R., 1977, Differentiation between *Clostridium acidiurici* and *Clostridium cylindrosporum* on the basis of specific metal requirements for formate dehydrogenase formation, *Arch. Microbiol.* **114:**219–224.

Wagner, R., and Andreesen, J. R., 1979, Selenium requirement for active xanthine dehydrogenase from *Clostridium acidiurici* and *Clostridium cylindrosporum, Arch. Microbiol.* **121:**255–260.

Wang, G. and Wang, D., 1984, Elucidation of growth inhibition and acetic acid production by *Clostridium thermoaceticum, Appl. Env. Microbiol.* **47:**294–298.

Whitehead, T. R., and Rabinowitz, J. C., 1986, Cloning and expression in *Escherichia coli* of the gene for 10-formyltetrahydrofolate synthetase from *Clostridium acidiurici* ("*Clostridium acidiurici*"), *J. Bacteriol.* **167**:205–209.

Wiegel, J., and Garrison, R., 1985, Utilization of methanol by *Clostridium thermoaceticum*, Abstr. *Annu. Meet. Am. Soc. Microbiol.* **I115**, p. 165.

Wiegel, J., and Ljungdahl, L. G., 1986, The importance of thermophilic bacteria in biotechnology, *CRC Crit. Rev. Biotechnol.* **3**:39–108.

Wiegel, J., Braun, M., and Gottschalk, G., 1981, *Clostridium thermoautotrophicum* species novum, a thermophile producing acetate from molecular hydrogen and carbon dioxide, *Curr. Microbiol.* **5**:255–260.

Wiegel, J., Kuk, S., Kohring, G. W., 1989, *Clostridium thermobutyricum*, spec. nov. a moderate thermophile isolated from a cellulolytic culture producing butyrate as major product from glucose *Int. J. Syst. Bacteriol.* **39**:199–204.

Wieringa, K. T., 1940, The formation of acetic acid from carbon dioxide and hydrogen by anaerobic spore-forming bacteria, *Antonie van Leeuwenhoek, J. Microbiol. Serol.* **6**:251–262.

Winters-Ivey, D. K., 1987, Metabolism of methanol in acetogenic bacteria, Dissertation, Univ. of Georgia, Athens.

Wise, D. L., ed., 1983, *Organic Chemicals from Biomass*, Benjamin Cummings, Menlo Park, CA.

Wise, D. L., ed., 1983 and 1984, *CRC Series in Bioenergy Systems*, five volumes entitled: *Fuel Gas Systems; Fuel Gas Developments, Liquid Fuel Systems, Liquid Fuel Developments; Bioconversion Systems*, CRC, Boca Raton.

Wood, H. G., 1952a, A study of carbon dioxide fixation by mass determination of the types of C^{13}-acetate, *J. Biol. Chem.* **194**:905–931.

Wood, H. G., 1952b, Fermentation of 3,4-C^{14}- and 1-C^{14}-labeled glucose by *Clostridium thermoaceticum*, *J. Biol. Chem.* **199**:579–583.

Wood, H. G., 1982, The discovery of the fixation of CO_2 by heterotrophic organisms and metabolism of the propionic acid bacteria, in: *Of Oxygen, Fuels, and Living Matter*, Part 2 (G. Semenza, ed.), John Wiley and Sons, New York, pp. 173–250.

Wood, H. G., and Kumar, G. K., 1985, Transcarboxylase. Its quaternary structure and the role of the biotinyl subunit in the assembly of the enzyme and in catalysis, *Ann. N.Y. Acad. Sci.* **447**:1–22.

Wood, H. G., O'Brien, W. E., and Michaels, G., 1977, Properties of carboxytransphosphorylase; pyruvate phosphate dikinase; pyrophosphate-phosphofructokinase and pyrophosphate-acetate kinase and their role in the metabolism of inorganic pyrophosphate, *Adv. Enzymol.* **45**:85–155.

Wood, H. G., Ragsdale, S. W., and Pezacka, E., 1986a, The acetyl-CoA pathway: a newly discovered pathway of autotrophic growth, *Trends Biochem. Sci.* **11**:14–18.

Wood, H. G., Ragsdale, S. W., and Pezacka, E., 1986b, The acetyl-CoA pathway of autotrophic growth, *FEMS Microbiol. Rev.* **39**:345–362.

Wood, W. A., 1961, Fermentation of carbohydrates and related compounds. in: *The Bacteria*, Vol. 2, *Metabolism* (I. C. Gunsalus and R. Y. Stanier, eds.), Academic Press, New York, pp. 59–149.

Wu, Z., Daniel, S. L., and Drake, H. L., 1988, Characterization of a CO-dependent O-demethylating enzyme system from the acetogen *Clostridium thermoaceticum*, *J. Bacteriol.* **170**:5747–5750.

Yamamoto, I., Saiki, T., Liu, S.-M., and Ljungdahl, L. G., 1983, Purification and properties of NADP-dependent formate dehydrogenase from *Clostridium thermoaceticum*, a tungsten-selenium-iron protein, *J. Biol. Chem.* **258**:1826–1832.

Yang, S.-S., Ljungdahl, L. G., DerVartanian, D. V., and Watt, G. D., 1980, Isolation and characterization of two rubredoxins from *Clostridium thermoaceticum*, *Biochim. Biophys. Acta*, **590**:24–33.

Yates, R. A., 1981, Removal and concentration of lower molecular weight organic acids from dilute solutions, U.S. Patent 4,282,323.

Young, L. Y., and Frazer, A. C., 1987, The fate of lignin and lignin-derived compounds in anaerobic environments, *Geomicrobiol. J.* **5:**261–293.

Zeikus, J. G., 1980, Chemical and fuel production by anaerobic bacteria, *Ann. Rev. Microbiol.* **34:**423–464.

Zindel, U., Freudenberg, W., Rieth, M., Andreesen, J. R., Schnell, J. and Widdel, F., 1988, *Eubacterium acidaminophilum* sp. nov., a versatile amino acid-degrading anaerobe producing or utilizing H_2 or formate. Description and enzyme studies, *Arch. Microbiol.* **150:**254–266.

Bioconversions 6

J. GARETH MORRIS

1. INTRODUCTION

Among the characteristic properties of clostridia are several that make these bacteria particularly useful agents of anaerobic bioconversions. Despite being obligate anaerobes they are not so aero-intolerant that in culture, or in washed cell suspension, they cannot survive the occasional encounter with oxygen, in this sense behaving as moderate anaerobes. Even so, they are in general highly reducing organisms capable of developing and sustaining a low redox potential (E_h) in their environment. The required reducing power is generated by their fermentative metabolism, though many species are also equipped with an uptake hydrogenase capable of utilizing H_2 gas as a supplemental electron donor. Almost exclusively, they acquire their free energy from fermentation processes and different species can utilize different substrates (carbohydrates, amino acids, purines, and pyrimidines) by a variety of fermentation pathways. In consequence, the genus is a particularly rich mine of unfamiliar biochemistry and novel enzymology.

Even their ability to sporulate can prove helpful to those who might wish to exploit *Clostridium* spp. Certainly, the durability of spore suspensions and the readiness whereby their germination can be elicited and vegetative growth reestablished makes for ease of selection, long-term storage, and transport of cultures. Indeed in several instances, when immobilized cells of a *Clostridium* sp. are required to effect a certain bioconversion, it has been recommended that the immobilization might beneficially be performed using a suspension of spores which are thereafter germinated *in situ*. Yet sporulation can also be a source of problems, while loss of the capacity to sporulate, which is highly selected for under conditions of chemostat culture, can be associated with concurrent loss of other useful metabolic trait(s) and so contribute to strain "degeneration."

In this chapter we shall not be concerned with the more usual fermen-

J. GARETH MORRIS ● Department of Biological Sciences, University College of Wales, Penglais, Aberystwyth, Dyfed SY23 3DA, Wales.

tations accomplished by clostridia nor with the extracellular enzymes whose secretion enables many species to extend the range of utilizable substrates; these topics have been fully considered earlier in this volume. Instead, we shall consider, for exemplary purposes, a few unusual fermentative bioconversions and the manner in which more usual fermentations can often be subverted to effect desirably novel biotransformations. Single-step bioconversions may be accomplished *in vitro* using isolated enzyme(s); if the required enzyme normally plays a key role in major fermentative pathways, it is likely that it would be present in the source bacterium in such high specific activity as to facilitate its extraction and purification. Yet often, and especially when the bioconversion is dependent on recycling of cofactor(s), the same biotransformation can be undertaken even more simply using whole organisms (semipermeabilized if necessary) in suspension or immobilized in, or on, a suitable support material. Since many of their most characteristic (and unusual) metabolic reactions exploit low potential reductants (i.e., so-called high-energy electrons), it is not surprising that the predominant exploitation of clostridia to date has been as agents of reductive bioconversions. We shall therefore consider examples of reductive biotransformations in some detail. In so doing, we run the risk of constructing a tedious catalogue of disparate reactions which have fortuitously been reported to be accomplished by identified species of *Clostridium* but are otherwise little related. In compensation we shall finally survey the various bioconversions wrought by clostridia within a single "family" of related compounds, i.e., bile acids and steroids.

2. CLOSTRIDIAL FERMENTATIONS: NATURAL AND MODIFIED

2.1. Natural Fermentations

The organic substrates of clostridial fermentations are many and varied, but the end products of these fermentations are generally very similar, i.e., fatty acids of short chain length, corresponding alcohols, occasional amines, etc. This is due mainly to the fact that relatively few substrate level phosphorylation (SLP) reactions are available for exploitation (Thauer *et al.*, 1977). Consequently, the primary substrate of the fermentation has to be converted during the course of its metabolism into one or more of a very limited number of SLP metabolites. It is this constraint, together with the additional necessity that a redox fermentation must maintain perfect oxidation–reduction balance, that calls for some quite extraordinary feats of biochemistry to be accomplished during the early preparatory stages of several clostridial fermentations. A good example is provided by the route of fermentation of 5-aminovalerate by *C. aminovalericum* (Fig. 1). The substrate is converted to glutaric semialdehyde and thence to 5-hydroxyvaler-

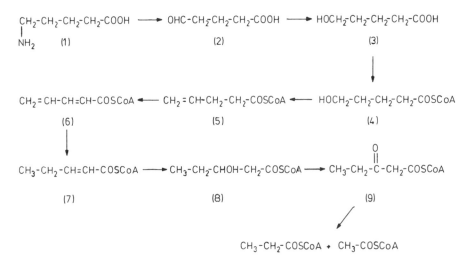

Figure 1. Initial steps in the fermentation of 5-aminovalerate (1) by *C. aminovalericum.* The intermediates are (2) glutaric semialdehyde, (3) 5-hydroxyvalerate, (4) 5-hydroxyvaleryl-CoA, (5) 4-pentenoyl-CoA, (6) 2,4,-pentadienoyl-CoA, (7) *trans*-2-pentenoyl-CoA, (8) L-3-hydroxyvaleryl-CoA, and (9) 3-oxovaleryl-CoA.

ate before this intermediate is activated by CoA thioesterification. Dehydration of the 5-hydroxyvaleryl-CoA gives 4-pentenoyl-CoA, which is dehydrogenated to yield 2,4-pentadienoyl-CoA. This is then reduced to *trans*-2-pentenoyl-CoA which is hydrated to yield L-3-hydroxyvaleryl-CoA, which is then oxidized to give 3-oxovaleryl-CoA (Barker *et al.*, 1987). This sequence of some eight bioconversions, which effects removal of the initially terminal amino group of the 5-aminovalerate and the introduction of a crucial carbonyl group at C-3, is merely the prelude to the ATP-yielding segment of the fermentation pathway. This effects the thiolytic cleavage of 3-oxovaleryl-CoA to yield acetyl-CoA and propionyl-CoA whose conversion to the terminal free acids is accomplished via SLP reactions. Thus while in this and similar "unusual" fermentations it is these end reactions that are of crucial import to the organism and the *raison d'être* of all that has gone before, it is often among the catalysts of the preliminary reactions that we find some of the most serviceable enzymes with which to effect analogous bioconversions.

2.1.1. Catabolism of Homocyclic and Heterocyclic Compounds

Particular interest attaches to whether and how the anaerobic clostridia are able to circumvent the requirement for molecular oxygen that is displayed by aerobic microbes in their catabolism of aromatic organic compounds. The extent of anaerobic degradation of such compounds has in the past been considerably underestimated, but it is now recognized that many are subject to bacterial attack in the complete absence of air. Sometimes the biotransformation stops short of ring fission; in other circumstances ring breakage and further catabolism is accomplished. In the latter case, the reaction pathway generally consists of preliminary ring hydrogenation followed by a ring hydration/ring cleavage sequence, the substitution with an aptly situated hydroxyl group (originating in water) facilitating the subsequent ring cleavage (Evans, 1977). It follows that prehydroxylated aromatic compounds as well as those bearing substituent carboxylic groups are more vulnerable to anaerobic bacterial attack.

Much of the investigation of the nature and extent of aromatic degradation by fermentative anaerobes has been performed with methanogenic consortia of very mixed composition and little is known of the part specifically played by the clostridial members of such consortia (Sleat and Robinson, 1984; Young, 1984; Berry *et al.*, 1987). However, there are indications that species of *Clostridium* could be of some importance therein, if only in accomplishing initial reactions which remove side chain substituents prior to ring cleavage. Thus it is known, for example, that *C. pfennigii* utilizes the methoxyl group of monobenzenoids to give hydroxybenzenoids plus butyrate. Vanillate is thereby converted to protocatechuate plus protocatechuate aldehyde, ferulate yields caffeate plus hydrocaffeate, and syringate is converted to gallate (Krumholtz and Bryant, 1985). One characteristic property of *C. difficile* is its formation of p-cresol from p-hydroxyphenylacetate (D'Ari and Barker, 1985), and *C. tetanomorphum* produces phenol from tyrosine (Brot and Weissbach, 1970), whilst *C. aerofoetidum* can form toluene from phenylacetate or phenylalanine (Pons *et al.*, 1984).

The ability of various species of *Clostridium* to degrade heterocyclic compounds is much better documented, especially in the case of those organisms that ferment purines and pyrimidines. Uracil, which at the outset is a highly oxidized molecule, is utilized by *C. uracilicum* via initial reduction followed by hydrolytic ring cleavage (Fig. 2, and see Campbell, 1957). *Clostridium cylindrosporum* and *C. acidiurici* are both capable of fermenting the hydroxylated purines, guanine, uric acid, or xanthine, but not the unhydroxylated molecules of adenine or purine (Dürre and Andreesen, 1983). However, *C. purinolyticum* can ferment purine or adenine via preliminary conversion to hypoxanthine and xanthine (Fig. 3). Xanthine is subsequently converted to formiminoglycine, a process in which both rings of the purine molecule are sequentially cleaved by hydrolytic reactions (Fig. 4).

A particularly intriguing fermentation of nicotinic acid is undertaken by *C. barkerii* (Tsai *et al.*, 1966). The initial attack on the pyridine ring is by hydroxylation, which yields 6-hydroxynicotinate in a reaction catalyzed by a NADP-dependent nicotinate hydroxylase, which contains both selenium and molybdenum (Holcenberg and Stadtman, 1969; Imhoff and An-

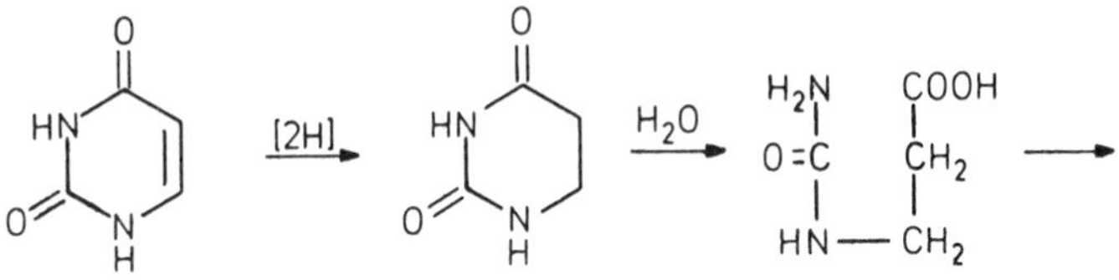

Figure 2. Ring cleavage in the fermentation of uracil by *C. uracilicum.*

dreesen, 1979; Dilworth, 1983). Ring cleavage, in order to yield α-methylene glutarate, occurs subsequent to reduction of the 6-hydroxynicotinate to 1,4,5,6-tetrahydro-6-oxonicotinate (Fig. 5).

2.2. Modified Fermentations and Anaerobic Cometabolism

Although a species of *Clostridium* may be noted for a particular fermentation yielding characteristic products, modification of the conditions of its culture can in several instances encourage its usage of an alternative metabolic pathway. Thus although *C. thermosaccharolyticum* is generally noted for its homoacetic fermentation of glucose, under certain culture conditions it can form substantial quantities of R(−)-1,2-propanediol and acetol from various sugars, probably as a consequence of the routing of intermediates via the methylglyoxal bypass (Cameron and Cooney, 1986;

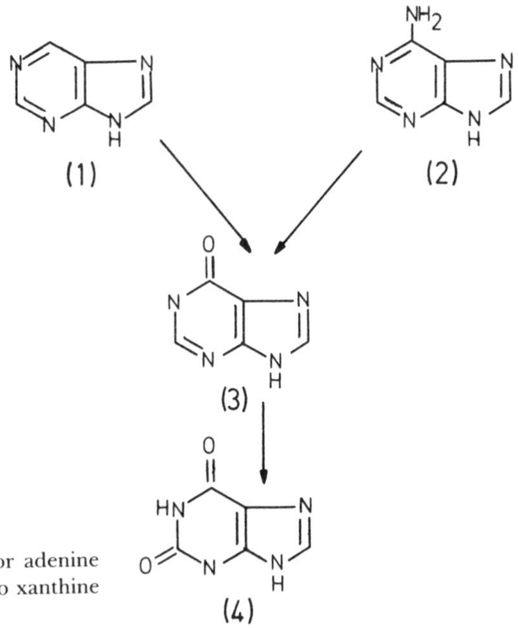

Figure 3. Conversion of purine (1) or adenine (2) to hypoxanthine (3) and thence to xanthine (4) by *C. purinolyticum.*

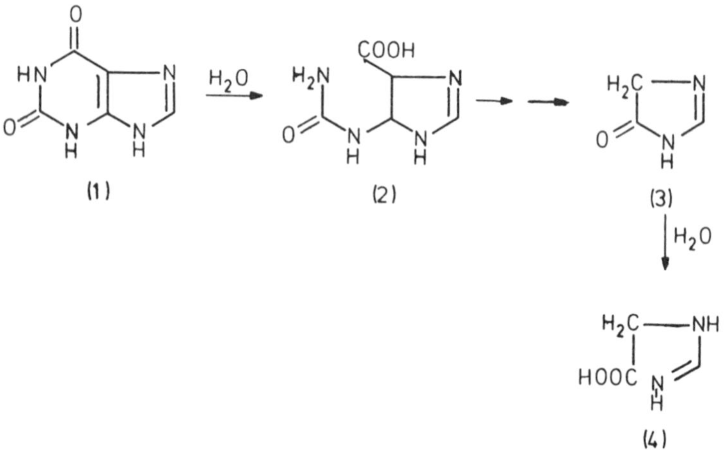

Figure 4. Ring cleavage reactions in fermentations accomplished by purinolytic clostridia.

Sanchez-Riéra et al., 1987). Both R(−)-1,2-propanediol and D(−)-lactate are formed from glucose by C. sphenoides when it is grown under conditions of phosphate limitation (Tran-Din and Gottschalk, 1985). Again, though noted for its acetone-butanol fermentation of glucose, C. acetobutylicum produces 1,3-propanediol from glycerol (Forsberg, 1987) and, even when fermenting glucose, it can cometabolize certain aldehydes and acids, e.g., propionate, valerate, and 4-hydroxybutyrate, reducing them to their corresponding alcohols (Blanchard and MacDonald, 1935; Jewell et al., 1986).

Knowledge of the usual fermentation pathway can sometimes suggest

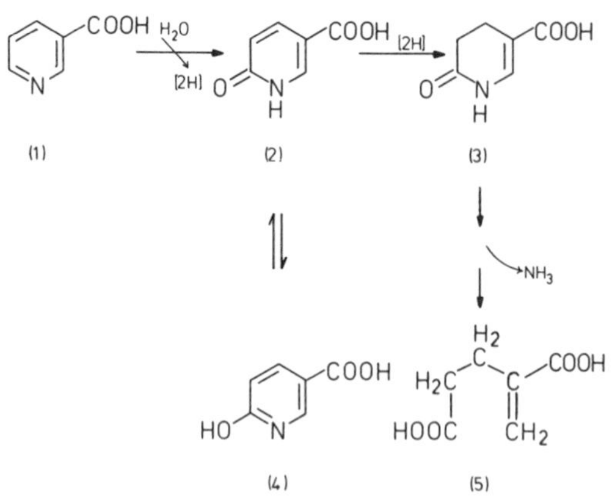

Figure 5. Ring cleavage during fermentation of nicotinic acid by C. barkeri. The intermediates are (1) nicotinate, (2) and (4) 6-hydroxynicotinate, (3) 1,4,5,6-tetrahydro-6-oxonicotinate, and (5) methyleneglutarate.

how it may be "distorted" so that the yield of a particularly desirous end product may be favored at the expense of some other. For example, by blockading the normal route of electron flow, the reducing power generated by a fermentation may be channeled into some alternative reductive process. One can instance the inhibition by carbon monoxide of the hydrogenase of *C. acetobutylicum* which has the effect of diverting more reducing equivalents to the formation of butan-1-ol and consequently decreasing the yield of acetone (Datta and Zeikus, 1985; Meyer *et al.*, 1986). The same purpose is served, though by a different means, by incorporating viologen dyes into the growth medium (Rao and Mutharasan, 1987), a procedure that also causes *C. thermoaceticum* to produce methanol from carbon monoxide (White *et al.*, 1987).

In an alternative strategy, fermentative electron flow might be diverted to an added alien, but preferred, organic electron acceptor whose product of reduction then accumulates at the same time that the organism forms more than usually oxidized fermentation end product(s). On occasion this has (sometimes unwittingly) been accomplished by supplying cultures of clostridia with an inorganic oxidant such as nitrate or sulfite or a nonlethal concentration of oxygen. In a more deliberate application of this strategy, *C. propionicum*, which normally ferments lactate to form propionate, was caused to synthesize acrylate (Sinskey *et al.*, 1981). A washed suspension of the organism provided with lactate (25 mM) in the presence of oxygen and methylene blue and excess propionate (200 mM) accumulated 18 mM of acrylate.

Organic compounds that are foreign to the normal metabolism of an organism may be used as the alternative "electron sinks" so that their specific bioreduction is accomplished. Specific examples of such reductive cometabolism will be given in Section 5.2, but it may here be noted that in exploiting clostridia for such purposes one may merely be taking advantage of enzymes whose broad substrate specificities are probably advantageous to the organisms in their natural habitats. As Schink (1986) pointed out, the reductive elimination reactions undertaken by various anaerobes (dehydroxylations, dehalogenations, and deaminations) "may mainly represent means of disposing of excess electrons for the fermenting bacteria involved." If this is so, then anaerobic cometabolism would appear to have a promising future as a ready means of effecting specific bioreductions. Nor should controlled anaerobic oxidations be neglected. Here a single species could link the oxidation of a natural or alien substrate X to the reduction of a natural electron acceptor Y, as is most obviously accomplished in the Stickland reactions between pairs of amino acids (Seto, 1980) or in mixed Stickland reactions when one of the substrates is not an amino acid. Alternatively, the oxidation could be accomplished by a *Clostridium* with the transfer of reducing equivalents occurring in the form of H_2 gas, which is consumed by a companion hydrogenotrophic anaerobe. It is this device which sustains growth of *C. bryantii* on short-chain fatty acids (C_4 to C_{11}) when the even-chain substrates yield acetate and those substrates with an odd number of carbon atoms yield propionate plus acetate (Stieb and Schink, 1985). Again, while *C. sporogenes* in pure culture scarcely ferments isoleucine, in coculture with a H_2-scavenging methanogen it quantitatively converts this amino acid to 2-methylbutyrate (Wildenauer and Winter, 1986).

3. INTERESTING AND EXPLOITABLE ENZYMES FROM CLOSTRIDIA

3.1. Intracellular Enzymes

Though the clostridia are obligate anaerobes, it does not follow that their intracellular enzymes are necessarily oxygen-sensitive. In fact, the majority are probably as robustly oxyduric as any derived from aerobic bacteria, although it is true that certain others, especially those involving cobamide coenzymes or reduced folate cofactors, plus some that operate at a low E_h, are so aero-intolerant that they must be purified and used under strictly anaerobic conditions. The existence of a comparatively large number of thermophilic species of *Clostridium* (e.g., *C. hydrosulfuricum, C. thermocellum, C. thermosaccharolyticum*) means that several usefully thermoduric enzymes are also obtainable from the genus.

Among the more unusual clostridial enzymes are several that catalyze C–C rearrangements. They include a number of mutases which demonstrate a requirement for a B_{12} coenzyme. Such an enzyme is the β-methylaspartate glutamate mutase from glutamate-grown *C. tetanomorphum* (Barker, 1985) which forms (2S,3S)-3-methyl-L-asparate from L-glutamate (Hartrampf and Buckel, 1984), the stereochemical course of this reaction resembling that catalyzed by the adenosylcobalamin-dependent 2-methyleneglutarate mutase of *C. barkerii* (Hartrampf and Buckel, 1986). Such reactions (Fig. 6) are not just of academic interest, but could prove useful in the synthesis of appropriately [14]C- or ^{13}C-labeled chiral synthons, e.g., ^{13}C-labeled β-methylaspartate for β-lactam synthesis.

Other rearrangements are also specifically catalyzed by enzymes of clostridial origin, including C–N rearrangements that feature in the fermentations of several amino acids. Both the L-β-lysine mutase (β-lysine to 3,5-diaminohexanoate) and the D-lysine mutase (D-lysine to 2,5-diaminohexanoate) of *C. sticklandii* are cobamide-dependent enzymes (Stadtman and Grant, 1971) as is the leucine-2,3-aminomutase of *C. sporogenes* (Poston, 1976). Yet the lysine-2,3-aminomutase of *C. subterminale* SB4 (Chirpich and Barker, 1971) displays no requirement for B_{12} but is instead dependent on Fe(II) ions, pyridoxal phosphate, and S-adenosylmethionine

Figure 6. Two mutase reactions catalyzed by B_{12}-dependent enzymes from clostridia. (1) Catalyzed by the β-methylaspartate-glutamate mutase of *C. tetanomorphum*. (2) Catalyzed by the 2-methyleneglutarate-3-methylitaconate mutase of *C. barkeri*. In both cases, substituent R is shifted from the βC to the αC (with inversion of configuration at the αC) and a H atom is moved in the opposite direction.

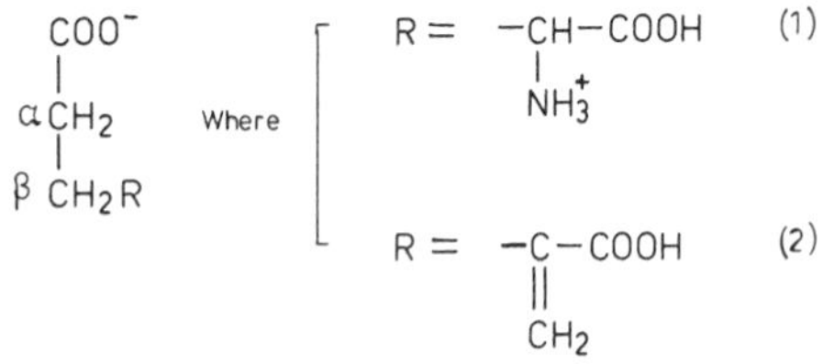

$$
\begin{array}{l}
\text{COO}^- \\
|\ \\
\alpha\,\text{CH}_2 \quad \text{Where} \\
|\ \\
\beta\ \text{CH}_2\text{R}
\end{array}
\qquad
\begin{array}{ll}
\text{R} = & -\text{CH}-\text{COOH} \quad (1) \\
 & \quad\ |\ \\
 & \quad \overset{+}{\text{NH}_3} \\
 & \\
\text{R} = & -\text{C}-\text{COOH} \quad (2) \\
 & \quad\ \| \\
 & \quad \text{CH}_2
\end{array}
$$

(Aberhart *et al.*, 1983). The reaction which it catalyzes, i.e., (2S)-α-lysine to (3S)-β-lysine, proceeds with inversion of configuration at both C-2 and C-3. It follows, interestingly enough, the same stereochemical course as that catalyzed by the B_{12}-dependent β-lysine mutase in which the 6-amino group of β-lysine replaces the 5(proS)H to form (3S,5S)-3,5-diaminohexanoate.

Elimination reactions are a feature of several clostridial fermentations and can again provide the means of synthesis of a specific chiral product. The majority of such biological elimination reactions lead to removal of a proton that is α to a carboxyl (or other activating group) plus a leaving nucleophile that is β to the carbonyl group. This is exemplified by the reversible dehydration of L(+)-3-hydroxyacyl-CoA thioesters of C_4–C_6 chain length to the corresponding 2-*trans*-enoyl-CoA thioesters that is catalyzed by the crotonase purified from *C. acetobutylicum* (Waterson and Conway, 1981). However, the lactyl-CoA dehydratase of *C. propionicum* (Kuchta and Abeles, 1985) accomplishes an unusual elimination reaction in its production of acrylyl-CoA, for the H atom that is removed is not activated and the α-hydroxyl is a very poor leaving group. A similarly unusual *syn* elimination of water is catalyzed by the phenyl lactate dehydratase of *C. sporogenes* when from (2R)-phenyl lactate it forms (E)-cinnamate (Pitsch and Simon, 1982; Machacek-Pitsch *et al.*, 1985), and by the reversible dehydration of (R)-2-hydroxyglutarate to (E)-glutaconate accomplished by *C. microsporum* (Buckel, 1980), which again involves stereospecific elimination of the (proS)3-H (Fig. 7). Other dehydratases purified from clostridia include the diol dehydratase of *C. glycolicum* (Hartmanis and Stadtman, 1986) and the D-gluconate dehydratase obtained from gluconate-grown *C. pasteurianum* (Gottschalk and Bender, 1982). Analogous in its mode of action to the crotonase (i.e., L-3-hydroxyacyl-CoA hydrolyase, EC 4.2.1.17) of butyric *Clostridia* is the enzyme from lysine-fermenting *C. subterminale* which eliminates NH_3 from L-3-aminobutyryl-CoA to form crotonyl-CoA (Jeng and Barker, 1974). Another interesting elimination of NH_3 is catalyzed by the ethanolamine deaminase of a choline-fermenting *Clostridium* (Kaplan and Stadtman, 1971), which, unlike the superficially analogous

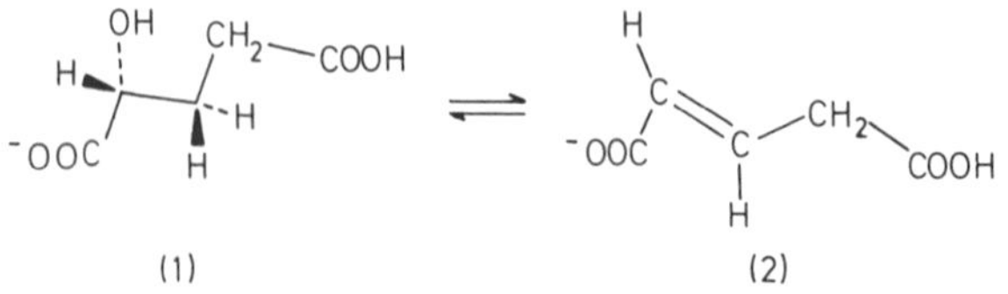

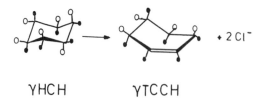

Figure 7. Reversible dehydration of (R)-2-hydroxyglutarate (1) to (E)-glutaconate (2) accomplished by *C. microsporum.*

diol dehydratase of *C. glycolicum,* is a B_{12}-dependent enzyme. The β-methylaspartase of *C. tetanomorphum* (Hsiang and Bright, 1969) catalyzes an elimination of NH_3 that leads to the formation of mesaconate (i.e., 3-carboxyl-3-methyl-2-propenoic acid). Elimination of two chlorine atoms from hexachlorocyclohexane to yield tetrachloro-1-cyclohexane (Fig. 8) is accomplished by several clostridia including *C. butyricum* and *C. pasteurianum* (Jagnow *et al.,* 1977), *C. sphenoides* (Heritage and MacRae, 1977), and *C. rectum* (Ohisa *et al.,* 1980).

Cleavage of C–C bonds is of course a necessary feature of catabolism and several useful enzymes which catalyze such lytic reactions have been purified from clostridial sources. Among them are several decarboxylases, including the acetoacetate decarboxylase of *C. acetobutylicum* (Westheimer, 1969), which has been found to be useful in several organic syntheses, and the L-lysine decarboxylase of *C. cadaveris,* which has been made the basis of an enzyme electrode employed to assay L-lysine (Tran *et al.,* 1983). Other amino acid decarboxylases including the L-glutamate decarboxylase of *C. perfringens* (Cozzani *et al.,* 1975), the L-aspartate decarboxylase and L-histidine decarboxylase of the same organism, and the L-ornithine decarboxylase of *C. septicum,* were all formerly used in the specific manometric assays devised for their substrate amino acids (Najjar, 1957). An ornithine decarboxylase has also been obtained from *C. thermohydrosulfuricum* (Paulin and Pösö, 1983) and a phosphatidylserine decarboxylase has been isolated from *C. butyricum* (Verma and Goldfine, 1985). Other lytic enzymes from clostridia have also been put to practical uses. Thus the L-methioninase of *C. sporogenes* has been used to destroy the methionine present in natural media (Kreis and Hession, 1973) and an L-threonine aldolase is obtainable from *C. pasteurianum* (Stöcklein and Schmidt, 1985). Several useful and quite stable enzymes which catalyze CoA-dependent thiolytic reactions

YHCH   YTCCH $+ 2 Cl^-$

Figure 8. Clostridial dechlorination of γ-hexachlorohexane to yield tetrachloro-1-cyclohexane. For simplicity, the Cl atoms are represented as ○ and H atoms as ●.

have been purified from clostridia, though their utility for large-scale bio-conversions is limited by the high cost of the coenzyme. They include the β-ketothiolase of *C. pasteurianum* (Berndt and Schlegel, 1975) and of *C. kluyveri* (Sliwkowski and Hartmanis, 1984), and the enzyme from *C. sticklandii* which catalyzes the CoA-dependent cleavage of 2-amino-4-oxopentanoate to yield alanine plus acetyl-CoA (Jeng *et al.*, 1974).

The formation of acyl-CoA thioesters is common to many clostridial fermentations. It is frequently accomplished by CoA transfer from some donor CoA thioester, but acyl-CoA thioesters can also be made from acyl phosphates, e.g., as in the reaction catalyzed by the purified phosphotrans-acetylase of *C. kluyveri* (Klotzsch, 1969), and several kinases which catalyze the production of acyl phosphates from ATP plus the free acid have been isolated from various clostridia, e.g., the butyrate kinase of *C. acetobutylicum* (Hartmanis, 1987). Quite serendipitously it was found that a crude cell-free extract of *C. kluyveri* when supplied with acetyl phosphate in the presence of excess cyanide would acetylate most amino acids (Lieberman and Barker, 1955). *Clostridium kluyveri* is also renowned as the source of a most effective diaphorase which links the oxidation of NADH, and, though less well, NADPH, to the reduction of dichlorophenolindophenol, or ferricyanide (Kaplan *et al.*, 1969). This enzyme finds a place in many "diagnostic kits" whose specificity resides in the enzymic component which generates the NAD(P)H.

The dehydrogenases of clostridial origin that have been subjected to the most study are again those that participate in their fermentations and which may therefore be synthesized in relatively large quantities by the organism. For example, the NAD-dependent glutamate dehydrogenase of *C. symbiosum* is a major component of the cellular protein of this organism (Rice *et al.*, 1985). Other potentially useful dehydrogenases include the xanthine dehydrogenase of *C. acidiurici* (Wagner *et al.*, 1984), the formate dehydrogenase of *C. pasteurianum* (Liu and Mortenson, 1984), and the formate-NADP oxidoreductase of *C. thermoaceticum*, which is a selenium- and tungsten-containing protein which operates best at 45°C and is capable of reacting with methylviologen and benzylviologen in place of the normal pyridine nucleotide (Ljundahl and Andreesen, 1978). Other clostridial dehydrogenases include the NAD-dependent γ-hydroxybutyrate dehydrogenase of *C. aminobutyricum* (Hardman, 1962) and the NADP-dependent 3-hydroxybutyryl-CoA dehydrogenase from *C. kluyveri* (Sliwkowski and Hartmanis, 1984).

Enzymes which display strict substrate specificity and which generate a product that is readily quantifiable are much sought after for assay purposes. Among several clostridial enzymes which could fall in this category are the guanase of *C. acidiurici* (Rakosky *et al.*, 1955), and the creatinine desimidase of *C. paraputrificum* (Szulmajster, 1958) and other clostridia, several of which also release trimethylamine from choline or betaine (Möller *et al.*, 1986). Arginine deiminase is obtainable from *C. perfringens,* especially after growth in the presence of caffeine (Sacks, 1985).

The biosynthetic capacities of clostridia have been less exploited than have their fermentative abilities, but *C. kluyveri* has been employed to synthesize radioactively labeled flavin nucleotides (Decker and Hamm, 1980), while radioactive nicotinic acid was synthesized from ^{14}C-labeled *N*-formyl-L-aspartate using cell-free extracts of *C. acetobutylicum* (Scott *et al.,* 1969).

3.2. Cloned Enzymes

To date, the relatively few clostridial enzymes that have been cloned have been synthesized in *E. coli,* though clostridial genes are expressible in numerous bacterial genera (see Chapter 3). Aside from the obvious desirability of cloning, for example, the several cellulases of *C. thermocellum* (Knowles *et al.,* 1987), there seems to be little rationale for some of the other choices that have been made. In the main, the cloned enzymes have been of some special interest to the investigators, who have often been equally concerned with cloning the controlling elements together with the structural genes so as to be better able to comprehend the mechanisms of *in vivo* regulation of the clostridial enzyme product. Early genes to be cloned in *E. coli* were those specifying the hydrogenase and β-isopropylmalate dehydrogenase of *C. butyricum* (Karube *et al.,* 1983; Ishii *et al.,* 1983). Cloning of the ferredoxin (Graves *et al.,* 1985) and galactokinase (Daldal and Applebaum, 1985) of *C. pasteurianum* was followed by cloning of the 10-formyltetrahydrofolate synthetase of *C. acidiurici* (Whitehead and Rabinowitz, 1986), the leucine dehydrogenase of *C. thermoaceticum* (Shimoi *et al.,* 1987), and the glutamine synthetase and NADP-dependent alcohol dehydrogenase of *C. acetobutylicum* (Usdin *et al.,* 1986; Youngleson *et al.,* 1988). One is therefore encouraged to believe that there will prove to be no insuperable obstacle to the successful cloning of most of the clostridial enzymes likely to be of practical value in bringing about desirable biotransformations.

4. IMMOBILIZED CLOSTRIDIAL CELLS AS AGENTS OF BIOCONVERSION

Suspensions of clostridial cells can be maintained at high biomass density immobilized within polymer beads (gel-entrapped suspensions) or adherent to the surface(s) of an inert support material. Calcium alginate or acrylic acid-derived photopolymerizable prepolymers have been most frequently employed for the former purpose and a simple conductimetric method for the assessment of the biomass content of gel-entrapped cell suspensions has been described with *C. pasteurianum* as the test organism (Lovitt *et al.*, 1986b). Inert support materials of a great many kinds varying from bone ash to clays have been used to create immobilized films of clostridia. Like other anaerobic bacteria such immobilized suspensions of clostridia lend themselves to exploitation by column filter procedures since no provision has to be made for aeration. The requisite low E_h is readily maintained by the activity of the cells themselves and the only problem is caused by gas production (e.g., H_2 and CO_2), which can bring about the splitting of polymer beads. Yet immobilized growing cells of *C. butyricum* have been employed to generate H_2 gas (Karube *et al.*, 1982) and to regenerate NAD(P)H when supplied with H_2 (Matsunaga *et al.*, 1985). Complete fermentations may be accomplished by immobilized clostridial cells; for example, there are many reports of acetone and butanol production by immobilized *C. acetobutylicum* (Taya *et al.*, 1986). Single-step bioconversions can also be undertaken as, for example, the hydrogenation of aldehydes or enoates by polymer-entrapped cells of *C. tyrobutyricum* La1 (Egerer and Simon, 1982), or the conversion of chenodeoxycholic acid to ursodeoxycholic acid by immobilized cells of *C. absonum* (Kole and Altosaar, 1985).

5. BIOREDUCTIONS CATALYZED BY CLOSTRIDIAL CELLS AND ENZYMES

5.1. Miscellaneous Reductive Capacities of Clostridia

It is in the nature of any redox fermentation that one or more terminal bioreductions are carried out to yield distinctive products. In the case of the saccharolytic clostridia, these may be alcohols (e.g., ethanol, butanol), acids (lactate, propionate, butyrate, hexanoate, etc.), or H_2 produced by the reduction of protons. More unusual reductions are accomplished by some species; an example is provided by the synthesis of acetyl-CoA from CO_2 and H_2 with CoA and methyltetrahydrofolate that is accomplished by *C. thermoaceticum* (Pezacka and Wood, 1984; Lebertz *et al.*, 1987). The reductive branches of the various Stickland reactions in amino acid-fermenting clostridia are also a source of several reductive enzymes that could, by virtue of their broad substrate specificities, prove useful in catalyzing related bioconversions and/or of several highly specific terminal reductases. The 2-oxoacid reductases and enoate reductases of organisms such as *C. sporogenes* are examples of the former group (Bader *et al.*, 1982; Giesel and Simon, 1983) while the glycine reductase and proline reductase of *C. sticklandii* illustrate the latter (Tanaka and Stadtman, 1979; Seto and Stadtman, 1976). *Clostridium kluyveri* contributes a potentially useful acryloyl-CoA reductase (Sedlmaier and Simon, 1985) and among biosynthetic enzymes those that employ reduced ferredoxin to effect a chain-lengthening carboxylation have attracted considerable interest. Thus *C. kluyveri* is one of several clostridia that contain a pyruvate synthase which catalyzes the production of pyruvate from acetyl-CoA plus CO_2 (Andrew and Morris, 1965), while *C. sporogenes* contains a 2-oxoacid synthase which brings about reductive carboxylation of propionyl-CoA to 2-oxobutyrate or of 2-methylbutyryl-CoA to 3-methyl-2-oxopentanoate (Monticello *et al.*, 1984; Lovitt *et al.*, 1987b).

Several inorganic oxidants can also be reduced to clostridia. Indeed species of *Clostridium* can be expected to contribute to the anaerobic dissolution of metal oxides in aquatic sediments, both by virtue of their creation of localized acidic conditions and also, as has been reported in the case of one clostridial isolate, by direct reduction of Fe_2O_3 and MnO_2 (Francis and Dodge, 1988). An inducible, dissimilatory sulfite reductase is produced by *C. pasteurianum* (Harrison *et al.*, 1984) and both a nitrate reductase (Seki *et al.*, 1987) and a nitrite reductase which yields hydroxylamine (Sekiguchi *et al.*, 1983) have been purified from *C. perfringens*. Indeed, although nitrite is potentially toxic to most clostridia in varying degrees, the ability to reduce nitrate to ammonia is displayed by several species. Many clostridia also have the ability to reduce aliphatic and aryl nitro compounds. Reduction of the nitro aliphatic compounds would appear to be attributable to reduced ferredoxin, but specific reductase enzymes seem to be required to effect reduction of nitroaryl compounds (Angermaier and Simon, 1983). Thus complete reduction of the aryl nitro group of chloramphenicol was the cause of its destruction by *C. acetobutylicum* (O'Brien and Morris, 1971) while stepwise reduction of the nitro group of metronidazole, via a lethal free radical intermediate, is the probable cause of this drug's effectiveness against all clostridia and other obligate anaerobes (O'Brien and Morris, 1972; Edwards *et al.*, 1982).

5.2. Clostridial Reductions of Unusual Chemicals

5.2.1. Background to Chiral Reductions

The lax specificity of many microbial redox enzymes, both dehydrogenases and reductases, means that they can be taken advantage of by organic chemists and used to catalyze single-step reactions which are difficult or impossible to undertake by normal chemical procedures. This is particularly the case when a stereospecific conversion of a prochiral synthon to one of several alternatively possible enantiomeric products is required. Yeasts and a wide range of aerobic filamentous fungi have long been used for this purpose even when reduction is involved. It is logical, however, to suppose that obligately anaerobic bacteria which generate excess reducing power at low E_h values should be even better equipped to catalyze a wide range of bioreductions. Much of the work so far undertaken with these organisms is fragmentary, but evidence is accumulating the clostridia could be particularly rewarding agents of directed reductive syntheses. Although in some instances the oxidant may simply be added to growing cultures, it is more usual to employ dense suspensions of washed cells (whole or semipermeabilized, and immobilized if this is advantageous). The normal fermentation substrate might be supplied as the source of reducing power, though in some instances H_2 can serve as the chief reductant, possibly with methylviologen being additionally supplied as a mediator of electron transport. In organisms which happen to possess a potent NAD-dependent formate dehydrogenase, provision of formate could conveniently assure regeneration of NADH.

Several factors have to be taken into account when deciding which of possibly several species capable of carrying out the desired bioconversion should best be employed for that purpose. To the chemist, the ease with which the organism can be grown in culture is likely to be a major consideration, but thereafter one has to be concerned with the rate of biotransformation, the percentage yield of product, and the enantiomeric excess (ee) in which the desired stereoisomer is produced. The specific rate of bioreduction of a prochiral substrate is generally expressed as a productivity number (PN), a term defined by Simon and his colleagues (1985) as the ratio:

$$\frac{\text{Amount of product (mmoles)}}{\text{Dry weight of catalyst (kg)} \times \text{time (hr)}}$$

5.2.2. Reduction of C=O Groups

Enantioselective reduction of ketones to chiral alcohols was obtained using yeasts and other microbes (Sih and Chen, 1984). The specificity evidenced in the excess in which one of the possible enantiomers accumulates occurs because the responsible (chiral) enzyme interacts with the prochiral substrate to form diastereomeric transition states which differ in free energy content; the magnitude of this difference determines the enantiomeric excess. Thus if a prochiral ketone displays the enantiotopic faces *Re* and *Si*, complexation of the enzyme with the *Re* face will yield one enantiomeric form of the chiral alcohol product whereas complexation with the *Si* face will yield the other enantiomer. Pragmatic observation suggests (Prelog's rule) that it is most usual for the stereochemical course of reduction of the prochiral ketone to proceed via hydrogen transfer to the *Re* face when the disposition of larger R_L and small R_s substituent groups is as shown in Figure 9. Yet this is not invariably the case in all of the clostridial reductions of ketones that have been reported. Poor enan-

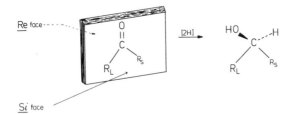

Figure 9. Reduction of a prochiral ketone by hydrogen transfer to the *Re* face. (Modified from Sih and Chen, 1984.)

tiomeric selectivity could result (1) from a reductive enzyme not favoring either face of the ketone, or (2) when whole cells or crude cell extracts are employed, through competition for the substrate between two or more reductases which differ in their selectivity and relative reaction rates (i.e., in K_m and V_{max} values with the given substrate).

Bühler *et al.* (1980) reported the presence in *C. sporogenes* of a NADH-dependent phenylpyruvate reductase which catalyzed the formation of (R)-phenyl lactate. This 2-oxoacid reductase also accepted as a substrate 4-methyl-2-oxopentanoate (from leucine) but was far less active on 3-methyl-2-oxopentanoate derived from isoleucine (Bader *et al.*, 1982). A potentially most useful NADPH-dependent alcohol–ketone oxidoreductase capable of effecting the reduction of methyl ketones has been obtained from the thermophilic *C. thermohydrosulfuricum* (Lamed *et al.*, 1981). Both *C. kluyveri* and *C. tyrobutyricum* La1 were found by Simon *et al.* (1985) to reduce a range of ketones to chiral secondary alcohols. With *C. kluyveri* acetoacetic ethyl ester was reduced with a PN of 1500 to yield the (S) alcohol in 95% yield, and 4-chloro-3-oxobutyrylethyl ester was reduced with a PN of 350 to give the (R)-alcohol product in 99% yield (Fig. 10). When (R,S)-2-methyl-3-oxobutyrylethyl ester (a 50 : 50 mixture of both enantiomers) was introduced to a cell suspension of *C. kluyveri* maintained in an atmosphere of H_2, (2R,3S)-3-hydroxyl-2-methylbutyrylethyl ester was formed in 81% yield. This was deemed to be the consequence of rapid reduction of the (2R) form of the substrate and spontaneous racemization of the (2S) form (Simon and Günther, 1983).

It is evident, therefore, that it is difficult to predict the outcome of a reduction even by a single species of *Clostridium* if it has been pregrown on different media. Thus Belan *et al.* (1987) reported microbial reduction of

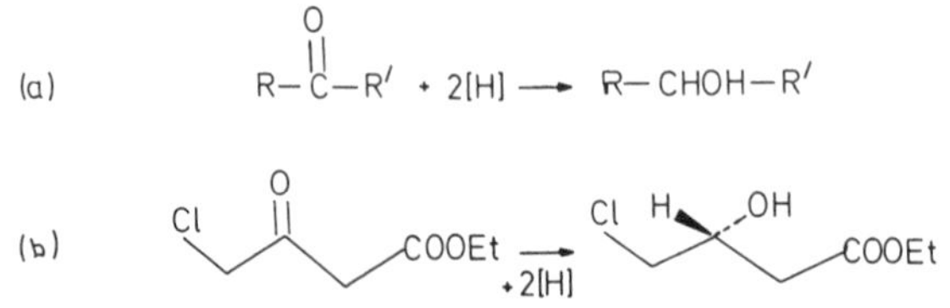

(a)
$$R-\overset{\overset{\displaystyle O}{\|}}{C}-R' + 2[H] \longrightarrow R-CHOH-R'$$

(b)

Figure 10. Reduction of ketones to secondary alcohols by *C. kluyveri*. (a) Model reaction, (b) reduction of 4-chloro-3-oxobutyrylethyl ester to give the (R) alcohol.

the ketone sulcatone (i.e., 6-methylhept-5-ene-2-one) to the enantiomeric forms of sulcatol (Fig. 11). By Prelog's rule, the likely product would be the S(+) enantiomer of sulcatol (which is the aggregation pheromone of the Ambrosia beetle) and, indeed, washed cells of *C. tyrobutyricum* La1 pregrown in crotonate medium when supplied with methylviologen and held under H_2 did produce S(+)-sulcatol in 88% ee and 100% yield. Yet the same organism pregrown in glucose medium under the same conditions gave R(−)-sulcatol in 80% ee and 76% yield. It was presumed that the organism possessed two alcohol dehydrogenases of opposite stereospecificity, the proportions in which they were synthesized being determined by the nature of the carbon and energy source in the growth medium.

The range of carbonyl compounds susceptible to reduction by various clostridia and their cell-free extracts is very extensive. Yet even if the cell-free extract of a H_2-utilizing *Clostridium* does not of itself perform the desired bioreduction, it can usefully complement the activity of an extract derived from another bacterium by supplying reductant for the reaction. Thus a mixture of cell-free extracts of *C. kluyveri* and of aerobically grown *Enterobacter agglomerans* was successfully employed to hydrogenate 3-hydroxy-3-methyl-butan-2-one to yield (S)-2,3-dihydroxy-2-methylbutane with a PN of 3300 (Simon *et al.*, 1985). The *C. kluyveri* extract alone was incapable of reducing this ketone, while that of *E. agglomerans* could not utilize H_2 but could effect the desired reduction using the NADH which was constantly regenerated by the *C. kluyveri* extract plus H_2. For the same reason, the reductive ability of an extract derived from a *Clostridium* species that possesses only weak hydrogenase activity can often be much improved by mixing it with *C. kluyveri* extract and maintaining the mixture under H_2.

5.2.3. Reduction of C=C Double Bonds

There is every reason to suppose that clostridia should prove to be effective agents of hydrogenation of unsaturated organic compounds.

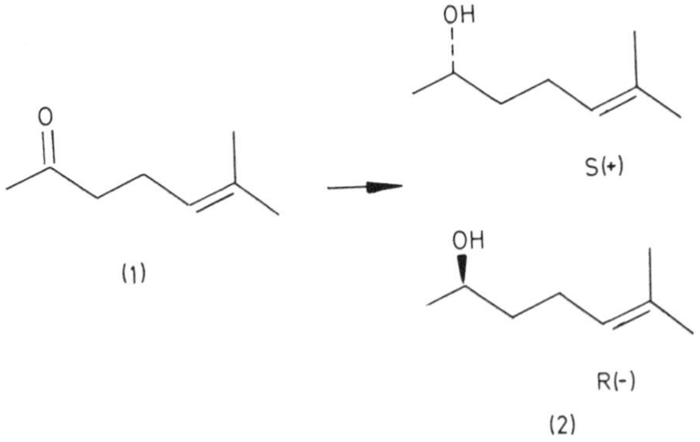

Figure 11. Reduction of sulcatone (1) to the enantiomeric forms of sulcatol (2), which is accomplished by *C. tyrobutyricum* La1.

Thus biohydrogenation of linoleic acid (9-*cis*,12-*cis*-octadecadienoic acid) to yield *trans*-vaccenic acid (11-*trans*-octadecanenoic acid) was accomplished by *C. sporogenes, C. bifermentans,* and *C. sordellii* (Verhulst *et al.,* 1985). The reduction of the 9-*cis* double bond followed isomerization of the linoleic acid to give 9-*cis*,11-*trans*-octadecadienoic acid. Yet the most informative studies and the most useful applications to date have derived from the discovery by Simon and his colleagues (e.g., Simon *et al.,* 1985) that several species of clostridia contain enoate reductases which are sometimes exceptionally permissive in their substrate specificities.

In *C. sporogenes* a 2-enoate reductase plays a key role in the reductive branch of the Stickland fermentations practiced by the organism. The enzyme catalyzes a NADH-dependent reduction of (E)-cinnamate (from phenylalanine) to phenylpropionate and also the reduction of 4-hydroxycinnamate (from tyrosine) and of 4-methyl-2-pentenoate (from leucine). It does not, however, reduce the 3-methyl-2-pentenoate that would arise from isoleucine nor does it reduce crotonate (i.e., butan-2-enoate). Interestingly, this enoate reductase-catalyzed step in the fermentation, when accomplished by whole cells, may be associated with conservation of free energy (Bader and Simon, 1983). Though the mechanism of any associated ATP formation has not been disclosed, it is noteworthy that in the same organism reduction of proline is associated with transmembrane export of protons (Lovitt *et al.,* 1986a).

However, the most attention has been given to the enoate reductases of *C. kluyveri* and *C. tyrobutyricum* La1, which display broader substrate specificity than the enzyme from *C. sporogenes.* Though similar, the enzymes are not immunologically identical, and it is that from *C. tyrobutyricum* La1 that has been most intensively studied. It is a dodecameric, iron-sulfur flavoprotein which in certain growth conditions can be produced in large quantities, sometimes accounting for 5–10% of the total cell protein. Though normally NADH-dependent, the enzyme can also accept electrons from reduced methylviologen (MV) (1,1'-dimethyl-4,4'-bipyridinium cation), the K_m for NADH (0.018 mM), however, being smaller than that for MV^+ (0.4 mM). The enzyme catalyzes the reduction of a great variety of nonactivated, unsaturated carboxylic acids and aldehydes (Fig. 12). It re-

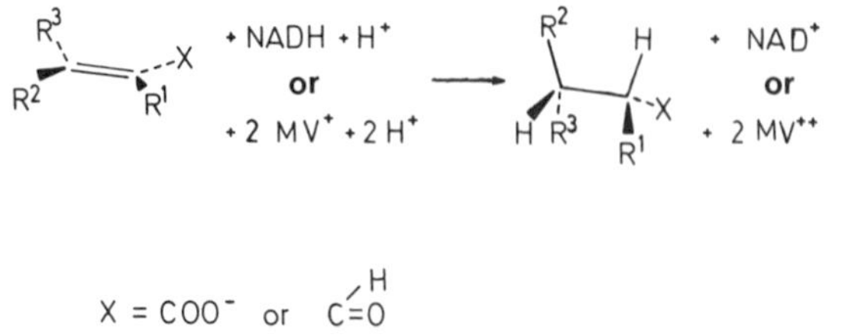

Figure 12. Reduction of 2-ene carboxylic acids and aldehydes that is catalyzed by the 2-enoate reductase of *C. tyrobutyricum* La1 with NADH as natural electron donor or with methylviologen cation as artificial electron donor.

quires of the substrate that the R_1 substituent not be too large but the choice of R_2 groups is less restrictive. Halogenation of the α-C is tolerated but if the β-C atom of the substrate (i.e., C-3) carries a halogen atom, then this is reductively eliminated in the course of the reaction. The attack on C-3 is always *trans* to that on C-2 so that if R_2 and R_3 groups are interchanged in the substrate (i.e., E and Z isomers are employed), then different enantiomers of the product are obtained. Thus, (E)-gerianate was readily hydrogenated by cells of *C. tyrobutyricum* La1 supplied with H_2 plus MV to give enantiomerically pure (R)-citronellate. (Z)-Geraniate was however hydrogenated much more slowly. With purified enoate reductase, the product was (S)-citronellate but whole cells still gave (R)-citronellate (60–85% ee) due to prior isomerization of (Z)-geraniate to (E)-geraniate. Whole cells of either *C. kluyveri* or *C. tyrobutyricum* La1 also reduced allyl alcohol derivatives (i.e., 2-alken-1-ols). These substrates were first dehydrogenated to the corresponding aldehydes which were then reduced by the enoate reductase, the aldehyde group in the product subsequently being reduced to an alcohol group (Bader *et al.*, 1978). Several chiral alcohols were produced by this procedure; for example, cells of *C. kluyveri* supplied with ethanol as the primary source of reducing power produced (R)-2-methyl-butan-1-ol, (R)-3-methylpentan-1-ol, and (2R,3S)-2-methyl-3-phenyl[2,3-^{2}H]propan-1-ol. Methylviologen actually inhibited reduction of the allyl alcohol substrates, presumably because in its presence no NAD was available for the initial oxidation of the alcohol to the aldehyde, since enoate reductase catalyzes reduction of NAD by MV^+ as follows:

$$2MV^+ + NAD^+ + H^+ \rightarrow NADH + 2MV^{2+}$$

It is presumed that the natural substrate for the enoate reductases of *C. kluyveri* and *C. tyrobutyricum* La1 is crotonate (2-butenoic acid), though both acrylate and 2-methylacrylate are more rapidly hydrogenated by *C. tyrobutyricum* La1, and the K_m for cinnamate of its enoate reductase is 0.01 mM, while that for (E)-2-methyl-2-butenoate is 1.5 mM.

The selectivity displayed by the enoate reductase of whole cells of *C. tyrobutyricum* La1 is well evidenced by the fact that such cells when given H_2 plus MV could hydrogenate allene carboxylates, but only with reduction of the α,β double bond. Also with an enoate which carried a nitrophenyl substituent on its β-C the double bond was reduced without accompanying reduction of the aryl nitro group.

It would be a mistake, however, to assume that it is only the bio-hydrogenation of 2-enoate derivatives that is rapidly accomplished by *C. kluyveri* and *C. tyrobutyricum*. The ability of these organisms to reduce carbonyl groups was mentioned earlier, and other reductive reactions have also been reported, e.g., both organisms reductively split the N–O bond of dihydroxazines (Fig. 13; Klier *et al.*, 1987).

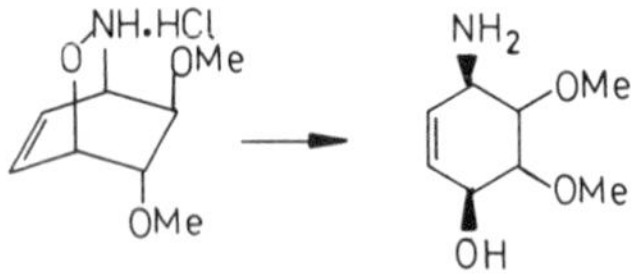

Figure 13. Reductive splitting, by *C. kluyveri* and *C. tyrobutyricum* La1, of the N–O bond in a dimethoxylated dihydrooxazine (1) to yield 1-amino-2,3-dimethoxy-4-hydroxy-5-cyclohexane (2).

5.2.4. Reduction of C=C Double Bonds in Cyclic Compounds

Clostridial hydrogenation of cyclic compounds is encountered in several of the fermentations of such substrates that are accomplished by specialist members of the genus. We have already noted the hydrogenation of 6-hydroxynicotinate to 1,4,5,6-tetrahydro-6-oxonicotinate that occurs in the course of the fermentation of nicotinic acid by *C. barkeri* (Fig. 5). The first step in uracil fermentation by *C. uracilicum* consists of the reduction of a single-ring double bond in a reaction catalyzed by a NADH-specific uracil reductase (Hunninghake and Grisolia, 1967). Yet *Clostridia* can often hydrogenate cyclic compounds that are not substrates of complete growth-supportive fermentations. As shown in Figure 14, cyclohexenones are co-reduced in growing cultures of *C. tyrobutyricum* La1 (Bostmembrun-Desrut *et al.*, 1983). The substrate was *trans*-hydrogenated and cyclohexenone, when added to a culture of strain La1 growing in crotonate medium and sparged with H_2, was converted in 100% yield to cyclohexanone. Methylcyclohexenone was 30% converted into methylcyclohexanone, again with no concurrent production of the corresponding alcohol.

5.3. Electromicrobial Reductions with Clostridia

The principle underlying this procedure is very straightforward. Electrical potential at a suitable cathode is used to reduce a redox mediator which in turn reduces intracellular electron donors. The oxidized mediator then returns to the cathode where it is re-reduced. Methylviologen possesses the required properties of a suitable mediator, and Simon *et al.* (1984, 1985) reported on how they were so able to exploit the ability of

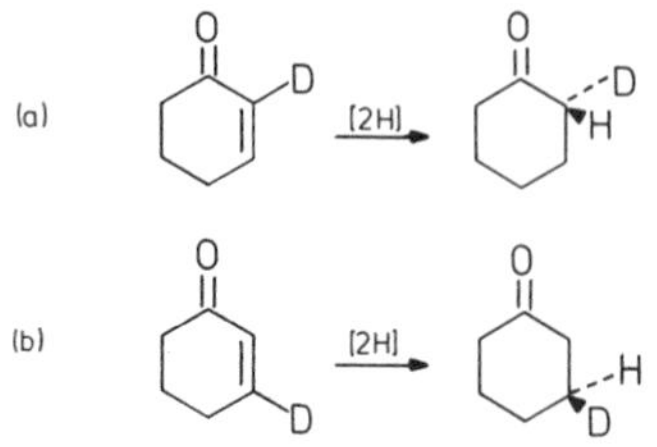

Figure 14. Reduction by *C. tyrobutyricum* La1 of (a) 2-deuterio-2-cyclohexene-1-one and (b) 3-deuterio-2-cyclohexene-1-one, demonstrating the *trans* addition of hydrogen.

enoate reductase to use the MV^+ cation as reductant. By employing an electrochemical cell with a mercury cathode, they could effect the preparative reduction of 2-enoate derivatives using a cell suspension of *C. tyrobutyricum* La1 (about 3 mg dry wt/ml) with 3 mM MV in 100 mM phosphate buffer, pH 6.3, under anaerobic conditions. In the same laboratory (Simon *et al.*, 1984), suspensions of *C. ghonii* or *C. sordellii* were used to prepare (4S)-[4-^{2}H]NADH by electromicrobial reduction of NAD in ^2H_2O buffer. Suspensions of *C. kluyveri* were similarly used to prepare (4S) [4-^{2}H]NADPH. Using suspensions of permeabilized cells of *C. sporogenes*, Lovitt *et al.* (1987a) developed polarographic assays for 2-oxoacid synthase and MV-NAD reductase and, with MV as mediator, accomplished the synthesis of several 2-oxoacids by bioelectrical reductive carboxylation. A similar procedure was employed as a convenient method of assay of the proline reductase of this organism. In earlier work, Fernandez (1983) described a H_2 electrode for continuous monitoring of the hydrogenase activity of *C. pasteurianum*, and in several laboratories *C. butyricum* was employed as the reductive element in an electrochemical fuel cell (e.g., Karube *et al.*, 1977; Ardeleanu *et al.*, 1984).

6. CLOSTRIDIAL BIOCONVERSIONS OF BILE ACIDS AND STEROIDS

6.1. Bile Acids

6.1.1. Release of Bile Acids from Conjugate Bile Acids

In man, the cholic acid and chenodeoxycholic acid (Fig. 15) that are synthesized in the liver are then conjugated therein with glycine or taurine. Several species of *Clostridium*, including *C. perfringens*, that are present in the intestine, however, are able to release the primary bile acids from these conjugates by a hydrolytic cleavage reaction catalyzed by a conjugated bile acid hydrolase (Nair *et al.*, 1965; Masuda, 1981). An interesting organism (*Clostridium* S_1) isolated from rat feces and displaying a growth requirement for taurine was found to possess a bile salt sulfatase that hydrolyzed the 3α-

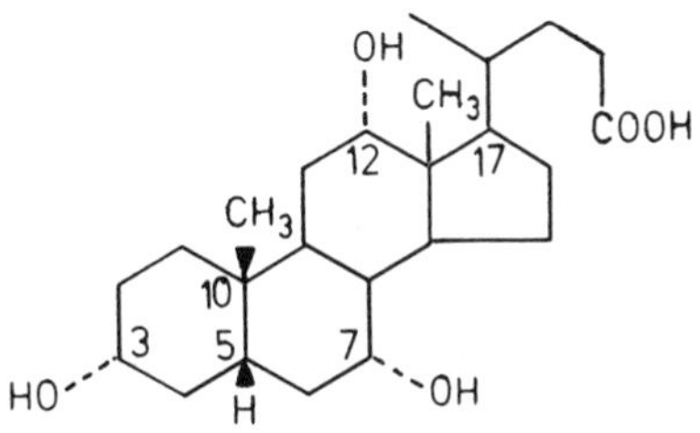

Figure 15. Primary bile acids. The molecule illustrated is that of cholic acid (3α, 7α, 12α-trihydroxy-5β-cholan-24-oic acid). Note that the A/B ring junction is *cis*. Deoxycholic acid lacks the 7-hydroxyl group of cholic acid. Chenodeoxycholic acid lacks the 12-hydroxyl group of cholic acid. Ursodeoxycholic acid is the 7β epimer of chenodeoxycholic acid. Lithocholic acid is 3α-hydroxy-5β-cholan-24-oic acid.

sulfates of lithocholic and chenodeoxycholic acids. A free C-24 or C-26 carboxyl group was required in the substrate so that the organism would not hydrolyze cholesterol-3α-sulfate (Huijghebaert and Eyssen, 1982). A similar 3α-sulfatase was obtained from another organism (*Clostridium* S_2) present in rat intestine which showed a growth requirement not only for taurine but also for vitamin K (Robben *et al.*, 1986).

6.1.2. Epimerization of Hydroxyl Groups

The major microbial bioconversion of primary bile acids results from hydroxysteroid dehydrogenase (HSDH) activities which effect epimerization of substituent hydroxyl groups via oxobile acid intermediates (Hylemon and Glass, 1983). Clostridia are well known for their HSDH activities (Edenharder and Deser, 1981; Edenharder and Knaflic, 1981), which have been the subject of scrutiny because of some suspicion that bile acid and steroid degradation could play a role in the etiology of large-bowel cancer.

The reduction of 3-oxobile acids by cell-free extracts of *C. perfringens* supplied with a NADPH regeneration system yielded 3α-hydroxybile acids (Aries and Hill, 1970) and the reverse production by this organism of 3-oxo and 3β-hydroxybile acids from 3α-hydroxybile acids has also been reported (Macdonald *et al.*, 1976; Hirano *et al.*, 1981). Although *C. perfringens* may contain 7α- and 12α-HSDH activities, its constitutive 3α-HSDH is its major enzyme of this type (Macdonald *et al.*, 1983a). It is NADP-linked and demonstrates a broad substrate specificity with 3-hydroxy-conjugated or free bile acids. Growing cultures supplied with 3-oxochenodeoxycholic acid gave chenodeoxycholic acid (84%) and its 3β epimer (18%).

Epimerization of the 7α hydroxy group of bile acids (via the 7-oxo bile acids) was demonstrated with *C. absonum* which contains both 7α-HSDH and 7β-HSDH enzymes (Macdonald and Roach, 1981; Sutherland and Macdonald, 1982; Macdonald and Sutherland, 1983). The NADP-dependent 7α-HSDH was gratuitously induced by deoxycholic acid but its synthesis was not provoked by ursodeoxycholic acid despite the fact that this compound was utilized as a substrate (Hylemon and Glass, 1983; Macdonald *et al.*, 1983b). On the other hand, the 7β-HSDH of *C. absonum* was made the basis of a method of assay of ursodeoxycholic acid (Macdonald *et al.*, 1983c). A NADP-dependent 7α-HSDH, optimally induced by 7-oxolithocholic acid, was discovered in *C. bifermentans* (Sutherland *et al.*, 1987). It catalyzed interconversion (1) of chenodeoxycholic acid and 7-oxolithocholic acid and (2) of cholic acid and 7-oxodeoxycholic acid. It did not attack ursodeoxycholic acid or ursocholic acid, and was therefore specific for the 7α-hydroxy configuration. The strain of *C. bifermentans* examined by Sutherland *et al.* (1987) contained no 3α-HSDH or 12α-HSDH activities and no 7-dehydroxylase activity. Additional 7α-HSDH activities have been reported in other clostridia (Mahony *et al.*, 1977); that in *C. limosum*, unlike the enzyme of *C. bifermentans*, was not induced by 7-oxolithocholic acid (Sutherland and Williams, 1985). As an alternative to the intraspecies epimerization of 7-hydroxybile acids brought about by *C. absonum*, in anaerobic consortia the epimerization may be carried out by cooperation between two different organisms, one of which possesses only a 7α-HSDH and the other a 7β-HSDH, with 7-oxobile acid transfer serving to couple the process.

Epimerization of the 12α-hydroxyl group has thus far not been demonstrated even in those clostridia which contain a 12α-HSDH and may thus form 12-oxobile acids from 12α-hydroxybile acids. Such a NADP-specific 12α-HSDH was purified from *C. leptum* (Harris and Hylemon, 1978) and from *Clostridium* ATCC 29733 (Macdonald *et al.*, 1979). The enzyme from *C. leptum* showed the greatest affinity for conjugated bile acids and with deoxycholic acid as substrate operated best at an alkaline pH of 8.5–9. A 12α-HSDH which converted deoxycholic acid to 3α-hydroxy-12-oxo-5β-cholanoic acid was present in *Clostridium* S_2 (Robben *et al.*, 1986). The 12β dehydrogenation of bile acids was demonstrated with *C. difficile, C. paraputrificum*, and *C. tertium* (Edenharder and Schneider, 1985).

Finally, a 6β-HSDH of *Clostridium* R6x76 (Sacquet *et al.*, 1979) was capable of producing ω-muricholic acid (3α,6α,7β-trihydroxy-5β-cholanoic acid) from β-muricholic acid (3α,6β,7β-trihydroxy-5β-cholanoic acid). When incubated with 3α-hydroxy-6-oxo-5β-cholanoic acid the organism formed only the 6α-hydroxy product. Conversion in the reverse mode of β-muricholic acid to ω-muricholic acid was accomplished by *Clostridium* S_2, which possesses 6α-HSDH and 6β-HSDH activities (Robben *et al.*, 1986).

6.1.3. Dehydroxylation

The 7α dehydroxylation of primary bile acids is accomplished by *C. perfringens* (Aries and Hill, 1970) and by *C. bifermentans* and *C. sordellii* (Hayakawa and Hattori, 1970; Archer *et al.*, 1981). Whole cells of *C. leptum* also catalyzed 7α dehydroxylation of bile acids but the activity was not sustained in cell-free extracts of the organism (Stellwag and Hylemon, 1979). As to the mechanism of the reaction, the suggestion was made (Ferrari *et al.*, 1977) that there is diaxial elimination of water (involving 6β-H and 7α-OH) yielding a Δ^6 intermediate which is subsequently transhydrogenated at the 6 and 7 positions.

6.2. Other Steroids

Reductions

Clostridia have been found to play a significant part in the anaerobic bioconversions of steroids other than bile acids which also occur in the gut.

For example, intestinal clostridia can inactivate contraceptive steroids (Bok-kenheuser and Winter, 1983) and synthesize an androgen from cortisol (Bokkenheuser *et al.*, 1984). Although HSDH-catalyzed reactions occur with steroids of all types, reductive bioconversions tend to prevail. A good example is provided by the reduction of Δ^4-3-oxosteroids. This configuration is necessary for the hormonal activity of corticosteroids so that reduction of ring A of the steroid will destroy this activity. The reduction is actively undertaken by *C. paraputrificum* which, for example, readily reduces 16α-hydroxyprogesterone to 16α-hydroxypregnanolone. Indeed, this organism has been employed for preparative bioconversion of aldosterone and 18-hydroxycorticosterone to their tetrahydro and hexahydro derivatives (Harnick *et al.*, 1983). The Δ^4-3-oxo reduction begins with hydrogenation of the double bond which is then followed by the formation of a hydroxyl group at C-3 to give the 3α-hydroxy-5β configuration (Bokkenheuser *et al.*, 1976). The structure of rings A and B in the substrate molecule much influences the rate of reduction, an additional Δ^1 double bond making the substrate much less susceptible to reduction. Thus medroxyprogesterone acetate is not reduced by *C. paraputrificum*. However, with 17α-hydroxysteroid substrates, this organism will simultaneously effect side chain cleavage to yield a 17-one product (Bokkenheuser and Winter, 1983). Two other clostridia isolated from human feces, namely, *C. innocuum* and *Clostridium* J-1 (Bokkenheuser *et al.*, 1983), reduced Δ^4-3 oxosteroids to 3β-hydroxy-5β and 3β-hydroxy-5α configurations, respectively, and the Δ^4-3 oxosteroid-5β reductase of *C. innocum* together with its 3β-HSDH has been the subject of further study (Stokes and Hylemon, 1985).

Nuclear dehydrogenation of steroids and bile acids (Owen, 1985) introduces into ring A a double bond conjugated to a 3-oxo group yielding a product with a Δ^4-3-oxo configuration (i.e., a 4-ene-3-one structure). The ability is prevalent among lecithinase-negative intestinal clostridia but is rare in lecithinase-positive isolates. Thus the activity is to be found in some strains of *C. perfringens* and in *C. paraputrificum*. Oxidation of 3-oxobile acids is rather more efficient than oxidation of androstanes, and the presence of a single hydroxyl group at C-6 or C-7 or C-12 appears to enhance induction of synthesis of the enzyme and the rate of reaction (Fig. 16). Oxo

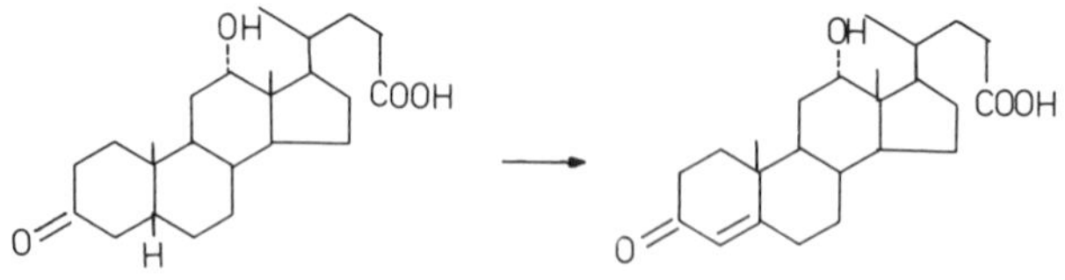

(1)   (2)

Figure 16. Nuclear dehydrogenation by *C. paraputrificum* of 12α-hydroxy-5β-cholan-3-oxo-24-oic acid (1) to yield 12α-hydroxychol-4-ene-3-one-24-oic acid (2).

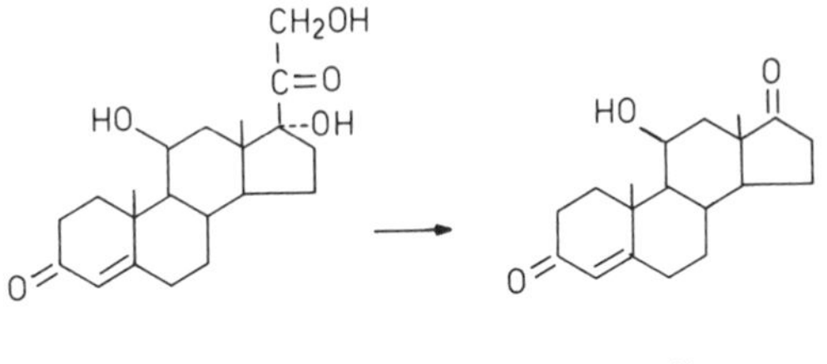

CH₂OH

Figure 17. Desmolytic activity of *C. cadaveris* and *C. scindens*. Conversion of cortisol (1) to 11β-hydroxyandrost-4-ene-3,17-dione (2).

groups at C-6 and C-7 render the substrate refractory to nuclear dehydrogenation, presumably because of the proximity of such groups to the enzyme binding site. In *C. paraputrificum* the dehydrogenation can be coupled to reduction of nitrate (Barnes *et al.*, 1975; Goddard *et al.*, 1975). A 3-oxo-5β-steroid-Δ¹-dehydrogenase has been reported in clostridia in addition to the more usual Δ⁴-dehydrogenase (Aries *et al.*, 1971) and, interestingly, clostridia that undertake nuclear dehydrogenation do not appear to be able to oxidize 3-hydroxyl groups to yield the appropriate 3-oxo configuration (Macdonald and Hill, 1978).

Side chain cleavage from steroids (so-called desmolytic activity) is a feature of *C. cadaveris* (Bokkenheuser *et al.*, 1986) and *C. scindens* (Morris *et al.*, 1985) as well as *C. paraputrificum*. In both of the former species, the cleavage enzyme is constitutively synthesized and requires the substrate to possess hydroxyl groups at C-17 and C-21 plus *either* an oxo group *or* a hydroxyl group at C-20. Thus cortisol is a good substrate which is converted to 11β-hydroxyandrost-4-ene-3,17-dione (Fig. 17). Krafft *et al.* (1987) reported on the cofactor requirements of both the steroid 17-20-desmolase of *C. scindens* and its 20α-HSDH.

7. CONCLUSION

The bulk fermentations accomplished by clostridia have attracted sustained academic and industrial interest over many years. Only recently has it been appreciated that these organisms can also be used to carry out a variety of preparative single-step anaerobic biotransformations, with the whole organism often proving as useful as purified enzymes. To exploit these abilities to the full, it will now be necessary to investigate the range of capacities possessed by different species of *Clostridium* in a far more systematic manner. Only when underlying patterns are discernible in the rich mosaic of currently "singular properties" will it be possible to harness the full catalytic potential of the genus.

REFERENCES

Aberhart, D. J., Gould, S. J., Lin, H-J., Thiruvengadam, T. K., and Weiller, B. H., 1983, Stereochemistry of lysine 2,3-aminomutase isolated from *Clostridium subterminale* strain SB4, *J. Am. Chem. Soc.* **105:**5461–5469.

Andrew, I. G., and Morris, J. G., 1965, The biosynthesis of alanine by *Clostridium kluyveri*, *Biochim. Biophys. Acta* **97:**176–179.

Angermaier, L., and Simon, H., 1983, On the reduction of aliphatic and aromatic nitro compounds by clostridia: The role of ferredoxin and its stabilization, *Hoppe-Seyler's Z. Physiol. Chem.* **364:**961–976.

Archer, R. H., Maddox, I. S., and Chong, R., 1981, 7α-dehydroxylation of cholic acid by *Clostridium bifermentans*, *Eur. J. Appl. Microbiol. Biotechnol.* **12:**46–52.

Ardeleanu, J., Margineanu, D-G., and Vais, H., 1984, Electrochemical conversion in biofuel cells using *Clostridium butyricum* or *Staphylococcus aureus* Oxford, *Bioelectrochem. Bioenerg.* **11:**273–278.

Aries, V., and Hill, M. J., 1970, Degradation of steroids by intestinal bacteria. II. Enzymes catalysing the oxidoreduction of the 3α-, 7α- and 12α-hydroxyl groups in cholic acid and the hydroxylation of the 7-hydroxyl group, *Biochim. Biophys. Acta* **202:**535–543.

Aries, V. C., Goddard, P., and Hill, M. J., 1971, Degradation of steroids by intestinal bacteria. III. 3-oxo-5β-steroid-Δ¹-dehydrogenase and 3-oxo-5β-steroid-Δ⁴-dehydrogenase, *Biochim. Biophys. Acta* **248:**482–488.

Bader, J., and Simon, H., 1983, ATP formation is coupled to the hydrogenation of 2-enoates in *Clostridium sporogenes*, *FEMS Microbiol. Lett.* **20:**171–175.

Bader, J., Günther, H., Rambeck, B., and Simon, H., 1978, Properties of two clostridia strains acting as catalysts for the preparative stereospecific hydrogenation of 2-enoic acids and 2 alken-1-ols with hydrogen gas, *Hoppe-Seyler's Z. Physiol. Chem.* **359:**19–27.

Bader, J., Rauschenbach, P., and Simon, H., 1982, On a hitherto unknown fermentation path of several amino acids by proteolytic clostridia, *FEBS Lett.* **140:**67–72.

Barker, H. A., 1985, β-Methylaspartate-glutamate mutase from *Clostridium tetanomorphum*, *Methods Enzymol.* **113:**121–132.

Barker, H. A., D'Ari, L., and Kahn, J., 1987, Enzymatic reactions in the degradation of 5-aminovalerate by *Clostridium aminovalericum*, *J. Biol. Chem.* **262:**8994–9003.

Barnes, P. J., Bilton, R. F., Mason, A. N., Fernandez, F., and Hill, M. J., 1975, The coupling of anaerobic steroid dehydrogenation to nitrate reduction in *Pseudomonas* NCIB 10590 and *Clostridium paraputrificum*, *Biochem. Soc. Trans.* **3:**299–300.

Belan, A., Bolte, J., Fauve, A., Gourcy, J. G., and Veschambre, H., 1987, Use of biological systems for the preparation of chiral molecules. 3. An application in pheromone synthesis: Preparation of sulcatol enantiomers, *J. Org. Chem.* **52:**256–260.

Berndt, A., and Schlegel, H. G., 1975, Kinetics and properties of β keto-thiolase from *Clostridium pasteurianum*, *Arch. Microbiol.* **10:**21–30.

Berry, D. F., Francis, A. J., and Bollag, J-M., 1987, Microbial metabolism of homocyclic and heterocyclic aromatic compounds under anaerobic conditions, *Microbiol. Rev.* **51:**43–59.

Blanchard, K. C., and MacDonald, J., 1935, Bacterial metabolism. 1. The reduction of propionaldehyde and of propionic acid by *Clostridium acetobutylicum*, *J. Biol. Chem.* **110:**145–150.

Bokkenheuser, V. D., and Winter, J., 1983, Biotransformation of steroids, in: *Human Intestinal Microflora in Health and Disease* (D. J. Hentges, ed.), Academic Press, New York, pp. 215–239.

Bokkenheuser, V. D., Winter, J., Dehazya, P., DeLeon, O., and Kelly, W. G., 1976, Formation and metabolism of tetrahydrodeoxycorticosterone by human fecal flora, *J. Steroid Biochem.* **7:**837–843.

Bokkenheuser, V. D., Winter, J., Cohen, B. I., O'Rourke, S. O., and Mosbach, E. H., 1983,

Inactivation of contraceptive steroid hormones by human intestinal clostridia, *J. Clin. Microbiol.* **18:**500–504.

Bokkenheuser, V. D., Morris, G. N., Ritchie, A. E., Holdeman, L. V., and Winter, J., 1984, Biosynthesis of androgen from cortisol by a species of *Clostridium* recovered from human fecal flora, *J. Infect. Dis.* **149:**489–494.

Bokkenheuser, V. D., Winter, J., Morris, G. N. and Locascio, S., 1986, Steroid desmolase synthesis by *Eubacterium desmolans* and *Clostridium cadaveris*, *Appl. Env. Microbiol.* **52:**1153–1156.

Bostmembrun-Desrut, M., Kergomard, A., Renard, M. F., and Veschambre, H., 1983, Microbiological reduction of cyclohexenones by growing *Clostridium* cells, *Agric. Biol. Chem.* **47:**1997–2000.

Brot, N., and Weissbach, H., 1970, Conversion of L-tyrosine to phenol *(Clostridium tetanomorphum)*, *Methods Enzymol.* **17A:**642–645.

Buckel, W. C., 1980, The reversible dehydration of (R)-2-hydroxyglutarate to (E)glutaconate, *Eur. J. Biochem.* **106:**439–447.

Bühler, M., Giesel, H., Tischer, W., and Simon, H., 1980, Occurrence and the possible physiological role of 2-enoate reductases. *FEBS Lett.* **109:**244–246.

Cameron, D. C., and Cooney, C. L., 1986, A novel fermentation: the production of R(−)-1,2-propanediol and acetol by *Clostridium thermosaccharolyticum*, *Biotechnology* **4:**651–654.

Campbell, L. L., 1957, Reductive degradation of pyrimidines, *J. Bacteriol.* **73:**220–224.

Chirpich, T. P., and Barker, H. A., 1971, Lysine-2,3-aminomutase *(Clostridium)*, *Methods Enzymol.* **17B:**215–222.

Cozzani, I., Barsacchi, R., Dibenedetto, G., Saracchi, L., and Falcone, G., 1975, Regulation of breakdown and synthesis of L-glutamate decarboxylase in *Clostridium perfringens*, *J. Bacteriol.* **123:**1115–1123.

Daldal, F., and Applebaum, J., 1985, Cloning and expression of *Clostridium pasteurianum* galactokinase gene in *Escherichia coli* K-12 and nucleotide sequence analyses of a region affecting the amount of the enzyme. *J. Mol. Biol.* **186:**533–546.

D'Ari, L., and Barker, H. A., 1985, pCresol formation by cell-free extracts of *Clostridium difficile*, *Arch. Microbiol.* **143:**311–312.

Datta, R., and Zeikus, J. G., 1985, Modulation of acetone-butanol-ethanol fermentation by carbon monoxide and organic acids, *Appl. Env. Microbiol.* **49:**522–529.

Decker, K., and Hamm, H. H., 1980, A convenient biosynthetic method for the preparation of radioactive flavin nucleotides using *Clostridium kluyveri*, *Methods Enzymol.* **66E:**227–235.

Dilworth, G. L., 1983, Occurrence of molybdenum in the nicotinic acid hydroxylase from *Clostridium barkeri*, *Arch. Biochem. Biophys.* **221:**565–569.

Dürre, P., and Andreesen, J. R., 1983, Purine and glycine metabolism by purinolytic clostridia, *J. Bacteriol.* **154:**192–199.

Edenharder, R., and Deser, H. J., 1981, The significance of the bacterial steroid degradation for the etiology of large bowel cancer. VIII. Transformation of cholic, chenodeoxycholic and deoxycholic acid by lecithinase-negative Clostridia. *Zentralbl. Bakteriol. Parasitenkd. Infektionskr. Hyg. Abt. 1: Orig. Reihe B* **174:**91–104.

Edenharder, R., and Knaflic, T., 1981, Epimerization of chenodeoxycholic acid to ursodeoxycholic acid by human intestinal lecithinase-negative clostridia. *J. Lipid Res.* **22:**652–658.

Edenharder, R., and Schneider, J., 1985, 12β-Dehydrogenation of bile acids by *Clostridium paraputrificum*, *Clostridium tertium*, and *Clostridium difficile* and epimerization of carbon-12 of deoxycholic acid by co-cultivation with 12α-dehydrogenating *Eubacterium lentum*, *Appl. Env. Microbiol.* **49:**964–968.

Edwards, D. I., Knox, R. J., Skolimowski, I. M., and Knight, R. C., 1982, Mode of action of nitroimidazoles, *Eur. J. Chemother. Antibiot.* **2:**65–72.

Egerer, P., and Simon, H., 1982, Hydrogenation with entrapped *Clostridium* spec. LA1 and observations on its stability, *Biotechnol. Lett.* **4:**501–506.

Evans, W. C., 1977, Biochemistry of the bacterial catabolism of aromatic compounds in anaerobic environments, *Nature* **270:**17–22.

Fernandez, V. M., 1983, An electrochemical cell for reduction of biochemicals: its application to the study of the effect of pH and redox potential on the activity of hydrogenases. *Anal. Biochem.* **130:**54–59.

Ferrari, A., Scolastinco, C., and Beretta, L., 1977, On the mechanism of cholic acid 7α-dehydroxylation by a *Clostridium bifermentans* cell-free extract, *FEBS Lett.* **75:**166–168.

Forsberg, C. W., 1987, Production of 1,3-propanediol from glycerol by *Clostridium acetobutylicum* and other clostridium species, *Appl. Env. Microbiol.* **53:**639–643.

Francis, A. J., and Dodge, C. J., 1988, Anaerobic microbial dissolution of transition and heavy metal oxides, *Appl. Env. Microbiol.* **54:**1009–1014.

Giesel, H., and Simon, H., 1983, On the occurrence of enoate reductase and 2-oxo-carboxylate reductase in clostridia, and some observations on the amino acid fermentation by *Peptostreptococcus anaerobius*, *Arch. Microbiol.* **135:**51–57.

Goddard, P., Fernandez, F., West, B., Hill, M. J., and Barnes, P., 1975, The nuclear dehydrogenation of steroids by intestinal bacteria, *J. Med. Microbiol.* **8:**429–435.

Gottschalk, G., and Bender, R., 1982, D-Gluconate dehydratase from *Clostridium pasteurianum*, *Methods Enzymol.* **90:**283–287.

Graves, M. C., Mullenbach, G. T., and Rabinowitz, J. C., 1985, Cloning and nucleotide sequence determination of the *Clostridium pasteurianum* ferredoxin gene, *Proc. Natl. Acad. Sci. USA* **82:**1653–1657.

Hardman, J. K., 1962, γ-Hydroxybutyrate dehydrogenase from *Clostridium aminobutyricum*, *Methods Enzymol.* **5:**778–783.

Harnick, M., Aharonowitz, Y., Lamed, R., and Kashman, Y., 1983, Tetra- and hexa- hydro derivatives of aldosterone and 18-hydroxycortico-sterone by chemical and microbial reductions, *Steroid Biochem.* **19:**1441–1450.

Harris, J. N., and Hylemon, P. B., 1978, Partial purification and characterization of NADP-dependent 12α-hydroxysteroid dehydrogenase from *Clostridium leptum*, *Biochim. Biophys. Acta* **528:**148–157.

Harrison, G., Curle, C., and Laishley, E. J., 1984, Purification and characterisation of an inducible dissimilatory type sulfite reductase from *Clostridium pasteurianum*, *Arch. Microbiol.* **138:**72–78.

Hartmanis, M. G. N., 1987, Butyrate kinase from *Clostridium acetobutylicum*, *J. Biol. Chem.* **262:** 617–621.

Hartmanis, M. G. N., and Stadtman, T. C., 1986, Diol metabolism and diol dehydratase in *Clostridium glycolicum*, *Arch. Biochem. Biophys.* **245:**144–152.

Hartrampf, G., and Buckel, W., 1984, The stereochemistry of the formation of the methyl group in the glutamate mutase-catalyzed reaction in *Clostridium tetanomorphum*, *FEBS Lett.* **171:**73–78.

Hartrampf, G., and Buckel, W., 1986, On the steric course of the adenosylcobalamin-dependent 2-methyleneglutarate mutase reaction in *Clostridium barkeri*, *Eur. J. Biochem.* **156:** 301–304.

Hayakawa, S., and Hattori, T., 1970, 7α-Dehydroxylation of cholic acid by *Clostridium bifermentans* strain ATCC 9714 and *Clostridium sordellii* strain NCIB 2629, *FEBS Lett.* **6:**131–133.

Heritage, A. D., and MacRae, I. C., 1977, Identification of intermediates formed during the degradation of hexachlorocyclohexanes by *Clostridium sphenoides*, *Appl. Env. Microbiol.* **33:** 1295–1297.

Hirano, S., Masuda, N., Oda, H., and Mukai, H., 1981, Transformation of bile acids by *Clostridium perfringens*, *Appl. Env. Microbiol.* **42:**394–399.

Holcenberg, J. S., and Stadtman, E. R., 1969, Nicotinic acid metabolism. III. Purification and properties of a nicotinic acid hydroxylase, *J. Biol. Chem.* **244:**1194–1203.

Hsiang, M. W., and Bright, H. J., 1969, β-Methylaspartase from *Clostridium tetanomorphum*. *Meth. Enzymol.* **13:**347–353.

Huijghebaert, S. M., and Eyssen, H. J., 1982, Specificity of the bile salt sulphatase from *Clostridium* S₁, *Appl. Env. Microbiol.* **44:**1030–1034.

Hunninghake, D., and Grisolia, S., 1967, Uracil and thymine reductases, *Methods Enzymol.* **12A:**50–59.

Hylemon, P. B., and Glass, T. L., 1983, Biotransformation of bile acids and cholesterol by the intestinal microflora, in: *Human Intestinal Microflora in Health and Disease* (D. J. Hentges, ed.), Academic Press, New York, pp. 189–213.

Imhoff, D., and Andreesen, J. R., 1979, Nicotinic acid hydroxylase from *Clostridium barkerii:* Selenium-dependent formation of active enzyme, *FEMS Microbiol. Lett.* **5:**155–158.

Ishii, K., Kudo, T., Honda, H., and Horikoshi, K., 1983, Molecular cloning of β-isopropylmalate dehydrogenase gene from *Clostridium butyricum* M588, *Agric. Biol. Chem. (Tokyo)* **47:** 2313–2318.

Jagnow, G., Haider, K., and Ellwardt, P. C., 1977, Anaerobic dechlorination and degradation of hexachlorocyclohexane isomers by anaerobic and facultatively anaerobic bacteria, *Arch. Microbiol.* **115:**285–292.

Jeng, I. M., and Barker, H. A., 1974, Purification and properties of a L-3-aminobutyryl CoA deaminase from a lysine-fermenting clostridium, *J. Biol. Chem.* **249:**6578–6584.

Jeng, I. M., Somack, R., and Barker, H. A., 1974, Ornithine degradation in *Clostridium sticklandii:* Pyridoxal phosphate and coenzyme A-dependent thiolytic cleavage of 2-amino-4-ketopentanoate to alanine and acetylCoA, *Biochemistry* **13:**2898–2903.

Jewell, J. B., Coutinho, J. B., and Kropinski, A. W., 1986, Bioconversion of propionic, valeric and 4-hydroxybutyric acids into the corresponding alcohols by *Clostridium acetobutylicum* NRRL 527, *Curr. Microbiol.* **13:**215–220.

Kaplan, B. H., and Stadtman, E. R., 1971, Ethanolamine deaminase (*Clostridium* sp.), *Methods Enzymol.* **17:**818–824.

Kaplan, F., Setlow, P., and Kaplan, N. O., 1969, Purification and properties of a DPNH-TPNH diaphorase from *Clostridium kluyveri, Arch. Biochem. Biophys.* **132:**91–98.

Karube, I., Matsunaga, T., Tsuru, S., and Suzuki, S., 1977, Biochemical fuel cell utilising immobilised cells of *Clostridium butyricum, Biotechnol. Bioeng.* **19:**1727–1733.

Karube, I., Urano, N., Matsunaga, T., and Suzuki, S., 1982, Hydrogen production from glucose by immobilised growing cells of *Clostridium butyricum, Eur. J. Appl. Microbiol. Biotechnol.* **16:**5–9.

Karube, I., Urano, N., Yamada, T., Hirochika, H., and Sakaguchi, K., 1983, Cloning and expression of the hydrogenase gene from *Clostridium butyricum* in *Escherichia coli, FEBS Lett.* **158:**119–122.

Klier, K., Kresze, G., Werbitsky, O., and Simon, H., 1987, The microbial reductive splitting of the N-O bond of dihydrooxazines; an alternative to the chemical reduction, *Tetrahedron Lett.* **28:**2677–2680.

Klotzsch, H. R., 1969, Phosphotransacetylase from *Clostridium kluyveri, Methods Enzymol.* **13:** 381–386.

Knowles, J., Lehtovaara, P., and Teeri, T., 1987, Cellulase families and their genes, *Trends Biotechnol.* **5:**255–261.

Kole, M. M., and Altosaar, I., 1985, Conversion of chenodeoxycholic acid to ursodeoxycholic acid by *Clostridium absonum* in culture and by immobilized cells, *FEMS Microbiol. Lett.* **28:** 69–72.

Krafft, A. E., Winter, J., Bokkenheuser, V. D., and Hylemon, P. B., 1987, Cofactor requirements of steroid-17-20-desmolase and 20α-hydroxysteroid dehydrogenase activities in cell extracts of *Clostridium scindens, J. Steroid Biochem.* **28:**49–54.

Kreis, W., and Hession, C., 1973, Isolation and purification of L-methionine-α-deamino-γ-mercaptomethane lyase (L-methioninase) from *Clostridium sporogenes, Cancer Res.* **33:** 1862–1865.

Krumholtz, L. R., and Bryant, M. P., 1985, *Clostridium pfennigii* sp. nov. uses methoxyl groups of monobenzenoids and produces butyrate, *Int. J. Syst. Bacteriol.* **35:**454–456.

Kuchta, R. D., and Abeles, R. H., 1985, Lactate reduction in *Clostridium propionicum*. Purification and properties of lactylCoA dehydratase, *J. Biol. Chem.* **260**:13181–13189.

Lamed, R., Keinan, E., and Zeikus, J. G., 1981, Potential applications of an alcohol/ketone oxidoreductase from thermophilic bacteria, *Enz. Microb. Technol.* **3**:144–148.

Lebertz, H., Simon, H., Courtney, L. F., Benkovic, S. J., Zydowsky, L. D., Lee, K., and Floss, H. G., 1987, Stereochemistry of acetic acid formation from 5-methyltetrahydrofolate by *Clostridium thermoaceticum*, *J. Am. Chem. Soc.* **109**:3173–3174.

Lieberman, I., and Barker, H. A., 1955, Amino acid acetylase of *Clostridium kluyveri*, *Methods Enzymol.* **1**:616–619.

Liu, C-L., and Mortenson, L. E., 1984, Formate dehydrogenase of *Clostridium pasteurianum*, *J. Bacteriol.* **159**:375–380.

Ljungdahl, L. G., and Andreesen, J. R., 1978, Formate dehydrogenase, a selenium-tungsten enzyme from *Clostridium thermoaceticum*, *Methods Enzymol.* **53**:360–372.

Lovitt, R. W., Kell, D. B., and Morris, J. G., 1986a, Proline reduction by *Clostridium sporogenes* is coupled to vectorial proton ejection, *FEMS Microbiol. Lett.* **36**:269–273.

Lovitt, R. W., Walter, R. P., Morris, J. G., and Kell, D. B., 1986b, Conductimetric assessment of the biomass content is suspensions of immobilised (gel-entrapped) microorganisms. *Appl. Microbiol. Biotechnol.* **23**:168–173.

Lovitt, R. W., James, E. W., Kell, D. B., and Morris, J. G., 1987a, Bioelectrochemical transformations catalyzed by *Clostridium sporogenes*, in: *Bioreactors and Biotransformations* (G. W. Moody and P. B. Baker, ed.), Elsevier, Amsterdam, pp. 263–278.

Lovitt, R. W., Morris, J. G., and Kell, D. B., 1987b, The growth and nutrition of *Clostridium sporogenes* NCIB 8053 in defined media, *J. Appl. Bacteriol.* **62**:71–80.

Macdonald, I. A., and Hill, M. J., 1978, The inability of nuclear dehydrogenating clostridia to oxidise bile salt hydroxyl groups, *Experientia* **35**:722–723.

Macdonald, I. A., and Roach, P. B., 1981, Bile salt induction of 7α- and 7β-hydroxysteroid dehydrogenases in *Clostridium absonum*, *Biochim. Biophys. Acta* **665**:262–269.

Macdonald, I. A., and Sutherland, J. D., 1983, Further studies on the bile salt induction of 7α- and 7β-hydroxysteroid dehydrogenases in *Clostridium absonum*, *Biochim. Biophys. Acta* **750**:397–403.

Macdonald, I. A., Meier, E. C., Mahony, D. E., and Costain, G. A., 1976, 3α-, 7α-, and 12α-hydroxysteroid dehydrogenase activities from *Clostridium perfringens*, *Biochim. Biophys. Acta* **450**:142–153.

Macdonald, I. A., Jellet, J. F., and Mahony, D. E., 1979, 12α-hydroxy-steroid dehydrogenase from *Clostridium* group P, strain C48-50 ATCC 29733; partial purification and characterization, *J. Lipid Res.* **20**:234–239.

Macdonald, I. A., Hutchison, D. M., Forrest, T. P., Bokkenheuser, V. D., Winter, J., and Holdeman, L. V., 1983a, Metabolism of primary bile acids by *Clostridium perfringens*, *J. Steroid Biochem.* **18**:97–104.

Macdonald, I. A., White, B. A., and Hylemon, P. B., 1983b, Separation of 7α- and 7β-hydroxysteroid dehydrogenase activities from *Clostridium absonum* ATCC 27555 and cellular responses of this organism to bile acid inducers, *J. Lipid Res.* **24**:1119–1126.

Macdonald, I. A., Williams, C. N., Sutherland, J. D., and MacDonald, A. C., 1983c, Estimation of ursodeoxycholic acid in human and bear biles using *Clostridium absonum* 7β-hydroxysteroid dehydrogenase, *Anal. Biochem.* **135**:349–354.

Machacek-Pitsch, C., Rauschenbach, P., and Simon, H., 1985, Observations on the elimination of water from 2-hydroxy-acids in the metabolism of amino acids by *Clostridium sporogenes*, *Biol. Chem. Hoppe-Seyler* **366**:1057–1062.

Mahony, D. E., Mier, C. E., Macdonald, I. A. and Holdeman, L. V., 1977, Bile salt degradation by nonfermentative clostridia. *Appl. Env. Microbiol.* **34**:419–423.

Masuda, N., 1981, Deconjugation of bile salts by *Bacteroides* and *Clostridium*, *Microbiol. Immunol.* **25**:1–11.

Matsunaga, T., Matsunaga, N., and Nishimura, S., 1985, Regeneration of NAD(P)H by immobilised whole cells of *Clostridium butyricum* under hydrogen high pressure, *Biotechnol. Bioeng.* **27**:1277–1281.

Meyer, C. L., Roos, J. W. and Papoutsakis, E. T., 1986, Carbon monoxide gassing leads to alcohol production and butyrate uptake without acetone formation in continuous cultures of *Clostridium acetobutylicum*, *Appl. Microbiol. Biotechnol.* **24**:159–167.

Möller, B., Hippe, H., and Gottschalk, G., 1986, Degradation of various amine compounds by mesophilic clostridia, *Arch. Microbiol.* **145**:85–90.

Monticello, D. J., Hadioetomo, R. S., and Costilow, R. N., 1984, Isoleucine synthesis by *Clostridium sporogenes* from propionate or α-methylbutyrate, *J. Gen. Microbiol.* **130**:309–318.

Morris, G. N., Winter, J., Cato, E. P., Ritchie, A. E., and Bokkenheuser, V. D., 1985, *Clostridium scindens* sp. nov., a human intestinal bacterium with a desmolytic activity on corticoids, *Int. J. Syst. Bacteriol.* **35**:478–481.

Nair, P. P., Gordon, M., Gordon, S., Reback, J., and Mendeloff, A. I., 1965, The cleavage of bile acid conjugates by cell-free extracts from *Clostridium perfringens*, *Life Sci.* **4**:1887–1892.

Najjar, V. A., 1957, Determination of amino acids by specific bacterial decarboxylases in the Warburg apparatus. *Methods Enzymol.* **3**:462–466.

O'Brien, R. W., and Morris, J. G., 1971, The ferredoxin-dependent reduction of chloramphenicol by *Clostridium acetobutylicum*, *J. Gen. Microbiol.* **67**:265–271.

O'Brien, R. W., and Morris, J. G., 1972, Effect of metronidazole on hydrogen production by *Clostridium acetobutylicum*, *Arch. Microbiol.* **84**:225–233.

Ohisa, N., Yamaguchi, M., and Kurihara, N., 1980, Lindane degradation by cell-free extracts of *Clostridium rectum*, *Arch. Microbiol.* **125**:221–225.

Owen, R. W., 1985, Biotransformation of bile acids by clostridia, *J. Med. Microbiol.* **20**:233–238.

Paulin, L., and Pösö, H., 1983, Ornithine decarboxylase activity from an extremely thermophilic bacterium, *Clostridium thermohydrosulfuricum*. Effect of GTP analogues on enzyme activity, *Biochim. Biophys. Acta* **742**:197–205.

Pezacka, E., and Wood, H. G., 1984, The synthesis of acetylCoA by *Clostridium thermoaceticum* from CO_2, H_2, CoA and methyltetrahydrofolate, *Arch. Microbiol.* **137**:63–69.

Pitsch, C., and Simon, H., 1982, The stereochemical course of the water elimination from (2R)-phenyllactate in the amino acid fermentation of *Clostridium sporogenes*, *Hoppe Seyler's Z. Physiol. Chem.* **363**:1253–1257.

Pons, J-L., Rimbault, A., Darbord, J-C., and Leluan, G., 1984, Biosynthesis of toluene by *Clostridium aerofoetidum* strain WS, *Ann. Microbiol.* **135B**:219–222.

Poston, J. M., 1976, Leucine, 2,3-aminomutase, an enzyme of leucine catabolism, *J. Biol. Chem.* **251**:1859–1863.

Rakosky, J., Zimmerman, L. N., and Beck, J. V., 1955, Guanine degradation by *Clostridium acidiurici*. II. Isolation and characterization of guanase, *J. Bacteriol.* **69**:566–570.

Rao, G., and Mutharasan, R., 1987, Altered electron flow in continuous cultures of *Clostridium acetobutylicum* induced by viologen dyes, *Appl. Env. Microbiol.* **53**:1232–1235.

Rice, D. W., Hornby, D. P., and Engel, P. C., 1985, Crystallisation of a NAD-dependent glutamate dehydrogenase from *Clostridium symbiosum*, *J. Mol. Biol.* **181**:147–149.

Robben, J., Parmentier, G., and Eyssen, H., 1986, Isolation of a rat intestinal *Clostridium* strain producing 5α- and 5β-bile salt, 3α-sulfatase activity, *Appl. Env. Microbiol.* **51**:32–38.

Sacks, L. E., 1985, Increased formation of arginine deiminase by *Clostridium perfringens* FD-1 growing in the presence of caffeine, *Experientia* **41**:1435–1436.

Sacquet, E. C., Raibard, P. M., Mejean, C., Riottot, M. J., Leprince, C., and Leglise, P. C., 1979, Bacterial formation of ω-muricholic acid in rats, *Appl. Env. Microbiol.* **37**:1127–1131.

Sanchez-Riéra, F., Cameron, D. C., and Cooney, C. L., 1987, Influence of environmental factors in the production of R(−)-1,2,propanediol by *Clostridium thermosaccharolyticum*, *Biotechnol. Lett.* **9:**449–454.

Schink, B., 1986, Environmental aspects of the degradation potential of anaerobic bacteria, in: *Biology of Anaerobic Bacteria* (H. C. Dubourguier et al., ed.), Elsevier, Amsterdam, pp. 2–15.

Scott, T. A., Bellion, E., and Matley, M., 1969, The conversion of N-formyl-L-aspartate into nicotinic acid by extracts of *Clostridium acetobutylicum*, *Eur. J. Biochem.* **10:**318–323.

Sedlmaier, H., and Simon, H., 1985, Purification and some properties of an acryloylCoA reductase of *Clostridium kluyveri, Biol. Chem. Hoppe-Seyler* **366:**953–962.

Seki, S., Hattori, Y., Hasegawa, T., Haraguchi, H., and Ishimoto, M., 1987, Studies on nitrate reductase of *Clostridium perfringens.* IV. Identification of metals, molybdenum cofactor and iron-sulfur cluster, *J. Biochem.* (Tokyo) **101:**503–510.

Sekiguchi, S., Seki, S., and Ishimoto, M., 1983, Purification and some properties of nitrite reductase from *Clostridium perfringens, J. Biochem.* (Tokyo) **94:**1053–1059.

Seto, B., 1980, The Stickland reaction, in: *Diversity of Bacterial Respiratory Systems*, Vol. 2 (C. J. Knowles, ed.), CRC, Boca Raton, pp. 50–64.

Seto, B., and Stadtman, T. C., 1976, Purification and properties of proline reductase from *Clostridium sticklandii, J. Biol. Chem.* **251:**2435–2439.

Shimoi, H., Nagata, S., Esaki, N., Tanaka, H., and Soda, K., 1987, Leucine dehydrogenase of a thermophilic anaerobe, *Clostridium thermoaceticum:* Gene cloning, purification and characterisation, *Agric. Biol. Chem. Tokyo* **51:**3375–3382.

Sih, C. J., and Chen, C-S., 1984, Microbial asymmetric catalysis-enantioselective reduction of ketones, *Angew. Chem. Int. Ed. Engl.* **23:**570–578.

Simon, H., and Günther, H., 1983, Chiral synthons by biohydrogenation or electroenzymatic reductions, in: *Biomimetic Chemistry* (Z. Yoshida and N. Ise, eds.), Elsevier, Amsterdam, pp. 207–227.

Simon, H., Bader, J., Günther, H., Neumann, S., and Thanos, J., 1984, Biohydrogenation and electromicrobial and electroenzymatic reduction methods for the preparation of chiral compounds, *Ann. N.Y. Acad. Sci.* **434:**171–185.

Simon, H., Bader, J., Günther, H., Neumann, S., and Thanos, J., 1985, Chiral compounds synthesised by biocatalytic reductions, *Angew. Chem. Int. Ed. Engl.* **24:**539–555.

Sinskey, A. J., Akedo, M., and Cooney, C. L., 1981, Acrylate fermentations, in: *Trends in the Biology of Fermentations for Fuels and Chemicals*, (A. Hollaender et al., eds.), Plenum Press, New York, pp. 473–492.

Sleat, R., and Robinson, J. P., 1984, The bacteriology of anaerobic degradation of aromatic compounds, *J. Appl. Bacteriol.* **57:**381–394.

Sliwkowski, M. X., and Hartmanis, M. G. N., 1984, Simultaneous single step purification of thiolase and NADP-dependent 3-hydroxybutyrylCoA dehydrogenase from *Clostridium kluyveri, Anal. Biochem.* **141:**344–347.

Stadtman, T. C., and Grant, M. A., 1971, Lβ-Lysine mutase (*Clostridium sticklandii*). Dα-Lysine mutase (*Clostridium*), *Methods Enzymol.* **17B:**206–215.

Stellwag, E. J., and Hylemon, P. B., 1979, 7α-Dehydroxylation of cholic acid and chenodeoxycholic acid by *Clostridium leptum, J. Lipid Res.* **20:**325–333.

Stieb, M., and Schink, B., 1985, Anaerobic oxidation of fatty acids by *Clostridium bryantii* sp. nov., a sporeforming obligately syntrophic bacterium, *Arch. Microbiol.* **140:**387–390.

Stöcklein, W., and Schmidt, H-L., 1985, Evidence for L-threonine cleavage and allothreonine formation by different enzymes from *Clostridium pasteurianum:* Threonine aldolase and serine hydroxymethyl transferase, *Biochem. J.* **232:**621–622.

Stokes, N. A., and Hylemon, P. B., 1985, Characterization of Δ⁴-3-ketosteroid-5β-reductase and 3β-hydroxysteroid dehydrogenase in cell extracts of *Clostridium innocuum, Biochim. Biophys. Acta* **836:**255–261.

Sutherland, J. D., and Macdonald, I. A., 1982, The metabolism of primary, 7-oxo- and 7β-hydroxy bile acids by *Clostridium absonum, J. Lipid Res.* **23:**726–732.

Sutherland, J. D., and Williams, C. N., 1985, Bile acid induction of 7α- and 7β-hydroxysteroid dehydrogenases in *Clostridium limosum*, *J. Lipid Res.* **26**:344–350.

Sutherland, J. D., Williams, C. N., Hutchison, D. M., and Holdeman, L. V., 1987, Oxidation of primary bile acids by 7α-hydroxysteroid dehydrogenase-elaborating *Clostridium bifermentans* soil isolate, *Can. J. Microbiol.* **33**:663–669.

Szulmajster, J., 1958, Bacterial degradation of creatinine. II. Creatinine desimidase, *Biochim. Biophys. Acta* **30**:154–163.

Tanaka, H., and Stadtman, T. C., 1979, Selenium-dependent clostridial glycine reductase; purification and characterization of the two membrane-associated protein components, *J. Biol. Chem.* **254**:447–452.

Taya, M., Yagi, Y., and Kobayashi, T., 1986, Production of acetone and butanol using immobilised growing cells of *Clostridium acetobutylicum*, *Agric. Biol. Chem. Tokyo* **50**:2141–2142.

Thauer, R. K., Jungermann, K., and Decker, K., 1977, Energy conservation in chemotrophic anaerobic bacteria, *Bacteriol. Rev.* **41**:100–180.

Tran, N. D., Romette, J. L., and Thomas, D., 1983, An enzyme electrode for specific determination of L-lysine: A real-time control sensor, *Biotechnol. Bioeng.* **25**:329–340.

Tran-Din, K., and Gottschalk, G., 1985, Formation of D(−)1,2 propanediol and D(−)lactate from glucose by *Clostridium sphenoides* under phosphate limitation, *Arch. Microbiol.* **142**:87–92.

Tsai, L., Pastan, I., and Stadtman, E. R., 1966, Nicotinic acid metabolism II. The isolation and characterisation of intermediates in the fermentation of nicotinic acid, *J. Biol. Chem.* **241**:1807–1813.

Usdin, K. P., Zappe, H., Jones, D. T., and Woods, D. R., 1986, Cloning, expression and purification of glutamine synthetase from *Clostridium acetobutylicum*, *Appl. Env. Microbiol.* **52**:413–419.

Verhulst, A., Semjen, G., Meerts, U., Janssen, G., Parmentier, G., Asselberghs, S., van Hespen, H., and Eyssen, H., 1985, Biohydrogenation of linoleic acid by *Clostridium sporogenes*, *Clostridium bifermentans*, *Clostridium sordelli* and *Bacteroides* sp., *FEMS Microb. Ecol.* **31**:255–259.

Verma, J. N., and Goldfine, H., 1985, Phosphatidylserine decarboxylase from *Clostridium butyricum*, *J. Lipid Res.* **26**:610–616.

Wagner, R., Cammack, R., and Andreesen, J. R., 1984, Purification and characterisation of xanthine dehydrogenase from *Clostridium acidiurici* grown in the presence of selenium. *Biochim. Biophys. Acta* **791**:63–74.

Waterson, R. M., and Conway, R. S., 1981, EnoylCoA hydratases from *Clostridium acetobutylicum* and *Escherichia coli*, *Methods Enzymol.* **71**:421–430.

Westheimer, F. H., 1969, Acetoacetate decarboxylase from *Clostridium acetobutylicum*, *Methods Enzymol.* **14**:231–241.

White, H., Lebertz, H., Thanos, I., and Simon, H., 1987, *Clostridium thermoaceticum* forms methanol from carbon monoxide in the presence of viologen dyes, *FEMS Microbiol. Lett.* **43**:173–176.

Whitehead, T. R., and Rabinowitz, J. C., 1986, Cloning and expression in *Escherichia coli* of the gene for 10-formyltetrahydrofolate synthetase from *Clostridium acidiurici*, *J. Bacteriol.* **167**:205–209.

Wildenauer, F. X., and Winter, J., 1986, Fermentation of isoleucine and arginine by pure and syntrophic cultures of *Clostridium sporogenes*, *FEMS Microbiol. Ecol.* **38**:373–379.

Young, L. Y., 1984, Anaerobic degradation of aromatic compounds, in: *Microbial Degradation of Organic Compounds* (D. T. Gibson, ed.), Marcel Dekker, New York, pp. 487–523.

Youngleson, J. S., Santangelo, J. D., Jones, D. T., and Woods, D. R., 1988, Cloning and expression of a *Clostridium acetobutylicum* alcohol dehydrogenase gene in *Escherichia coli*, *Appl. Env. Microbiol.* **54**:676–682.

Clostridial Enzymes   7

BADAL C. SAHA, RAPHAEL LAMED,
and J. GREGORY ZEIKUS

1. INTRODUCTION

Enzymes are catalytic proteins produced by living systems. In recent years, enzymes have gained wider applications in the biotechnology industry. Enzymes are grouped into six major classes: oxidoreductases, transferases, hydrolases, lyases, ligases, and isomerases. Table I lists some industrial enzymes that are used primarily in the food-related industries and in detergents. Enzymes are also finding applications in pharmaceutical synthesis, therapeutic contexts, and in clinical and chemical analysis. The majority of commercialized enzymes are hydrolases such as amylase, cellulase, pectinase, protease, lipase, and collagenase. Other important enzymes are glucose isomerase and glucose oxidase. *Clostridium* spp. or endospore-forming anaerobes may have a potential in enzyme technology for two general reasons. First, species are very diverse and they produce a wide variety of enzymes, although not oxygenases. Second, clostridia as anaerobes evolved early on earth under energy-limited conditions, which may have placed strong selection pressure on the evolution of very efficient catabolic enzymes. This hypothesis receives indirect support from the rapid growth rate of many anaerobes (e.g., *C. perfringens* grows with a doubling time of less than 15 min). Very little enzyme technology has exploited *Clostridium* spp. as the source of biocatalysts, largely because of their low growth yields. Nonetheless, with the development of genetic engineering techniques, cloning and expression of a gene for a superior enzyme from a *Clostridium* into a suitable producing host is certainly a future research direction with much promise. Most clostridial enzymes which do not perform low redox reactions are oxygen-stable. In this chapter, we will review both the characteristics of the most commonly used commercial enzymes found in *Clostridium* spp. and some of their very novel biocatalytic features not yet practiced.

BADAL C. SAHA and J. GREGORY ZEIKUS • Michigan Biotechnology Institute, Lansing, Michigan 48910, U.S.A. RAPHAEL LAMED • Center for Biotechnology, George S. Wise Faculty of Life Sciences, Tel Aviv University, Ramat Aviv, Israel.

Table I. Enzymes of Commercial Importance Present
in *Clostridium* sp.

Class—enzyme	Uses
Hydrolases	
alpha-Amylase	Starch liquefaction
beta-Amylase	Maltose production
Glucoamylase	Glucose production
Pullulanase	Debranching of starch in glucose and maltose syrup production
Protease	Detergent, meat, fish, dairy product processing, brewing, baking
Cellulases	Fruit and vegetable processing, ethanol production
Pectinases	Fruit juice clarification
Lactase	Lactose hydrolysis in milk, cheese, whey
Collagenase	Prevention and cure of keloids
Oxidoreductases	
Alcohol dehydrogenase	Synthesis of chiral alcohols
Isomerases	
Glucose isomerase	High-fructose syrups

2. AMYLOLYTIC ENZYMES

2.1. α-Amylase

α-Amylase (α-1,4-D-glucan glucanohydrolase, EC 3.2.1.1., endoamylase) hydrolyzes internal α-1,4 linkages of starch at random in an endo fashion, producing oligosaccharides of varying chain lengths. Generally, it cannot act on α-1,6 linkages of starch. α-Amylase activity is produced by a variety of microorganisms (Fogarty, 1983). The production of this amylase has been reported in a number of clostridia: *C. butyricum* (Whelan and Nasr, 1951; Hobson and MacPherson, 1952), *C. acetobutylicum* (French and Knapp, 1950; Hockenhull and Herbert, 1945; Ensley *et al.*, 1975; Chojecki and Blaschek, 1986), *C. thermohydrosulfuricum* (Melasneimi, 1987a; Plant *et al.*, 1987a), *C. thermoamylolyticum* (Katkocin *et al.*, 1985) and a new *Clostridium* isolate (Madi *et al.*, 1987; Antranikian, 1987a).

The effect of carbon source on the formation of extracellular α-amylase activity by *C. acetobutylicum* has been studied (Ensley *et al.*, 1975; Chojecki and Blaschek, 1986). When starch was the carbon source, α-amylase activity was excreted into the medium once the culture entered the stationary phase. Production of α-amylase activity was stimulated by cultivation of *C. acetobutylicum* on starch and biosynthesis of enzyme activity was affected by catabolic repression. Scott and Hedrick (1952) indicated that the β-limit dextrin in starch induced synthesis of α-amylase activity in *C. acetobutylicum,* but glucose or sorbitol did not stimulate the production of α-amylase activity.

Baker *et al.* (1950) first demonstrated the participation of a strain of *C. butyricum* in the breakdown of raw starch granules in the cecum of a pig. Recently, *C. butyricum* T-7 was isolated as a potent raw starch utilizing acidogenic anaerobe from acclimatized mesophilic methane sludge (Tanaka *et al.*, 1987a). The α-amylase activity was found in both supernatant and raw starch-binding forms, when the bacterium was grown in a nutrient medium with various types of raw starch granules (Tanaka *et al.*, 1987b). The raw starch-binding α-amylase was purified by chromatography on a column of Bio-Rex 70 after the enzyme had been liberated from potato starch granules.The molecular weight of the enzyme was 89,000. The enzyme displayed optima for pH at around 5.0 and temperature at 60°C. It was almost completely stable in the absence of Ca^{2+}, between pH 5.5 and 7.5 and temperatures below 35°C. The mode of maltooligosaccharide formation by the raw starch-binding α-amylase of *C. butyricum* T-7 was shown to be almost the same as that of the *Streptococcus bovis* raw starch-digesting α-amylase (Mizokami *et al.*, 1978) and α-amylase of *B. circulans* (Taniguchi *et al.*, 1983). The enzyme initially produced maltose, maltotriose, maltotetraose, and maltopentaose. Later maltotetraose and maltopentaose decreased with the concomitant formation of glucose, as well as increases in the levels of maltose and maltotriose. The iodine reaction apparently disappeared when 9% hydrolysis of the substrate was attained.

2.2. β-Amylase

β-Amylase (α-1,4-D-glucan maltohydrolase, EC 3.2.1.2, saccharogenic amylase) hydrolyzes alternate α-1,4-glucosidic linkages of starch in an exo fashion from the nonreducing end, producing β-maltose. It cannot act and bypass α-1,6 linkages in starch, so a high molecular weight β-limit dextrin remains. It occurs widely in many higher plants such as soybean, barley, sweet potato, and is also produced by microorganisms such as *Bacillus megaterium, B. circulans, B. cereus, B. polymyxa, Pseudomonas* spp., and *Streptomyces* spp. (Fogarty, 1983). So far only one anaerobic species, *C. thermosulfurogenes,* has been reported to produce an extracellular β-amylase (Hyun and Zeikus, 1985a).

The β-amylase was produced in high yield as a primary product during growth of *C. thermosulfurogenes* at 62°C. Hyun and Zeikus (1985b) studied the general mechanism for regulation of β-amylase synthesis in *C. thermosulfurogenes.* They found that it was inducible and subject to catabolic repression. A hyperproductive mutant was isolated which produced eightfold more β-amylase than the wild type. Synthesis of the enzyme was both constitutive and resistant to catabolite repression.

The β-amylase was easily recovered and purified by membrane ultra-filtration (100,000 mol. wt. cutoff), ethanol precipitation, DEAE-Sepharose CL-6B column chromatography, and Sephacryl S-200 gel filtration (Shen *et al.*, 1988). The pure enzyme was a glycoprotein and had a specific activity of 4215 units/mg protein (Table II). It was a tetramer of about 210,000 mol. wt. composed of a single type of subunit. In this respect, it is very similar to sweet potato β-amylase, with a molecular weight of 200,000 and monomer molecular weight of 47,500 (Balls *et al.*, 1946). The isoelectric point was at pH 5.1. The enzyme had a pH and temperature optimum at pH 5.5 and 75°C (respectively) and it was stable at pH 3.5–7.0 and temperatures up to 80°C. Like β-amylase from some other sources, the β-amylase from *C. thermosulfurogenes* is a sulfhydryl enzyme. It was inactivated by sulfhydryl reagents (*p*-chloromecuribenzoate (pCMB) and *N*-ethylmaleimide. It did not require metal ions for activity, but required Ca^{2+} for stability and was competitively inhibited by cyclodextrins. The K_m for soluble starch at pH 6.0 and 75°C was 1.68 mg/ml and K_{cat} was 440,000 min^{-1}.

The β-amylase from *C. thermosulfurogenes* was readily and strongly adsorbed onto insoluble starch (Saha *et al.*, 1987). *p*-Chloromercuribenzoate-treated β-amylase lost its activity toward insoluble or gelatinized starch, but the ability to be adsorbed onto insoluble starch was preserved. Adsorbed β-amylase was gradually released from starch in the liquid phase during hydrolysis at 75°C. The degradation of insoluble starch by β-amylase was greatly stimulated by pullulanase addition. Polyclonal antibodies raised against purified β-amylase did not cross-react with sweet potato β-amylase, suggesting that there is no significant homology between these two enzymes, although they have similar molecular weights. The β-amylase was rich in acidic and hydrophobic amino acids. The 20 N-terminal amino acid sequence of this β-amylase indicates only 45% homology with *B. polymyxa* β-amylase (Kawazu *et al.*, 1987). The properties of this oxygen-insensitive β-amylase are summarized in Table II.

The β-amylase also displayed industrial potential for the production of high-maltose syrups from raw or soluble starch at 75°C. Various maltose-containing syrups are used in the brewing, baking, canning, and confectionery industries. The β-amylase from *C. thermosulfurogenes*, with its superior thermoactivity and its environmental compatibility with current starch-processing conditions at high temperature and acidic pH, may have future industrial applications in various maltose-containing syrups production, if it can be produced economically (Zeikus and Saha, 1989).

2.3. Glucoamylase

Glucoamylase (1,4-α-D-glucan glucohydrolase, EC 3.2.1.3) is an exo-acting carbohydrase which cleaves glucose units consecutively from the nonreducing end of starch molecules. It is a microbial enzyme produced by

Table II. Biochemical Characteristics of Thermostable β-Amylase Purified from *C. thermosulfurogenes*[a]

Specific activity (units/mg protein)	4215
Chemical composition	Tetramer
Molecular weight	210,000 (subunit, 51,000)
Isoelectric point (pI)	5.1
Optimum pH	6.0
pH Stability	3.5–7.0
Optimum temperature (°C)	75
Thermostability (up to °C)	80
K_m for soluble starch (mg/ml)	1.68 (at 75°C)
K_{cat}	440,000/min (at 75°C)
Specificity	Hydrolyzes alpha-1,4 forming maltose (exo-acting)
Metal ion requirement:	
For stability	Calcium
For activity	None
Inhibitors	pCMB, cyclodextrins
Polyclonal antibody reaction	Does not cross-react with beta-amylase from some other sources

[a]From Shen *et al.*, 1988.

many molds and yeasts, e.g., *Aspergillus* spp., *Rhizopus* spp. and *Endomyces* spp. (Ueda and Saha, 1981). The production of this enzyme activity by some *Clostridium* spp. has been reported. Hockenhull and Herbert (1945) found that *C. acetobutylicum* produced a glucoamylase activity. This enzyme had a pH optimum at 4.5. Ensley *et al.* (1975) studied the production of glucoamylase activity by *C. acetobutylicum* and found that it was induced in the presence of glucose in the culture medium. Chojecki and Blaschek (1986) found that the biosynthesis of this enzyme was affected by catabolic repression. High levels of a thermostable glucoamylase activity was reported in *C. thermohydrosulfuricum* by Hyun and Zeikus (1985c), although it was later described as a maltase or α-glucosidase activity (Hyun *et al.*, 1985).

A thermostable extracellular glucoamylase activity from *C. thermoamylolyticum* (Katkocin *et al.*, 1985) was purified from the culture supernatant by precipitation with calcium chloride. The mixture of α-amylase and glucoamylase was separated by adsorbing the α-amylase on granular starch, with which it forms a complex. Further purification of the crude glucoamylase was accomplished by $(NH_4)_2SO_4$ precipitation, followed by several chromatographic separations. The purified enzyme had a molecular weight of 75,000 and a half-life of greater than 3 hr at pH 6.0 and 70°C, with a maximum glucoamylase activity at pH 5.0 and a temperature of about 70–75°C. The enzyme hydrolyzed maltodextrin and soluble starch. It also hydrolyzed maltose, but the rate of hydrolysis was less than half the rate of hydrolysis of maltodextrin.

Glucoamylase from fungal sources is a well-studied, industrially important enzyme in the starch bioprocessing and brewing industries. It is widely used in alcoholic fermentation of starchy materials and in the commercial production of glucose and high-glucose corn syrups from liquefied starch.

2.4. Pullulanase

Pullulanase (pullulan-6-glycanohydrolase, EC 3.2.1.41) is a debranching enzyme that specifically cleaves α-1,6 linkages in starch, amylopectin, pullulan, and related oligosaccharides (Abdullah *et al.*, 1966). It is an industrially important enzyme which is generally used in combination with saccharifying amylases such as glucoamylase, fungal α-amylase, or fungal β-amylase for the production of various sugar syrups because it improves saccharification and yield (Norman, 1982). Moreover, it has gained significant attention as a useful tool for structural studies of carbohydrates (Whelan, 1971).

Pullulanase is produced by mesophilic organisms such as *Aerobacter aerogenes* (Bender and Wallenfels, 1961), *Bacillus* sp. (Norman, 1982; Nakamura *et al.*, 1975), *Bacillus polymyxa* (Fogarty and Griffin, 1975), *B. macerans* (Adams and Priest, 1977), *B. cereus* var. *mycoides* (Takasaki, 1976), and *Streptococcus mitis* (Walker, 1968). Hyun and Zeikus (1985c) found that *C. thermohydrosulfuricum* produces highly thermoactive and thermostable cell-bound pullulanase. Enzyme synthesis was inducible and subjected to catabolic repression (Hyun and Zeikus, 1985d). They also developed a hyperproductive mutant which displayed improved starch metabolism features in terms of enhanced rates of growth, ethanol production, and starch consumption. In both wild-type and mutant strains, pullulanase was produced at high levels in starch-limited chemostats, but not in glucose- or xylose-limited chemostats. Hyun and Zeikus (1985e) reported the simultaneous and enhanced production of thermostable amylases and ethanol from starch by cocultures of *C. thermosulfurogenes* and *C. thermohydrosulfuricum*. Pullulanase has also been reported to be produced by other strains of *C. thermohydrosulfuricum* (Plant *et al.*, 1987a; Melasniemi, 1987a; Antranikian *et al.*, 1987a), *C. thermosaccharolyticum* (Koch *et al.*, 1987), and a new *Clostridium* isolate (Madi *et al.*, 1987). In the new isolate, the formation of pullulanase was dependent on growth and occurred predominantly in the exponential phase. The enzyme was largely cell-bound during growth of the organism with 0.5% starch. When a *C. thermohydrosulfuricum* strain DSM 567 was grown in continuous culture on a defined medium containing growth-limiting amounts of starch (Antranikian *et al.*, 1987b–d), pullulanase and α-amylase were overproduced and a partial disintegration of the cell surface layer occurred, which was associated with the formation of membrane "blebs" and extracellular vesicles.

The cell-bound pullulanase from *C. thermohydrosulfuricum* strain 39E was solubilized by treatment with detergent and lipase (Saha *et al.*, 1988). The solubilized pullulanase was purified 3511-fold to homogeneity (specific activity 481 units/mg protein) by treatment with streptomycin sulfate, ammonium sulfate, and by DEAE-Sephacel, octyl-Sepharose and pullulan-Sepharose chromatography. No multiplicity of solubilized pullulanase was detected during the course of purification, unlike the other pullulanase preparations (Eisele *et al.*, 1972; Mercier *et al.*, 1972; Ueda and Ohba, 1972; Ohba and Ueda, 1973); rather, it was a monomeric glycoprotein with an apparent molecular weight of about 136,500. The molecular weights of pullulanase from some other sources range from 56,000 to 150,000 (Fogarty, 1983). However, this pullulanase is very similar to that of *Aerobacter aerogenes* with respect to molecular weight (143,000) and possession of a single polypeptide chain (Eisele *et al.*, 1972). The glycoprotein nature of the *Clostridium* pullulanase is contrary to the finding that extracellular *Klebsiella pneumoniae* pullulanase is a lipoprotein (Pugsley *et al.*, 1986). The enzyme is an acidic protein with an isoelectric point of pH 5.9. The amino acid composition of the purified pullulanase showed a high degree of similarity with pullulanase from other sources (Eisele *et al.*, 1972; Brandt *et al.*, 1976; Ohba and Ueda, 1975), differing significantly only in its having less methionine, except that from *A. aerogenes* (Ohba and Ueda, 1975), which has much lower proportions of leucine and isoleucine and more hydrophobic amino acids.

The pullulanase from *C. thermohydrosulfuricum* was stable at temperatures up to 90°C and at pH 3.0–5.5. It was optimally active at 90°C and at pH 5.5 (Saha *et al.*, 1988). Melasniemi (1987b) reported that the temperature and pH optima of pullulanase from another strain of *C. thermohydrosulfuricum* were 85–90°C and pH 5.6, respectively. The pullulanase from *C. thermosaccharolyticum* was most active at 75°C and pH 5.0–6.0 (Koch *et al.*, 1987). The pH and temperature optima of most pullulanases are 5.0–7.0 and temperature 45–60°C, respectively (Fogarty, 1983). Schardinger dextrins inhibited the pullulanase activity from *C. thermohydrosulfuricum* strain 39E and β-dextrin was a stronger inhibitor than α-dextrin (Saha *et al.*, 1988). This is true with another pullulanase (Enevoldsen *et al.*, 1977). The pullulanase from *C. thermosaccharolyticum*, however, was weakly inhibited by β-cyclodextrin but not at all by α-cyclodextrin (Koch *et al.*, 1987).

The pullulanase from *C. thermohydrosulfuricum* was not inhibited by pCMB, although pullulanases from *B. cereus* (Takasaki, 1976) and *B. polymyxa* (Fogarty and Griffin, 1975) were inhibited. This indicates that there is no thiol group involvement in its catalytic activity. However, it was inhibited by EDTA, which indicates that some (divalent) metal ion is probably important for its stability. The enzyme was inhibited by *N*-bromosuccinimide, which indicates that tryptophan residues may be involved in catalysis. The pullulanase from *C. thermohydrosulfuricum*, like other reported pullulanase (Abdullah and French, 1970), produced condensation products. A summary of the properties of the purified pullulanase from *C. thermohydrosulfuricum* Z21-109 is given in Table III.

The initial reaction velocity for hydrolysis of various substrates by purified pullulanase from *C. thermohydrosulfuricum* was compared with that of commercial *Bacillus* pullulanase in order to assess the relative substrate specificities of these two enzymes. The *C. thermohydrosulfuricum* pullulanase displayed threefold higher specificity toward soluble starch and amylopectin and sevenfold higher specificity toward glycogen than the *Bacillus* enzyme. This pullulanase can cleave α-1,4 linkages in amylose and hence it is novel in activity when compared with other purified pullulanases described. Melasniemi (1987b) suggested that the α-amylase and pullulanase activities of *C. thermohydrosulfuricum* E101-69 are properties of the same protein, representing a novel, thermostable amylase. Similar findings were also reported with thermostable pullulanase from other thermoanaerobes. *Thermoanaerobium brockii* pullulanase has been cloned in *E. coli* and *B. subtilis*. The cloned enzyme cleaved all of the α-1,6-glycosidic linkages (and none of the α-1,4 bonds) in pullulan but it hydrolyzed mostly α-1,4 and very few of the α-1,6 linkages of starch (Coleman *et al.*, 1987). *Thermoanaerobium* Tok6-B1 pullulanase was also able to hydrolyze α-1,4-glycosidic bonds in starch (Plant *et al.*, 1987b). It is not clear why a pullulanase cleaves α-1,4 linkages in starch. It is most likely a new type of enzymatic mechanism for a pullulanase-amylase activity and we suggest the name amylopullulanase to distinguish it from normal pullulanase, which does not cleave α-1,4 linkages. The K_m for pullulan, the maximum initial reaction rate

Table III. Biochemical Characteristics of Thermostable Amylolytic Pullulanase Purified from *C. thermohydrosulfuricum* Z21-109

Specific activity (units/mg protein)	481
Chemical composition	Monomeric, glycoprotein
Molecular weight	136,500
Isoelectric point (pI)	5.9
pH optimum	5.0–5.5
pH stability	3.5–5.0
Temperature optimum (°C)	90
Thermostability (up to °C)	90
K_m for pullulan (mg/ml)	0.675
K_{cat}	Hydrolyzes both alpha-1,4 and alpha-1,6 (endo-acting)
Metal ion requirement for activity	None
Inhibitors	Cyclodextrins, EDTA, *N*-bromosuccinimide

*a*From Saha *et al.*, 1988.

(V_{max}), and K_{cat} of purified pullulanase from *C. thermohydrosulfuricum* Z21-109 were 0.675 mg/ml, 122.5 moles reducing sugar formed/min/mg protein, and 16,240 turnover number/min, respectively (Table III).

The potential industrial application of novel amylopullulanases from *Clostridium* spp. have not yet been examined. Taking advantage of the high thermoactivity and acidoactivity of these pullulanases, it may be assumed that these enzymes might effectively replace α-amylase and pullulanase in both starch liquefaction and saccharification processes.

2.5. α-Glucosidase

α-Glucosidase (EC 3.2.1.20, α-D-glucoside glucohydrolase) hydrolyzes terminal nonreducing α-1,4-linked glucose residues of various substrates, releasing α-D-glucose. It is generally considered to be a maltase, but has wide specificities—being able to cleave glucosides of nonsugars in addition to maltose, maltotriose, and other maltooligosaccharides, and transfer α-D-glucosyl residues of maltose and α-D-glucosides to suitable acceptors. It hydrolyzes smaller substrates faster than larger substrates and some α-glucosidases cannot attack starch at all. The enzyme is produced by a variety of microorganisms (Fogarty, 1983). Hyun and Zeikus (1985b) characterized a glucoamylase from *C. thermohydrosulfuricum*, whose activity was optimal at pH 5.0–6.0 and 75°C. Hyun *et al.* (1985) later concluded that the major enzyme responsible for glucose formation was maltase or α-glucosidase.

α-Glucosidase activity was also reported to be produced by *C. thermohydrosulfuricum* (Melasniemi, 1987a) and a new *Clostridium* isolate (Madi *et al.*, 1987; Antranikian *et al.*, 1987b). The glucosidase activity from the new isolate was active over a broad temperature range (40–85°C) and displayed a temperature optimum for activity at 60–72°C. The apparent K_m value of α-glucosidase for maltose was 25 mM. The maltase of *C. acetobutylicum* (French and Knapp, 1950; Hockenhull and Herbert, 1945) has almost identical initial rates of hydrolysis with starch, maltose, and amyloheptaose.

3. CELLULASES

3.1. Background

Cellulose, the major constituent of most plant material, is a β-1,4-linked linear polyglucopyranoside of very complex structure which is quite resistant to microbial attack. The use of microbial cellulases was recently the object of study as part of the renewed interest in the potential bioconversion of cellulosic biomass to useful chemical products (e.g., ethanol). In this context, thermophilic anaerobic fermentation may have significant advantages. The high specific activity of the degradation of crystalline cellulose by *C. thermocellum* is sufficient reason for basic research using this organism as the preferred model of a bacterial cellulase system.

The microbial degradation of insoluble cellulose cannot be accomplished by a single enzyme; a mixture of different cellulases acting synergistically is required for the successful solubilization of the crystalline substrate. The cellulase system in both bacteria and fungi comprises three different classes of enzymes: (1) endo-1,4-β-D-glucanases (EC 3.2.1.4), which cleave the internal β-1,4-cellulose bonds randomly; (2) exo-1,4-β-D-glucanases, including both 1,4-β-D-glucan cellobiohydrolases (EC 3.2.1.91) and 1,4-β-D-glucan glucohydrolases (EC 3.2.1.74), which liberate cellobiose and glucose moieties, respectively, from the nonreducing end of the cellulose chain; and (3) 1,4-β-D-glucosidases, also referred to as cellobiases (EC 3.2.1.21), which can hydrolyze cellobiose or the nonreducing ends of soluble oligocellodextrins to glucose. In addition, it is quite common for bacteria to possess cellobiose phosphorylase (EC 2.4.1.20) and/or cellodextrin phosphorylase (EC 2.4.1.49), which cleave the corresponding soluble oligosaccharide to yield glucose-1-phosphate.

3.2. Natural Occurrence of Cellulases

A wide variety of bacteria and fungi are known which produce cellulases (Ljungdahl and Eriksson, 1985). However, relatively few strains are capable of producing the high levels of the complementary set of endo- and exoglucanases, believed to be required for the extensive solubilization of the crystalline substrate. A great many strains in nature produce extracellular endoglucanases but cannot grow on cellulose. In other cases, a given microbe (particularly bacteria) grows quite well on the insoluble substrate, yet the cellulolytic enzymes are not readily detected in the extracellular medium.

Since the latter part of the last century, several species of cellulolytic clostridia have been described (Table IV). Although most of the work on

Table IV. *Clostridium* **Species Known to Produce Cellulase Which Degrades Crystalline Celluloses**

Species	Ref.
Mesophilic species	
C. acetobutylicum	Lee *et al.*, 1987
C. cellobioparum	Hungate, 1944
C. cellulovorans	Sleat *et al.*, 1984
Thermophilic species	
C. stercorarium	Madden, 1983
C. thermocellum	Viljoen, 1926

cellulases was accomplished using aerobic fungal systems, the anaerobic bacteria hold potential promise due to their intrinsic requirement for economy with respect to secretion of hydrolytic enzymes. The resultant cellulase system would thus be expected to exhibit higher specific activity. In anaerobic thermophilic ecosystems, *C. thermocellum* is by far the most abundant, almost universal cellulolytic species, occurring worldwide in a variety of autothermal, solar-heated, and volcanic habitats (Lamed and Bayer, 1988b). In view of its profuse distribution, it is interesting to note that its substrate utilization range is remarkably limited, mainly to cellulose and its degradation products.

3.3. The Cellulolytic System of *C. thermocellum*

3.3.1. Properties of the Crude Extracellular Enzyme System

One very puzzling early observation concerning the cellulase system of *C. thermocellum* (as well as for several other cellulolytic organisms) is that, although the cells grow very well on crystalline forms of cellulose, the crude extracellular cellulase preparations from the same organism exhibit relatively low levels of activity on the same substrate. Early work (Ng and Zeikus, 1981a) pointed out the remarkable comparative properties and high activities of *C. thermocellum* and *T. reesei* cellulases, as well as the sensitivity of the former to thiol inhibitors. Johnson *et al.* (1982a) coined the term "true" cellulase activity to describe the capacity of a given cellulase system essentially to solubilize the crystalline forms of the substrate. This activity was found to be dependent on the presence of Ca^{2+} and thiols, and to be different from the endoglucanase or carboxymethyl (CM) cellulase of *C. thermocellum* characterized previously (Ng *et al.*, 1977; Ng and Zeikus, 1981a,b). In addition, true cellulase activity was completely inhibited by the major end product, cellobiose, but not by glucose (Johnson *et al.*, 1982b). Interestingly, endoglucanase activity was unaffected by any of these three effectors (Ca^{2+}, thiols, or cellobiose). In a study by a different group (Ljungdahl *et al.*, 1983), extracellular cellulases which had been adsorbed onto cellulose were found to be released by lowering the ionic strength of the medium. As will be seen later, this finding proved to be very useful in correlating the properties of the crude system with a purified, functionally viable, cellulase complex. Johnson *et al.* (1985) studied the effect of carbon nutrition on the formation of cellulase. A direct relationship between repression of true cellulase production and growth rate was usually found. On the other hand, the endoglucanase or CM-cellulase activity of *C. thermocellum* was constitutive enzyme (Ng and Zeikus, 1988).

3.3.2. Purification of Cellulases

The purification of a single polypeptide which exhibits cellulolytic activity has proved extremely difficult. In one study, Ng and Zeikus

(1981b) purified an endoglucanase with an apparent molecular weight of 83,000. The enzyme was a glycoprotein with 11% carbohydrate. The enzyme catalyzed most actively the hydrolysis of CM cellulose. Avicel and acid-swollen cellulose were attacked, but the activity was very low. The enzyme produced cellobiose and a mixture of longer chain oligodextrins from noncrystalline cellulose.

In another study, Petre *et al.* (1981) purified a 56,000 mol. wt. polypeptide with endoglucanase activity by drastically different techniques. The procedure used involved ion exchange chromatography in the presence of 8 M urea. The purified enzyme was capable of converting acid-treated cellulose to cellotriose as the main product. Antibodies were prepared against this cellulase and were eventually used for the screening and identification if its gene, which was later cloned in *E. coli*.

In addition to these studies, it was recognized at a fairly early stage that many polypeptide bands exhibit endoglucanase activity (Beguin, 1983) and that some of this activity may be associated with higher order complexes (Ait *et al.*, 1979). Nevertheless, the overall contribution of the above purified enzymes to the intact cellulolytic apparatus in *C. thermocellum* went unrecognized.

3.3.3. Cloning of Cellulase Genes

Recombinant DNA technology may provide an alternative to conventional strain development for improving enzyme production. As a prerequisite for genetic manipulation, various *C. thermocellum* genes involved in cellulose degradation have been cloned in *E. coli* (Millet *et al.*, 1985; Schwartz *et al.*, 1987; Romaniec *et al.*, 1987). These achievements have somewhat eased the difficulties encountered in isolating to purity single polypeptide chains possessing cellulolytic activity, circumventing the requirement to separate them from multienzyme (cellulosome) complexes, which tend to be very tightly associated. However, cloning of cellulase genes has not yet contributed to the identity, localization, or contribution of the individual cellulases to the total cellulolytic process in the parent organism.

This approach to the cellulase system in *C. thermocellum* was developed mainly by the group at the Pasteur Institut (Millet *et al.*, 1985). In their studies, various cellulase genes of *C. thermocellum* were cloned in *E. coli* and the important clones were identified by either enzyme or immunochemical overlays. The first gene to be recognized was termed *celA* (Cornet *et al.*, 1983). The corresponding gene product, termed EGA, was the 56,000 mol. wt. polypeptide isolated by Petre *et al.* (1981). In this case, the clone was identified immunochemically. The *celA* gene has been sequenced (Beguin *et al.*, 1985).

Three other endoglucanases (termed EGB, EGC, and EGD; gene

products of *celB*, *celC*, and *celD*, respectively) have thus far been purified from extracts of *E. coli* which had been cloned with *C. thermocellum* genes. In these examples, the clones were identified by their expression of enzymatic activity. Polyclonal antibodies were raised against the respective proteins produced in *E. coli* and used to demonstrate that the corresponding target antigen is actually secreted by *C. thermocellum*. Using this approach, EGB was shown to be a polypeptide of about 66,000 mol. wt., EGC 40,000 mol. wt., and EGD about 65,000 mol. wt. The latter was produced in *E. coli* in sufficient quantities for the purified protein to be crystallized for X-ray diffraction studies (Jolif *et al.*, 1986a,b). These four endoglucanases isolated to date are different cellulases, since the antibodies failed to cross-react and the purified enzymes displayed different specificities toward different cellulodextrins.

It is necessary to produce active enzymes in large amounts to study the structure and function of cloned cellulases and to evaluate their biotechnological potential. In this regard, of the several endoglucanases cloned by the pioneering group at the Pasteur Institut, only EGD, the product of *celD* gene (and not EGA, EGB, or EGC), was obtained in substantial (above 10%) amounts. Recently, *celA* and *celC* were subcloned in a temperature-regulated *E. coli* expression vector containing the leftward heat-sensitive promoter PL of bacteriophage λ. Expression of these genes was improved to give genetically engineered proteins which composed 10–15% of the total cellular protein. Overproduction of EGA was toxic to the cells, since it accumulated in the membrane fraction and the periplasmic space, while EGC was well tolerated (Schwartz *et al.*, 1987).

It is important to note that the cellulase genes are not clustered and constitute not less than about one-third of the *C. thermocellum* genome. However, no cellulolytic system with the unique capacity to degrade crystalline cellulose (such as that of the parent organism) has yet been reconstructed employing cloned cellulases. In fact, despite the cloning of about 20 genes (so far), no polypeptide exhibiting "true" cellulase or cellobiohydrolase activity has yet been obtained (nor has such an enzyme been isolated from the parent organism).

3.3.4. Cellulosomes

During the course of studies on the cellulase system of *C. thermocellum*, Lamed *et al.* (1983a) purified a very high molecular weight, multifunctional, multienzyme complex which has been coined the cellulosome. The purified cellulosome was responsible for the adhesion of *C. thermocellum* to the insoluble cellulosic substrate, as well as for the majority of endoglucanase activity in the organism. Moreover, the cellulosome exhibited "true" cellulase activity as well as all of the previously established properties of the crude cellulase system of *C. thermocellum* (Lamed *et al.*, 1985).

3.3.4a. Cellulose-Binding Factor (CBF). The initial clue which led Bayer *et al.* (1983) to the cellulosome concept was based on the results of a combined immunochemical-genetic study, prompted by the observation that prior to cellulolysis, cells of *C. thermocellum* bind very strongly to the insoluble substrate. In order to investigate the existence of a cellulose-binding factor (CBF), which was postulated to mediate cell-substrate adhesion, an adhesion-defective mutant, which failed to adhere to cellulose following growth on cellobiose, was first isolated. Bayer *et al.* (1983) subsequently prepared antibodies against intact wild-type cell; following adsorption of the resultant antiserum onto large quantities of mutant cells, they were able to "subtract" the antibody species which were common to both wild-type and mutant cells. The residual antibodies were found to recognize a single major antigenic component which was associated with the surface of the wild type but absent from the mutant cell surface. Further work led to the isolation of the CBF, employing a procedure which included an efficient affinity chromatography step using cellulose as the adsorbent (Lamed *et al.*, 1983b). The isolation of the CBF was thus based on the fundamental characteristic of this surface component, i.e., its intrinsic recognition of the cellulosic substrate.

3.3.4b. Physical Properties. The purified CBF proved to be of very high molecular weight as evidenced by gel filtration, ultracentrifuge, and electron microscopic studies. The molecular weight was estimated at about 2.1 million.

Sodium dodecyl sulfate–polyacrylamide gel electrophoresis (SDS–PAGE) analysis revealed that the CBF (later termed cellulosome) is a complex composed of at least 14 different subunits, ranging in molecular weight from about 48,000 to 210,000 (Fig. 1). Interestingly, the highest molecular weight subunit (S1) appeared to react almost exclusively with the anti-CBF antiserum (prepared by a mutant adsorption procedure, as described in the previous section). The same subunit was also sensitive to saccharide-specific staining procedures (Lamed and Bayer, 1987, 1988a) and thus appears to be a glycopolypeptide. The saccharide portion of the S1 is subject to enzymatic oxidation with galactose oxidase and is reactive to the α-galactose-specific lectin from *Griffonia simplicifolia*. This lectin is the only one out of more than 20 tested which agglutinated cells of *C. thermocellum;* under similar conditions, the lectin fails to recognize the mutant (Lamed *et al.*, 1987).

The sugar components of S1 of the cellulosome have been elucidated (Morgenstern *et al.*, 1987). A novel oligosugar was found to constitute about 20% of this protein (Gerwig *et al.*, 1989). The cloning of S1 is impatiently awaited to obtain better understanding of this unique glycosylated protein and the role of sugars in its function.

The CBF complex is remarkably stable. Various conditions and treat-

Sugars Antigen CMCase Protein Subunit   Mr(K)

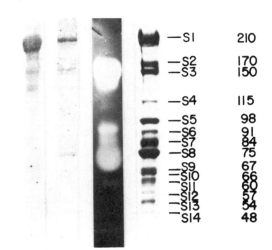

Subunit	Mr(K)
—S1	210
—S2	170
—S3	150
—S4	115
—S5	98
—S6	91
—S7	84
—S8	75
—S9	67
—S10	66
—S11	60
—S12	57
—S13	54
—S14	48

Figure 1. Subunit composition and biochemical characterization of the cellulosome isolated from cellulose-grown *C. thermocellum* YS. Following SDS–PAGE (6% gels), protein content was determined by Coomassie Brilliant blue staining, endoglucanase activity by CMC overlay, anticellulose specificity by immunoblotting, and sugar content by the enzyme hydrazide technique. (Lamed and Bayer, 1987.)

ments (including high salt concentrations, urea, guanidine-HCl, and detergents) known to be disruptive to noncovalent bonding failed to separate the CBF into its component polypeptide parts. Only by boiling in the presence of SDS it was possible to denature the CBF into its full complement of subunits. Moreover, as far as could be determined, none of the CBF subunits exist in the free state; they are all associated exclusively with the complex.

3.3.4c. Enzymatic Properties. One of the big surprises with the CBF came to light when a gel overlay technique was employed in search of endoglucanase activity in cell extracts. Following rocket immunoelectrophoresis, it became clear that the majority of endoglucanase activity in *C. thermocellum* was associated with the CBF. These results were further supported by examining the enzymatic properties of the purified CBF.

Using the identical overlay technique of SDS–PAGE-separated samples of the CBF, endoglucanase activity could be assigned to the majority of the CBF subunits (Fig. 1). Surprisingly, the highest molecular weight subunit (S1), which exhibited the majority of antigenic activity and contained the majority of saccharide-specific label, exhibited no detectable endoglucanase activity (Fig. 1). This is in marked contrast to nearly all of the other subunits. Additional experiments indicated that the various enzymatic activities associated with the various subunits represent cellulases of different specificities, as revealed by experiments which examined the product pattern of CBF subcomplexes. In one case, using monoclonal antibodies against one of the cloned endoglucanases (EGD) described by the group at the Pasteur Institute (Joliff *et al.*, 1986a), it could be demonstrated that this particular endoglucanase was equivalent to one of the minor subunits (S11) of the cellulosome (Lamed and Bayer, 1987). Consequently, in light of the multifunctional, multienzyme nature of the CBF, the term "cellulosome" was selected as a more appropriate descriptive term for the CBF complex (Lamed *et al.*, 1983a).

The localization of these enzymes in the form of a high molecular weight complex was envisaged as a major contributing factor in the synergistic action of the cellulase system of *C. thermocellum*.

Several additional studies have been performed with the goal of isolating components from the cellulosome after its disintegration by SDS (Lamed *et al.*, 1983a; Demain and Wu, 1987) or by limited proteolytic digestion (Lamed and Bayer, 1987). In these studies, there are strong indications for S8 (or Ss) being the target of cellobiose and Ca^{2+} sensitivity of the system and for S1 or (SL) being important for assembly of the complex and its tight binding to cellulose. However, more extensive and controlled experiments are required in order to establish these findings.

3.3.4d. The Cell Surface Cellulosome and Interaction with Cellulose. Histochemical and cytochemical evidence indicated that the cellulosome is localized abundantly on the cell surface of *C. thermocellum* on specialized protuberances (Bayer *et al.*, 1985; Bayer and Lamed, 1986; Lamed and Bayer, 1986). These exocellular polycellulosomal protuberances play a critical role in mediating the interaction between the cell and its cellulosic substrate (Fig. 2). The protuberances protract on interaction with cellulose, thus forming fibrillar extensions which connect the cell surface with the cellulosome particles. The latter have adsorbed onto the surface of the substrate. The cellulosome particles have the capacity to convert cellulose directly to cellobiose, which is then taken up by the cell via an appropriate transport apparatus (Ng and Zeikus, 1982). When grown on cellobiose, the adhesion-defective mutant lacks these protuberances (along with the cell surface cellulosome).

3.3.4e. Noncellulosomal Cellulases. The cellulosome is the major, but not the only, contribution to the cellulase system in *C. thermocellum*. There are additional cellulases which are not components of the cellulosome and fail to bind to cellulose on their own. Among these, it was recently demonstrated that the *C. thermocellum* equivalent of the cloned EGC is a separate (noncellulosomal) cellulase which appears in relatively high quantities both on the cell surface and as an extracellular cellulase (E. A. Bayer, personal communication). The cellulolytic properties of this endoglucanase have been extensively characterized (Petre *et al.*, 1986), although the contribution of this particular endoglucanase, as well as other noncellulosomal cellulases to the overall cellulase system of *C. thermocellum*, is still undefined.

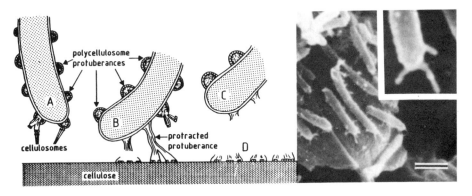

Figure 2. Interaction of *C. thermocellum* cells with cellulose mediated by protracted poly-cellulosomal protuberances. Left: cell A, prior to contact; cell B, following contact; cell C, following attachment. Right: SEM of cationized ferritin stained cells of *C. thermocellum* attached to cellulose. Bar, 1.0 μM. (Lamed and Bayer, 1987.)

3.3.4f. Extension of the Cellulosome Concept. Studies with at least five different strains of *C. thermocellum* indicated that the cellulosome is an integral part of this organism (Wu and Demain, 1985; Hon-nami *et al.*, 1985; Lamed and Bayer, 1988a). Although strain-specific alterations have been observed in the relative size of the cellulosome, in the composition of the subunits within the complex, and in the disposition of the protuberances on the cell surface, all of these strains (which were originally isolated from very different natural habitats) express a similar type of exocellular cellulosome. In addition, a variety of the cellulolytic bacteria of surprisingly divergent evolutionary and physiological states have been shown to possess cellulosome-like complexes (Lamed *et al.*, 1987). Despite the fact that many of these strains were mesophilic, some were gram-negative, and one was aerobic, all of the latter strains possessed cell surface cellulases and protuberance-like exo structures. Moreover, nearly all of the strains cross-reacted with the anticellulosome (CBF-specific) antibody and all but one interacted with the *Griffonia* lectin. The cellulosome may thus appear to be a quite general solution for bacterial cellulolysis in nature.

3.4. Cellulases of Other *Clostridium* spp.

Although the existence of several other cellulolytic clostridial species has been known for many years, the status of their respective cellulase systems is surprisingly obscure. To date, most of the work on clostridial cellulases described in the literature has involved the well-characterized strains of *C. thermocellum*, and relatively little has been published concerning mesophilic strains.

After *C. thermocellum*, the cellulase system which has attracted the most attention belongs to the other recognized thermophilic species, *C. stercorarium* (Madden, 1983). This bacterium appears to be quite different from *C. thermocellum*, especially in its wide substrate utilization pattern. Unlike *C. thermocellum* and most other cellulolytic species, *C. stercorarium* releases its endoglucanases only in the middle of the exponential phase of growth. An extracellular endoglucanase from this organism was isolated to apparent homogeneity (Creuzet and Frixon, 1983), exhibiting an isoelectric point of 3.85 and a molecular weight of 91,000 by SDS–PAGE (99,000 by gel filtration). Unlike the endoglucanase activity associated with the crude cellulase system and the purified cellulosome of *C. thermocellum*, this purified endoglucanase preparation was markedly inhibited by cellobiose. The enzyme activity was optimal at pH 6.4 and was inhibited by thiol reagents. The highest activity was measured with CMC as substrate, and the enzyme displayed very little activity on more ordered cellulose substrates. No activity could be detected using trinitrophenyl CMC as a substrate. In a second contribution (Creuzet *et al.*, 1983), the same research group isolated a second cellulase which had no action on either CMC or crystalline cellulose. On the basis of the product pattern resulting from partial hydrolysis of an amorphous form of cellulose by this enzyme, the authors concluded that this cellulase was a cellobiohydrolase. The enzyme was further purified by ion exchange chromatography; PAGE indicated a single protein band, but no values for molecular weight or any other physical or chemical characterizations were reported. The authors were able to demonstrate synergistic action among the cellulolytic components, by combining the purified endoglucanase and cellobiohydrolase preparations and measuring the combined effect on hydrolysis of different cellulosics. Additional synergistic action resulted from the addition of a crude β-glucosidase preparation from the same organism. Recently, Bronnenmeier and Staudenbauer (1988) reported the resolution of *C. stercorarium* cellulase by fast protein liquid chromatography (FPLC). Five components were obtained, which were termed avicelase I and II, β-cellobioase I and II, and a β-glucosidase. Only a vague correlation could be drawn between the newly isolated components (100,000, 87,000, 82,000, 128,000, and 85,000 mol. wt., respectively) and the endoglucanase and cellobiohydrolase reported earlier (Creuzet and Frixon, 1983; Creuzet *et al.*, 1983). However, one avicelase could be defined as an endoglucanase since it showed CMCase activity, while the second avicelase may be a cellobiohydrolase not showing this activity. The two avicelases were synergistic in their action, but the cellobiosidases (which split *p*-nitrophenyl cellobioside to give cellobiose) did not contribute to the activity on avicel and the β-glucosidase increased the activity only slightly. No evidence was obtained for oligomerization or complex formation between individual enzymes, but no data were shown to support this statement.

Another report, which appeared shortly after the work on *C. ster-*

corarium, described the isolation of another thermophilic anaerobe, identified as *Clostridium* sp. (Taya *et al.,* 1984). This strain exhibited a substrate utilization pattern remarkably similar to that of *C. stercorarium* and may well be the same species.

Clostridium acetobutylicum, a mesophilic organism, exhibits only low activity on crystalline cellulose and fails to grow on this substrate (Lee *et al.,* 1985). "True cellulases" *per se* have not been purified in this organism, and its extracellular endoglucanase and cell-associated β-glucosidase activities may very well be incidental to other activities (e.g., xylanase) (Lee *et al.,* 1987).

A very interesting, new, as yet unclassified, mesophilic clostridial isolate was recently shown to secrete a high-molecular-weight (approx. 700,000) cellulolytic complex, the integrity of which appears to be responsible for the efficient hydrolysis of crystalline cellulose (Cavedon *et al.,* 1987). The complex was shown to contain about 18 different subunits. A mutant was obtained which lacked cellulolytic activity on crystalline substrates but contained a single endoglucanase fraction of mol. wt. 130,000. This latter fraction contained several polypeptides which migrated on SDS–PAGE at rates identical to those of subunits derived from the wild-type strain. The authors postulated that the defective strain failed to form the high molecular weight complex due to the lack of a required component and that the intact complex is required for activity on insoluble substrates. Alternatively, the missing component(s) is an enzyme(s) which has a central role in the degradation of cellulose.

In a recent work, Lamed *et al.* (1987) analyzed the topographical disposition of cellulase-containing cell surface structures in a variety of cellulolytic organisms including several clostridial strains. It was found that *C. cellulovorans* and *C. cellobioparum* both contain cell-associated endoglucanase activity, as well as an abundance of protuberance-like structures on the cell surface. As mentioned earlier, many other characteristics (e.g., lectin binding and immunochemical cross-reactivity) suggest a similarity in the organization of cell-associated cellulases in *Clostridium.*

3.5. Comparative Aspects and Industrial Implications

It was previously determined that in *C. thermocellum* the specific activity of cellulolysis on highly crystalline substrates is superior to that of the well-characterized cellulolytic fungus *Trichoderma reesei* (Johnson *et al.,* 1982a). Due to the high specific enzymatic activity on the one hand and the high thermostability on the other, the cellulases from *C. thermocellum* have been considered for potential industrial utilization in direct alcohol fermentations, but not for saccharification *per se* (i.e., glucose production). Recently, endoglucanases have been used in laundry detergents. Unlike glucoamylases and β-amylases, which are not inhibited by glucose or maltose, glucanohydrolase and cellobiohydrolases are inhibited by their end products. Thus, enzymatic glucose production from cellulose does not seem to have industrial merit when compared to starch saccharification technology. Despite significant efforts invested in this direction, applicable processes have yet to be successfully implemented. One currently appealing approach for producing large amounts of a given enzyme would be to clone the gene(s) in an appropriate host organism. Indeed, it appears to be quite easy to clone the cellulases of *C. thermocellum* as well as those of other organisms. However, it seems to us that the potential to utilize clostridial cellulases industrially would be contingent on reconstruction of cellulosome-like complexes and not a simple accumulation of the complementary set of free cellulases (Lamed and Bayer, 1988b).

Another possibility, which has been attempted experimentally with quite encouraging results in some cases, is the coculture of a cellulolytic (polymer-degrading) strain with accompanying saccharolytic strains (Wang *et al.*, 1979; Ng *et al.*, 1981; Ljungdahl and Weigel, 1981). The latter organisms are usually included in attempts to alter the ensuing product pattern to enhance the yield of a desired product (Lamed and Bayer, 1988b). One advantage of this approach is that the rate of cellulolysis is often enhanced due to more rapid removal of inhibitory hydrolytic products such as cellobiose.

Another approach, which takes advantage of the higher activity of cell-bound cellulases, is to apply resting cell suspensions instead of using the classical approach, which involves extracellular cellulase preparations (Giuliano and Khan, 1985). Extracellular cellulases can be immobilized on solid supports in order to emulate the native cell-bound state.

4. COLLAGENASE AND PROTEINASES

4.1. Collagenase

Collagenases are endopeptidases which hydrolyze native insoluble fibrous collagen. They have been obtained from a variety of animal tissues and microorganisms such as *C. histolyticum*, *Achromobacter iophagus*, *Pseudomonas marinoglutinosa*, *Streptomyces* spp., and *Vibrio* B-30 (Siefter and Harper, 1971). Among collagenases of microbial origin, the one produced extracellularly by *C. histolyticum* has been studied extensively. It is available commercially. Recently, the production of extracellular collagenase activity by *C. proteolyticum* sp. nov. and *C. collagenovorans* sp. nov. was reported (Jain and Zeikus, 1988).

The basic collagen unit, the tropocollagen molecule, consists of a triple helix of three peptide chains (α-1, α-2, and a second α-1 or an α-3), each of approximately 95,000 mol. wt. (Seifter and Harper, 1970). Collagen comprises 33% of the total protein in mammalian organisms and is the main constituent of skin, tendon, and cartilage as well as organic constituents of teeth and bone (Mandl, 1972). The extracellular collagenase of *C. histolyticum* has been fractionated into different fractions that can be differentiated on the basis of substrate preference. Grant and Alburn (1959) first fractionated crude clostridial collagenase into three fractions and designated them as A, B, and C in order of elution from the column. The enzyme from *C. histolyticum* has been purified by several methods (Yoshida and Noda, 1965; Kono, 1968; Lee-Owen and Anderson, 1975; Lwebuga-Mukasa, 1976; Emod and Keil, 1977; Oppenheim and Franzblau, 1978). It is essential to have collagenase free of other proteolytic activity.

Collagenase A has a molecular weight of about 105,000 (Seifter and Harper, 1970); earlier values for collagenase A obtained by other groups were 112,000 (Mandl *et al.*, 1964), 95,000 (Yoshida and Noda, 1965), and 109,000 (Strauch and Grassmann, 1966). Collagenase B has a molecular weight of 79,000 (Yoshida and Noda, 1965). Levdikova *et al.* (1963) showed that collagenase A could be dissociated into four inactive subunits each of 25,000 mol. wt. Harper and Seifter (1965) suggested that collagenase B could be an active dimer (57,400 mol. wt.) of the 25,000 mol. wt. monomer of Levdikova *et al.* (1963) and that collagenase A could be the parent molecule of collagenase B. These two enzymes are similar in (1) amino acid composition, (2) immunological cross-reactivities of the purified enzymes with antisera prepared to either enzyme, (3) molecular weight relationship between A and B, and (4) under certain conditions, the binding of a specific inhibitor (cysteine) in the molecular ratios of 2 : 1 with collagen A and 1 : 1 with collagen B (Harper and Seifer, 1965).

The optimum pH range for collagenase is relatively wide, at least pH 6–8 for crude preparations, but is narrower within these limits for purified collagenases. Low pH values inactivate the enzyme irreversibly. Temperatures above 56°C inactivate the enzyme rapidly and completely. The presence of various fractions of collagenases appears to be required for optimal digestion of collagen to acid-soluble peptides, and this may be due to slight differences in their peptide sequence substrate specificity (Peterkofsky, 1982). Calcium ions are required for both the binding of the enzyme to the collagen substrate and for full catalytic activity. Ethylene diaminetetraacetic acid (EDTA) inhibits the activity of collagenase A and B by binding calcium (Gallop *et al.*, 1957).

Cysteine and mercaptoethanol inhibit collagenase activity and their action seems to be on a metal constituent of the protein *per se* (Seifter and Gallop, 1962). Evidence was also presented for the involvement of zinc in the active center of collagenase (Harper, 1972). Thus collagenase has a dual metal requirement. Both collagenase A and B are inhibited by a number of metal-chelating agents such as *o*-phenanthroline, α,α-bipyridyl, and 8-hydroxyquinoline (Seifter and Harper, 1970). Diisopropylphosphofluoridate (DFP) does not inhibit collagenase, eliminating this enzyme from the class of serine protease (Seifter *et al.*, 1959). Collagenase was activated to some degree by CO^{2+} or Mn^{2+} (Yagisawa *et al.*, 1965). Both collagenase A and B can catalyze the hydrolysis of native and denatured collagens to yield seemingly identical peptide mixtures (Seifter and Harper, 1970). They act against synthetic peptide substrates designed for screening collagenolytic activities, Z-Gly-Pro-Gly-Gly-Pro-Ala-OH (Wuensch, 1963) and PZ-Pro-Leu-Gly-Pro-D-Arg-OH (Wuensch and Heidrich, 1963). In general, the specificity resides in a sequence-P-X-Gly-P-Y, common to all collagen molecular, enzyme action occurring between X-Gly residues, to yield a peptide with C-terminal X and always peptides with N-terminal glycine (Seifter and Harper, 1970). This indicates that the minimum requirement for specificity for the enzyme is a peptide bond involving the amino acid glycine. A summary of some characteristics of *C. histolyticum* collagenase is given in Table V.

Pure collagenase can be applied as a sensitive probe for biosynthetic studies and sequence determinations. It is useful in prevention or cure of keloids. Collagenase is useful for the dispersal and dissociation of animal tissues in the laboratory. It is routinely used to separate cells from their parent tissues. Collagenase may be employed in the study of certain disease processes and of wound-healing mechanisms (Seifter and Harper, 1970).

4.2. Clostripain and Clostridial Aminopeptidase

Clostridium histolyticum also produces an extracellular sulfhydryl proteolytic enzyme called clostripain (clostridiopeptidase B, EC 3.4.4.20). This enzyme possesses amidase, esterase, and proteolytic activity which is directed toward the carboxyl peptide linkage of arginine (Mitchell and Harrington, 1971). It has been purified and characterized. It has maximum activity at pH 7.4–7.8. Calcium enhances esterase activity and depresses

Table V. Some Characteristics of *C. histolyticum* Collagenase

Characteristic	Collagenase A	Collagenase B
Molecular weight	105,000	57,400
Activator	Ca^{2+}	Ca^{2+}
	Zn^{2+} probably intrinsic	Zn probably intrinsic
Inhibitor	EDTA, cysteine, not inhibited by DFP	EDTA, cysteine; not inhibited by diisopropylfluorophosphonate
Product (collagen)	Peptides of N-terminal Gly residues	Same as for collagenase A

thermal inactivation. The enzyme may be useful as a protease in sequencing work to obtain large initial peptide fragments and also in studies of enzyme active sites possessing arginine specificity. For further details about this enzyme, see Mitchell and Harrington (1971).

Clostridial aminopeptidase (CAP) was isolated from the spert culture medium of *C. histolyticum* by Kessler and Yaron (1976). Later it was found that large quantities of the enzyme are cell-associated, located in the periplasmic space of the bacteria. Purification to homogeneity of the cell-associated enzyme showed that the two forms of the enzyme are identical. Clostridial aminopeptidase is a metal-requiring enzyme, requiring Mn^{2+} and CO^{2+} for activity (Fleminger and Yaron, 1984). Since CAP is a nonspecific aminopeptidase, cleaving all the N-terminal L-amino acid residues (except proline) at comparable rates, it is very applicable to enzymatic peptide sequencing. Such application, using immobilized CAP in combination with immobilized aminopeptidase P (which cleaves only N-terminal prolyl residues), has been demonstrated by Fleminger and Yaron (1987).

5. PECTINASES AND OTHER HYDROLASES

Pectin is a complex carbohydrate containing pectic acid (polygalacturonic acid) esterified with methyl alcohol. Pectin-degrading enzymes can be classified in the following groups:

I. Esterases catalyze the hydrolysis of methoxyl groups on pectin, forming free carboxyl groups (i.e., polygalacturonic acid) and methanol.
II. Depolymerases
   (a) Polymethylgalacturonase hydrolyzes pectin in a random fashion (endo enzyme) or in an exo fashion from the nonreducing end (exo enzyme).
   (b) Polymethylgalacturonate lyase causes cleavage in pectin by a transelimination process (endo and exo types).
   (c) Polygalacturonate lyase hydrolyzes pectic acid (endo and exo fashion).
   (d) Polygalacturonate lyase causes cleavage in pectic acid by a transelimination process (endo and exo types).
   (e) Oligogalacturonase hydrolyzes oligo-D-galactosiduronates.
   (f) Oligogalacturonate lyase causes cleavage of oligo-D-galactosiduronate by a transelimination process.

Pectic enzymes are produced by a variety of microorganisms including clostridia (Fogarty and Kelly, 1983). The distribution of various pectic enzymes in some *Clostridium* spp. is shown in Table VI. Miller and Macmillan (1970) purified pectinesterase and exopolygalacturonate lyase 178-fold

Table VI. Production of Pectinolytic Enzymes by Some *Clostridium* spp.

Organism	PE[a]	PG	PGL	Ref.
C. aurantibutyricum	+	ND	+	Fogarty and Kelly, 1983
C. felsineum	ND	+	+	Fogarty and Kelly, 1983
C. multifermentans	+		+	Macmillan *et al.*, 1964
C. thermosulfurogenes	+	+	ND	Schink and Zeikus, 1983
C. roseum	+	+	ND	Lund and Brocklehurst, 1978

[a]PE, pectinesterase; PG, polygalacturonase; PGL, polygalacturonate lyase. ND, not determined.

and 156-fold, respectively, from *C. multifermentans* by gel filtration on Sephadex G-200. Both enzymes coeluted as a single protein peak. Further, the two activities could not be separated by DEAE-cellulose chromatography and zonal centrifugation. Both activities were most stable at pH 6.0. The esterase was inactivated rapidly below pH 5 or above pH 7. Lyase preparations were freed of pectinesterase activity by heating at pH 7 for 30 min at 38°C. The pectinesterase, which is complexed with exopolygalacturonate lyase, attacks highly esterified pectin at the reducing end only (Lee *et al.*, 1970; Miller and Macmillan, 1971; Sheiman *et al.*, 1976). *Clostridium thermosulfurogenes* produces an active thermostable polygalacturonate hydrolase and pectin methylesterase (Schink and Zeikus, 1986). The optimum temperatures for polygalacturonate hydrolase and pectin methylesterase activities were 75°C and 70°C, respectively. Both enzyme activities were completely stable on heating to 60°C for 30 min. The pH optima for polygalacturonate hydrolase and pectin methylesterase were 5.5 and 6.5, respectively. These thermostable pectinolytic activities may have application in fruit juice clarification and for processing food or agricultural/forestry products.

The thermophilic *clostridial* species isolated from thermophilic methane sludge possessed cell-bound xylanase and β-xylosidase activities, which were involved in the degradation of xylan (Tanaka *et al.*, 1986). An arabanase activity has been characterized from *C. felsineum* (Kaji *et al.*, 1963).

6. OXIDOREDUCTASES

Oxidoreductases are a vast group of enzymes involved in electron transfer reactions. These enzymes have been considered less applicable than the various kinds of hydrolase because they usually require the participation of electron transfer mediators, which tend to be expensive and, in many cases, unstable. In spite of these limitations, several reductases have been examined seriously and means for regeneration of the reduced or oxidized cofactors are in various stages of development. Most notable are the methods in which the coenzyme is regenerated by a cosubstrate recognized by the same oxidoreductase (e.g., alcohol dehydrogenase) or by a different enzyme (e.g., formate dehydrogenase, glucose dehydrogenase). Many chiral compounds can now be synthesized by microbial hydrogenation using hydrogen (or formate) and hydrogenase-containing microorganisms, as well as by chemical, electrochemical, electroenzymatic, or light-induced regeneration of reduced cofactors. Reductases, which do not require reduced pyridine nucleotides and can accept electrons directly from reduced viologens, are of particular potential value.

Methods for assaying both analytes and enzymes by coupling to nicotinamide dinucleotide-dependent oxidoreductases are widely used because of the ease of continuous spectrophotometric or fluorometric determination of the formation or disappearance of the reduced form of these coenzymes. These enzymes will not be dealt with in this chapter. Potentially useful reductive enzymes from clostridia are listed in Table VII (Lovitt *et al.*, 1987), few of which are described in further detail below.

There are several general strategies for application of an enzyme, such as an oxidoreductase, in a stereoselective catalysis. These include whole-cell

Table VII. Novel Oxidoreductases in *Clostridium* Species[a]

Reduction	Electron donor	Organism
Aldehyde/ketone dehydrogenase		
steroids	Unknown	*C. paraputrificum*
		C. bifermentans
methyl ketones	NADPH	*C. thermohydrosulfuricum*
ketones	Unknown	*C. pasteurianum*
	Unknown	*C. tyrobutyricum*
2-Oxoacid synthase		
fatty acids	Ferredoxin	*C. sporogenes*
acetate	Ferredoxin	*C. kluyveri*
Linoleic reductase		
linoleic acid	Unknown	*C. sporogenes*
Enoate reductase		
cinnamic acid	NADH	*C. sporogenes*
crotonic acid	NADH	*C. tyrobutyricum*
2-Oxoacid reductase		
phenylpyruvic acid	NADH	*C. sporogenes*
Nitroaryl reductase		
chloramphenicol	Ferredoxin/flavodoxin	*C. acetobutylicum*
metronidazole	Ferredoxin/flavodoxin	
paro-nitrobenzoate	Ferredoxin/flavodoxin	
2-nitrobenzene	Ferredoxin/flavodoxin	
Lipoamide hydrogenase		
NAD/lipoamide	Lipoamide/NADH	*C. kluyveri*

[a]From Lovitt *et al.*, 1987.

fermentation (which avoids the need for coenzyme regeneration, but may be subject to interference by competing enzymes), use of a crude enzyme in batch or continuous flow syntheses using an immobilized enzyme column, or the application of a purified enzyme in similar batch and continuous flow techniques.

6.1. Aldehyde Ketone Dehydrogenase

Perhaps the best known thermoactive alcohol dehydrogenase is that present in both *C. thermohydrosulfurcium* and other thermoanaerobes, the most notable being *Thermoanaerobium brockii*, which is nonsporulating. This NADP-linked secondary alcohol (aldehyde/ketone) dehydrogenase, found in thermoanaerobes, exhibits a wide substrate specificity toward linear and cyclic secondary alcohols (Lamed and Zeikus, 1981; Lamed *et al.*, 1981; Keinan *et al.*, 1986a). The assymmetrical reduction of small aliphatic ketones remains a major challenge in organic chemistry because the direct stereoselective reduction of acyclic ketones with well-known horse liver or yeast alcohol dehydrogenases, previously used for laboratory scale preparation of assymmetrical alcohols, are not suitable for linear secondary alcohols. In addition, the practical use of the above commercially available enzymes is somewhat limited by their limited temperature and solvent stability. The secondary alcohol dehydrogenase of *T. brockii* was detected and purified (Lamed and Zeikus, 1980, 1981) and then examined as a reduction catalyst employing a co-substrate-linked, NADPH-regenerating system (Lamed *et al.*, 1981; Keinan *et al.*, 1986a,b). The effect of reaction conditions on the stereoselectivity was studied in detail and described recently (Keinan *et al.*, 1986a). In addition to three monofunctional alcohols, alcohols bearing a second functional group to be used as a "handle" (e.g., chloride, nitrilo, olephinic, acetylenic, phenyl) were prepared and used for the synthesis of more complex chiral molecules (Keinan *et al.*, 1986b, 1987). A most striking feature of the enzyme is the reported size-induced reversal of stereoselectivity. This has been explained elegantly (but not conclusively) by assuming both small and large subsites at the active site of the enzyme. The small site is preferred for the large alkyl group, provided it is not too big (propyl) to be accommodated in it (Keinan *et al.*, 1986a).

Another somewhat different NADP-linked secondary alcohol dehydrogenase was found in several strains of *C. thermosaccharolyticum*. The enzyme differs from *T. brockii* alcohol dehydrogenase in its somewhat lower thermostability and its quite different stereoselectivity (R. Lamed, unpublished). Various alcohol dehydrogenase reactions have been assayed in *C. pasteurianum*, *C. tyrobutyricum* (Holt *et al.*, personal communication), and *C. thermocellum* (Lamed and Zeikus, 1980), but no detailed study of these enzymes is yet available. Ketones can also be reduced to chiral secondary alcohols by whole cells of *C. tyrobutyricum* or *C. kluyveri*. Excellent enantiomeric excess can be obtained in a number of cases, but low values are also encountered. Of special interest are various phenyl methyl ketones, reduced with high stereoselectivity to give the R alcohols employing whole clostridial cells and ethanol as an electron donor. The enzyme responsible for these reactions is not known (Simon et al., 1985).

6.2. Enoate Reductase

Reduction of the C=C double bond of α,β-unsaturated aldehydes and carboxylic acids was shown to take place in several clostridia and was studied in detail in some strains (Simon et al., 1985). Many α,β-unsaturated carboxylic acids or aldehydes can be hydrogenated by C. tyrobutyricum C. sp. strain Lal (DSM 1460) and C. kluyverii (DSM 555). These reactions were found to result from the presence of the enzyme enoate reductase (EC 1.3.1.31), an iron sulfur flavoprotein, catalyzing the general reaction:

$$
\begin{array}{c}
R^3 \quad\quad N \quad + \text{NADH} + H^\oplus \\
\diagdown\!\!\diagup \\
\diagup\!\!\diagdown \quad\quad\quad \text{or} \\
R^2 \quad\quad R^1 \quad + 2\,MV^{\oplus\oplus} + 2\,H^\oplus
\end{array}
\longrightarrow
\begin{array}{c}
R^2 \quad\quad H \quad + \text{NAD}^\oplus \\
\diagup\!\!\!\!\diagdown\!\!\!\!\diagup \\
H \quad R^3 \quad\quad R^1 \quad X \quad + 2\,MV^{\oplus\oplus}
\end{array}
$$

$$
\overset{\displaystyle H}{\underset{\displaystyle |}{}}
$$

X = COO$^\oplus$, C = 0; R^1, R^2, R^3 (Simon et al., 1985)

The reaction with reduced MV proceeds 1.5 times faster than with NADH. Enoate reductase can react with a very wide range of substrates. Various rules have been found to apply and determine the permissible R_1, R_2, and R_3 moieties and the sterochemistry of the resulting products (Simon et al., 1985). Several reactions were carried out on a preparative scale, e.g., the reduction of the following cyclic diketone using whole cells:

For more details the reader is directed to the original papers and to the cited review of Simon et al. (1985).

6.3. 2-Oxoacid Synthase

This section deals with reductive carboxylation of acyl-CoA catalyzed by 2-oxoacid synthases. Bachofen et al. (1964) obtained the first clue that reduced ferredoxin can drive the reductive synthesis of pyruvate from acetyl-CoA and CO_2 in cell-free extracts of *C. pasteurianum*. This is essentially the reversal of the "phosphoroclastic" splitting of pyruvate. Later work demonstrated that reduced ferredoxin can promote the synthesis of a ketoacid other than pyruvate according to the reaction:

$$\text{Acyl CoA} + \text{ferredoxin}_{red} + CO_2 \rightarrow \alpha\text{-ketoacid} + \text{CoA} + \text{ferredoxin}_{red}$$

For an early review, see Buchanan, 1972. All α-ketoacids shown to be synthesized via the ferredoxin-linked carboxylation are important intermediates in the biosynthesis of amino acids, and the reaction constitutes an important mechanism of CO_2 fixation. The synthetic and exchange reaction appear to be separate activities catalyzed by a single protein which is specific for each ketoacid. Thiamine pyrophosphate is required for each activity. A triphenyltetrazolium dye, flavin mononucleotide (FMN) or flavin adenine dinucleotide (FAD), can replace ferredoxin, but not NAD or NADP. α-Ketobutyrate synthase appears to be important for the synthesis of α-aminobutyrate and isoleucine (Buchanan, 1972).

Clostridium sporogenes performs the Stickland reaction in which pairs of amino acids are fermented, one amino acid acting as an electron donor (e.g., valine, leucine, isoleucine), and the other acting as an electron acceptor (e.g., proline, glycine). It has been shown that *C. sporogenes* synthesizes many of its growth requirements from deaminated and decarboxylated fatty acid analogs. The enzymes involved are highly active and show a broad specificity. The first step is reductive carboxylation to produce the 2-oxoacid, which is followed by further reduction to hydroxyl or amino acids (Lovitt *et al.*, 1988a). Studies of the synthesis of pyruvate and other 2-oxoacids from acyl phosphate derivatives were performed with permeabilised cells of *C. sporogenes* (Lovitt *et al.*, 1987). Regeneration of NAD was employed with MV as a mediator between an electrode and MV-linked NAD reductase present in the same cells. The reaction shows promise, but the authors conclude that for scale-up the relative area of the working electrode must be greater than the dropping mercury electrode used.

6.4. Miscellaneous Oxidoreductases

Several additional oxidoreductases of importance will be briefly mentioned here.

6.4.1. NADPH–NADH Diaphorase

A diaphorase from *C. kluyveri* is commercially available and has found wide application in enzyme-linked assays. Reduced di- or triphosphopyridine nucleotides are determined by diaphorase-catalyzed dye reduction yielding a colored product. The enzyme contains a FMN as a cofactor and has a molecular weight of 24,000. It does not catalyze hydrogen exchange between pyridine nucleotides (Kaplan *et al.*, 1969).

6.4.2. Glycine Reductase

The clostridial enzyme (Tanaka and Stadtman, 1979) is a complex of three or more dissimilar protein components, one of which is a selenoprotein. The enzyme complex catalyzes a coupled oxidation–reduction reaction, in which electrons from a dithiol are utilized for the reduction of glycine to acetate, together with esterification of ADP, to give ATP. Ferredoxin participates as one of the electron transfer components.

6.4.3. Nitroacryl Reductase

Crude extracts of *C. kluyveri, C. sp.* La1 (*C. tyrobutyricum*), *C. sporogenes,* and *C. pasteurianum* catalyze NADH-dependent reduction of the nitro group of *p*-nitrobenzene (Angermaier and Simon, 1983). The enzyme has been partially purified and characterized. Nitroalkyl derivatives do not serve as substrates for this enzyme. Based on chromatographic behavior, separation pattern, yields, stability, pH optima, molecular masses, and electron paramagnetic spin resonance (EPR) studies, it was determined that the nitroacryl reductase activity differed from aldehyde dehydrogenase, 2-hydroxybutyryl-CoA dehydrogenase, enoate reductase, ferredoxin NAD(P) oxidoreductases, and other enzymes. However, the physiological roles of nitroacryl reductase were not indicated.

7. CONCLUSIONS

Clostridium spp. and other spore-forming anaerobes produce a wide variety of enzymes of potential commercial importance. Many of the common classes of commercial enzymes found in aerobes are present in spore-forming clostridia. In particular, some thermophilic spore-forming anaerobes possess unique cellulases and amylases of high activity and stability that can be used directly as crude preparations (or even the organisms themselves) for production of alcohol from cheap substrates. Alternatively, the genes for these enzymes could be cloned and the activities overexpressed in an aerobic species for production at lower costs. One problem with using clostridia as enzyme sources is that the costs of production are relatively high. Although a few potentially useful *Clostridium* spp. or their proteins are pathogenic and harmful, most species and their proteins are not and provide safe enzyme production.

Clostridia also produce many novel enzymes, especially certain kinds of oxidoreductases not present in aerobes. These enzymes in particular (e.g., NADP-linked alcohol dehydrogenase of *C. thermohydrosulfuricum*) require more study before commercial uses can be realized. At present, studies of enzymes from clostridia are largely academic, but with the increasing need for novel biocatalysts and the advent of genetic engineering, enzymes from *Clostridium* may find their way into the marketplace.

ACKNOWLEDGMENTS. The help of Dr. E. A. Bayer, Department of Biophysics, The Weizman Institute of Science, Rehovot, Israel, in writing the cellulase section is greatly appreciated.

REFERENCES

Abdullah, M., and French, D., 1970, Substrate specificity of pullulanase, *Arch. Biochem. Biophys.* **137**:483–493.

Abdullah, M., Catley, B. J., Lee, E. Y. C., Robyt, J., Wallenfels, K., and Whelan, W. J., 1966, The mechanism of carbohydrase action. II Pullulanase, an enzyme specific for the hydrolysis of alpha-1,6 bonds in amylacious oligo- and polysaccharides, *Cereal Chem.* **43**:111–118.

Adams, K. R., and Priest, F. G., 1977, Extracellular pullulanase synthesis in *Bacillus macerans*, *FEMS Lett.* **1**:269–273.

Ait, N., Creuzet, N., and Forget, P., 1979, Partial purification of cellulase from *C. thermocellum*, *J. Gen. Microbiol.* **113**:399–400.

Angermaier, L., and Simon, H., 1983, On nitroaryl reductase activities in several clostridia. *Z. Physiol. Chem.* **354**:1653–1663.

Antranikian, G., Herzberg, C., and Gottschalk, G., 1987a, Production of thermostable α-amylase, pullulanase, and α-glucosidase in continuous culture by a new *Clostridium isolate*, *Appl. Environ. Microbiol.* **53**:1668–1673.

Antranikian, G., Herzberg, C., and Gottschalk, G., 1987b, Production of thermostable α-amylase, pullulanase and α-glucosidase in continuous culture by a new *Clostridium* isolate, *Appl. Environ. Microbiol.* **53**:1668–1673.

Antranikian, G., Herzberg, C., Mayer, F. and Gottschalk, G., 1987c, Changes in the cell envelope structure of *Clostridium* sp. strain EM1 during massive production of α-amylase and pullulanase, *FEMS Microbiol. Letters.* **41**:193–197.

Antranikian, G., Zablowski, P., and Gottschalk, G., 1987d, Conditions for the overproduction and excretion of thermostable α-amylase and pullulanase from *Clostridium thermohydrosulfuricum* DSM 567, *Appl. Microbiol. Biotechnol.* **27**:75–81.

Bachofen, R., Buchanan, B. B., and Arnon, D. I., 1964, Ferredoxin as a reductant in pyruvate synthesis by a bacterial extract. *Proc. Natl. Acad. Sci. USA* **51**:690–694.

Baker, F., Nasr, H., Morrice, F. and Bruce, J., 1950, Bacterial breakdown of structural starches and starch products in the digestive tract of ruminant and non-ruminant mammals, *J. Path. Bacteriol.* **62**:617–638.

Balls, A. K., Thompson, R. P., and Walden, M. K., 1946, Crystalline protein with beta-amylase activity, prepared from sweet potatoes, *J. Biol. Chem.* **163**:571–572.

Bayer, E. A., and Lamed, R., 1986, Ultrastructure of the cell surface cellulosome of *Clostridium thermocellum* and its interaction with cellulose, *J. Bacteriol.* **167**:828–836.

Bayer, E. A., Kenig, R., and Lamed, R., 1983, Studies on the adherence of *Clostridium thermocellum* to cellulose, *J. Bacteriol.* **156**:818–827.

Bayer, E. A., and Setter, E., and Lamed, R., 1985, Organization and distribution of the cellulosome in *Clostridium thermocellum*, *J. Bacteriol.* **163**:552–559.

Beguin, P., 1983, Detection of cellulase activity in polyacrylamide gels using Congo red-stained agar replicas, *Anal. Biochem.* **131**:333–336.

Beguin, P., Cornet, P., and Aubert, J.-P., 1985, Sequence of a cellulase gene of the thermophilic bacterium *C. thermocellum*, *J. Bacteriol.* **162**:102–105.

Bender, H. and Wallenfels, K., 1961, Untersuchungen an Pullulan II. Spezifischer Abban durch ein bakterielles, *Enzym. Biochem. Z.* **334**:79–95.

Brandt, C. J., Catley, B. J., and William, M. A., Jr., 1976, Extracellular and protease-released pullulanase, *J. Bacteriol.* **125**:501–508.

Bronnenmeier, K., and Staudenbauer, W. L., 1988, Resolution of *Clostridium stercorarium* cellulase by fast protein liquid chromatography, *Appl. Microbiol. Biotechnol.* **27**:432–436.

Buchanan, B. B., 1972, Ferredoxin-linked carboxylation reactions, in: *The Enzymes*, Vol. 4 (P. D. Boyer, ed.), Academic Press, New York, pp. 193–216.

Cavedon, K., Leschine, S. B., and Canale-Parola, E., 1987, A high molecular weight protein complex is responsible for the degradation of crystalline cellulose by a mesophilic *Clostridium, Ann. Meet. Am. Soc. Microbiol.*, Abst. K127, p. 223.

Chojecki, A. and Blaschek, H. P., 1986, Effect of carbohydrate source on α-amylase and glucoamylase formation by *Clostridium acetobutylicum* SA-1, *J. Ind. Microbiol.* **1**:63–67.

Coleman, R. D., Yang, S.-S., and McAlister, M. P., 1987, Cloning of the debranching enzyme gene from *Thermoanaerobium brockii* into *Escherichia coli* and *Bacillus subtilis, J. Bacteriol.* **169**:4302–4307.

Cornet, P., Millet, J., Beguin, P., and Aubert, M.-P., 1983, Characterization of two cell (cellulose degradation) genes of *C. thermocellum* coding for endoglucanases, *Bio/Technology* **1**: 589–594.

Creuzet, N., and Frixon, C., 1983, Purification and characterization of an endoglucanase from a newly isolated thermophilic anaerobic bacterium, *Biochimie* **65**:149–156.

Creuzet, N., Berenger, J.-F., and Frixon, C., 1983, Characterization of exoglucanase and synergistic hydrolysis of cellulose in *Clostridium stercorarium, FEMS Microbiol. Lett.* **20**:347–350.

Demain, A. L., and Wu, J. H. D., (1987), Proteins of the *Clostridium thermocellum* cellulase complex responsible for degradation of crystalline cellulose, in: *Biochemistry and Genetics of Cellulase Degradation* (J.-P. Aubert, P. Beguin, and J. Millet, eds.), Academic Press, London, pp. 117–131.

Eisele, E., Rashed, I. R., and Wallenfels, K., 1972, Molecular characterization of pullulanase from *Aerobacter aerogenes. Eur. J. Biochem.* **26**:62–67.

Emod, I., and Keil, B., 1977, Five sepharose-bound ligands for the chromatographic purification of *Clostridium* collagenase. *FEBS Lett.* **77**:51–56.

Enevoldsen, B. S., Reimann, L., and Hansen, N. L., 1977, Biospecific affinity chromatography of pullulanase, *FEBS Lett.* **79**:121–124.

Ensley, B., McHugh, J. J., and Barton, L. L., 1975, Effect of carbon source on formation of α-amylase and glucoamylase by *Clostridium acetobutylicum, J. Gen. Appl. Microbiol.* **21**:51–59.

Fleminger, G., and Yaron, A., 1984, Soluble and immobilized Clostridial aminopeptidase and aminopeptidase P as metal requiring enzymes, *Biochim. Biophys. Acta* **789**:245–256.

Fleminger, G., and Yaron, A., 1987, Application of immobilized aminopeptidases for sequential hydrolysis of proline containing peptides, *Methods Enzymol.* **136**:170–178.

Fogarty, W. M., 1983, Microbial amylases, in: *Microbial Enzymes and Biotechnology* (W. M. Fogarty, ed.), Applied Science, London, pp. 1–92.

Fogarty, W. M., and Griffin, P. J., 1975, Purification and properties of β-amylase produced by *Bacillus polymyxa, J. Appl. Chem. Biotechnol.* **25**:229–238.

Fogarty, W. M., and Kelly, C. T., 1983, Pectic enzymes, in *Microbial Enzymes and Biotechnology*, (W. M. Fogarty, ed.), Applied Science Publishers, London, pp. 131–182.

French, D., and Knapp, D. W., 1950, The maltase of *Clostridium acetobutylicum*. Its specificity range and mode of action. *J. Biol. Chem.* **187**:463–471.

Gallop, P. M., Seifter, S., and Meilman, E., 1957, Collagen. I. Partial purification, assay, and mode of activation of bacterial collagenase. *J. Biol. Chem.* **227**:891–906.

Gerwig, G. J., Waard, P. D., Kamerling, J. P., Vliegenthart, F. G., Morgenstern, E., Lamed, R., and Bayer, E. A., 1989, Novel O-linked carbohydrate chains in the cellulase-complex (cellulosome) of *Clostridium thermocellum, J. Biol. Chem.* **264**:1027–1035.

Giuliano, C., and Khan, A. W., 1985, Conversion of cellulose to sugars by resting cells of a mesophilic anaerobe, *Bacteroides succinogenes, Can. J. Microbiol.* **27**:517–530.

Grant, N. H. and Alburn, H. E., 1959, Collagenase of *Clostridium histolyticum, Arch. Biochem. Biophys.* **82**:245–255.

Harper, E., and Seifter, S., 1965, Mechanism of action of collagenase. Inhibition by cysteine, *Fed. Proc.* **24:**359.

Higashihara, M., and Okada, S., 1974, Studies on β-amylase of *Bacillus megaterium* strain No. 32, *Agric. Biol. Chem.* **38:**1023–1029.

Hobson, P. N., and MacPherson, M., 1952, Amylases of *Clostridium butyricum* and a *Streptococcus* isolated from the rumen of the sheep, *Biochem. J.* **52:**671–679.

Hockenhull, D. J. D., and Herbert, D., 1945, The amylase and maltase of *Clostridium acetobutylicum*, *Biochem. J.* **39:**102–106.

Hon-nami, K., Coughlan, M. P., Hon-nami, H., Carreira, L. H., and Ljungdahl, L. G., 1985, Properties of the cellulolytic enzyme system of *C. thermocellum, Biotechnol. Bioeng. Symp.*, **15:**191–205.

Hoshino, M., Hirose, Y., Sano, K., and Mitsuyi, K., 1975, Adsorption of microbial β-amylase on starch, *Agric. Biol. Chem.* **39:**2415–2416.

Hungate, R. E., 1944, Studies on cellulose fermentation, *J. Bacteriol.* **48:**499–513.

Hyun, H. H., and Zeikus, J. G., 1985a, General biochemical characterization of thermostable extracellular β-amylase from *Clostridium thermosulfurogenes, Appl. Environ. Microbiol.* **49:** 1162–1167.

Hyun, H. H., and Zeikus, J. G., 1985b, Regulation and genetic enhancement of β-amylase production in *Clostridium thermosulfurogenes, Appl. Environ. Microbiol.* **164:**1162–1170.

Hyun, H. H., and Zeikus, J. G., 1985c, General biochemical characterization of thermostable pullulanase and glucoamylase from *Clostridium thermohydrosulfuricum, Appl. Environ. Microbiol.* **49:**1168–1173.

Hyun, H. H., and Zeikus, J. G., 1985d, Regulation and genetic enhancement of glucoamylase and pullulanase production in *Clostridium thermohydrosulfuricum, J. Bacteriol.* **164:**1162–1170.

Hyun, H. H. and Zeikus, J. G., 1985e, Simultaneous and enhanced production of thermostable amylases and ethanol from starch by cocultures of *Clostridium thermosulfurogenes* and *Clostridium thermohydrosulfuricum, Appl. Environ. Microbiol.* **49:**1174–1181.

Hyun, H. H., Shen, J.-G., and Zeikus, J. G. 1985, Differential amylosaccharide metabolism of *Clostridium thermosulfurogenes* and *Clostridium thermohydrosulfuricum. J. Bacteriol.* **164:**1153–1161.

Jain, M. K., and Zeikus, J. G. (1988), Taxonomic distinction of two new protein specific, hydrolytic anaerobes: isolation and characterization of *Clostridium proteolyticum* sp. nov. and *Clostridium collagenovorans* sp. nov., *System. Appl. Microbiol.* **10:**134–141.

Johnson, E. A., Sakajoh, M., Halliwell, G., Madia, A., and Demain, A. L., 1982a, Saccharification of complex cellulosic substrates by the cellulase system from *C. thermocellum, Appl. Environ. Microbiol.* **43:**1125–1132.

Johnson, E. A., Reese, E. T., and Demain, A. L., 1982b, Inhibition of *C. thermocellum* cellulase by end products of cellulolysis, *J. Appl. Biochem.* **4:**64–71.

Johnson, E. A., Bouchot, F., and Demain, A. L., 1985, Regulation of cellulase formation in *C. thermocellum, J. Gen. Microbiol.* **131:**2302–2308.

Joliff, G., Beguin, P., Juy, M., Millet, J., Ryter, A., Poljak, R., and Aubert, J.-P., 1986a, Isolation, crystallization and properties of a new cellulase of *Clostridium thermocellum* overproduced in *Escherichia coli, Bio/Technology* **4:**896–900.

Joliff, G., Beguin, P., Millet, J., Aubert, J.-P., Alzari, P., Juy, M., and Poljak, R., 1986b, Crystallization and preliminary X-ray diffraction study of an endoglucanase from *Clostridium thermocellum, J. Mol. Biol.* **189:**249–250.

Kaji, A., Anabuki, Y., Taki, H., Oyama, Y., and Okada, T., 1963, Studies on the enzymes acting on arabans. Action and separation of arabanase produced by *Clostridium felsineum* (Eubacteriales). *Kagawa Daigaku Nogakubu Gabuzyuta Kokuku.* **15:**40–44.

Kaplan, F., Setlow, P. and Kaplan, N. O., 1969, Purification and properties of a DPNH-TPNH diaphorase from *Clostridium kluyverii, Arch. Biochem. Biophys.* **132:**91–98.

Katkocin, D., Word, N. S. and Wang, S. S., 1985, Thermostable glucoamylase and method for its production, U.S. Patent 4,536,477.

Kawazu, T., Nakanishi, Y., Uozumi, N., Sasaki, T., Yamagata, H., Tsukagoshi, N., and Udaka, S., 1987, Cloning and nucleotide sequence of the gene coding for enzymatically active fragments of the *Bacillus polymyxa* beta-amylase, *J. Bacteriol.* **169:**1564–1570.

Keinan, E., Hafeli, K. H., Seth, K. K., and Lamed, R., 1986a, Thermostable enzymes in organic synthesis II. Asymetric reduction of ketones with alcohol dehydrogenase from *Thermoanaerobium brockii, J. Am. Chem. Soc.* **108:**162–169.

Keinan, E., Kamal, K. K. and Lamed, R., 1986b, Organic synthesis with enzyme, 3. TBADH-catalyzed reduction of chloroketones. Total synthesis of (+)-(S,S)-cis-6-methyltetrahydro-pyran-2yl) acetic acid a civet constituent *J. Am. Chem. Soc.* **108:**3474–3480.

Keinan, E., Seth, K. K. and Lamed, R., 1987, Synthetic applications of alcohol dehydrogenase in *Thermoanaerobium brockii, 8th Symp. Enzyme Engineering*, Vol. 501, New York Academy of Sciences, pp. 130–149.

Kessler, E., and Yaron, A. (1976), An extracellular aminopeptidase from *Clostridium histolyticum, J. Biochem.* **63:**271–287.

Koch, R., Zablowski, P., and Antranikian, G., 1987, Highly active and thermostable amylases and pullulanases from various anaerobic thermophiles, *Appl. Microbiol. Biotechnol.* **27:** 192–198.

Kono, T., 1968, Purification and partial characterization of collagenolytic enzymes from *Clostridium histolyticum, Biochemistry* **7:**1106–1114.

Lamed, R., and Zeikus, J. G., 1980, Glucose fermentation pathway of *Thermoanaerobium brockii, J. Bacteriol.* **141:**1225.

Lamed, R., and Zeikus, J. G., 1981, Novel NADP-linked alcohol-aldehyde/ketone oxidoreductase in thermophilic ethanologenic bacteria, *Biochem. J.* **195:**183.

Lamed, R., and Bayer, E. A., 1986, Contact and cellulolysis in *Clostridium thermocellum* via extensive surface organelles, *Experientia* **42:**72–73.

Lamed, R., and Bayer, E. A., 1987, The cellulosome concept: exocellular/extracellular enzyme reactor centers for efficient binding and cellulolysis, in: *Biochemistry and Genetics of Cellulose Degradation* (J.-P. Aubert, P. Beguin, and J. Millet, eds.), Academic Press, London, pp. 101–116.

Lamed, R., and Bayer, E. A., 1988a, The cellulosome of *Clostridium thermocellum, Adv. Appl. Microbiol.* **33:**1–46.

Lamed, R., and Bayer, E. A., 1988b, Cellulose degradation by thermophilic anaerobic bacteria, in: *Biosynthesis and Biodegradation of Cellulose and Cellulose Materials* (P. J. Weimer and D. A. Haigler, eds.), Marcel Dekker, New York (in press).

Lamed, R., Keinan, E., and Zeikus, J. G., 1981, Potential applications of an aldose/ketose oxidoreductase from thermophilic bacteria, *Enzyme Microb. Tech.* **3:**144–148.

Lamed, R., Setter, E., Kenig, R., and Bayer, E. A., 1983a, The cellulosome - A discrete cell surface organelle of *Clostridium thermocellum* which exhibits separate antigenic, cellulose-binding and various cellulolytic activities, *Biotechnol. Bioeng. Symp.* **13:**163–181.

Lamed, R., Setter, E., and Bayer, E. A., 1983b, Characterization of a cellulose-binding, cellulase-containing complex in *Clostridium thermocellum, J. Bacteriol.* **156:**828–836.

Lamed, R., Kenig, R., Setter, E., and Bayer, E. A., 1985, The major characteristics of the celluloytic system of *Clostridium thermocellum* coincide with those of the purified cellulosome, *Enzyme Microb. Technol.* **7:**37–41.

Lamed, R., Naimark, J., Morgenstern, E., and Bayer, E. A., 1987, Specialized cell surface structures in cellulolytic bacteria, *J. Bacteriol.* **169:**3792–3800.

Lee, M., Miller, L. and Macmillan, J. D., 1970, Similarities in the action patterns of exopolygalacturonate lyase and pectinesterase from *Clostridium multifermentans, J. Bacteriol.* **103:**595–600.

Lee, S. F., Forsberg, C. W., and Gibbons, L. N., 1985, Cellulolytic activity of *Clostridium acetobutylicum, Appl. Environ. Microbiol.* **50:**220–228.

Lee, S. F., Forsberg, C. W., and Rattray, J. B., 1987, Purification and characterization of two endoxylanases from *Clostridium acetobutylicum* ATCC 824, *Appl. Environ. Microbiol.* **53**: 644–650.

Lee-Owen, V. and Anderson, J. C., 1975, Preparation of a bacterial collagenase containing negligible non-specific protease activity, *Prep. Biochem.* **5**:229–245.

Levdikova, G. A., Orekhovich, V. N., Soloveva, N. I. and Shpikiter, V. O., 1963, Dissociation of collagenase molecules into subunits, *Dokl. Akad. Nauk S.S.S.R.* **153**:725–727.

Ljundahl, L. G., and Wiegel, J. K. W., 1981, Anaerobic thermophilic culture system, U.S. Patent 4,292,406.

Ljungdahl, L. G., and Eriksson, K.-E., 1985, Ecology of microbial cellulose degradation, *Adv. Microbial.Ecol.* **8**:237–299.

Ljungdahl, L. G., Peterson, B., Eriksson, K.-E., and Wiegel, J., 1983, A yellow affinity substance involved in the cellulolytic system of *C. thermocellum, Current Microbiol.* **9**:195–200.

Lovitt, R. W., James, E. W., Kell, D. B., and Morris, J. G., 1987, Bioelectrochemical transformations catalyzed by *Clostridium sporogenes.* In: Bioreactors and Biotransformations (Moody, G. W., and Baker, P. B., eds.), Elsevier Appl. Sci. Publishers, London, pp. 263– 276.

Lovitt, R. W., Kell, D. B., and Morris, J. G., 1988a, Characterization of the carbon and electron flow pathways involved in the Stickland reaction of *C. sporogenes*, NC1B 8053. *J. Appl. Bacteriol.* (in press).

Lovitt, R. W., Kim, B.-H., Shen, G.-J., and Zeikus, J. G., 1988b, Solvent production by microorganisms, *Crit. Rev. Biotechnol.* **7**:107–186.

Lund, B. M., and Brocklehurst, T. F., 1978, Pectic enzymes of pigmented strains of *Clostridium, J. Gen. Microbiol.* **104**:59–66.

Lwebuga-Mukasa, J. S., Harper, E., and Taylor, P., 1976, *Clostridium:* Characterization of individual enzymes, *Biochemistry* **15**:4736–4741.

Macmillan, J. D., Phaff, H. J., and Vaughn, R. H., 1964, The pattern of action of an exopolygalacturonic acid trans-eliminase from *Clostridium multifermentans, Biochemistry* **3**: 572–578.

Madden, R. M., 1983, Isolation and characterization of *Clostridium stercorarium* sp. nov., cellulolytic thermophile, *Int. J. Systm. Bacteriol.* **32**:837–840.

Madi,E., Antranikian, G., Ohmiya, k. and Gottschalk, G., 1987, Thermostable amylolytic enzymes from a new *Clostridium* isolate, *Appl. Environ. Microbiol.* **53**:1661–1667.

Mandl, I., 1972, Collagenase comes of age, in *Collagenase* (I. Mandl, ed.), Gordon and Breach, New York, pp. 1–16.

Mandl, I., Keller, S., and Monahan, J., 1964, Multiplicity of *Clostridium hystolyticum* collagenases, *Biochemistry* **3**:1737–1741.

Melasniemi, H., 1987a, Effect of carbon source of production of thermostable α-amylase, pullulanase, and α-glucosidase by *Clostridium thermohydrosulfuricum, J. Gen. Microbiol.* **133**: 883–890.

Melasniemi, H., 1987b, Characterization of α-amylase and pullulanase activities of *Clostridium thermohydrosulfuricum, Biochem. J.* **246**:193–197.

Mercier, C., Frantz, B. M. and Whelan, W. J., 1972, An improved purification of cell-bound pullulanase from *Aerobacter aerogens, Eur. J. Biochem.* **26**:1–9.

Miller, L. and Macmillan, J. D., 1970, Mode of action of pectic enzymes. II. Further purification of exopolygalacturonate lyase and pectinesterase from *Clostridium multifermentans, J. Bacteriol.* **102**:72–78.

Millet, J., Petre, D., Beguin, P., Raynaud, O., and Aubert, J.-P., 1985, Cloning of ten distinct DNA fragments of *C. thermocellum* coding for cellulases, *FEMS Microbiol. Lett.* **29**:145– 149.

Mitchell, W. M., and Harrington, W. F., 1971, Clostripain, in: *The Enzymes*, Vol. 3 (P. D. Boyer, ed.), Academic Press, New York, pp. 699–719.

Mizokami, K., Kozaki, M., and Kitahara, K., 1978, Crystallization and properties of raw starch hydrolyzing enzyme produced by *Streptococcus bovis, J. Jpn. Soc. Starch Sci.* **25**:132–139.

Morgenstern, E., Bayer, E. A., Vliegenthart, J. F. G., Gerwig, G., Kamerling, H., and Lamed, R., 1987, Structure and possible function of glycosylated polypeptides in *C. thermocellum, Proc. FEMS Symp.,* 43:59.

Murao, S., Ohyama, K., and Arai, M., 1979, β-Amylases from *Bacillus polymyxa, No. 72, Agric. Biol. Chem.* **43**:719–726.

Nakamura, N., Watanabe, K. and Horikoshi, K., 1975, Purification and some properties of alkaline pullulanase from a strain by Bacillus no. 202-1, an alkaline microorganism, *Biochim. Biophys. Acta* **397**:188–193.

Ng, T. K. and Zeikus, J. G., 1981a, Comparison of extracellular cellulase activities of *Clostridium thermocellum* LQRI and *Trichoderma reesei* QM 9414, *Appl. Environ. Microbiol.* **42**:231–240.

Ng, T. K., and Zeikus, J. G., 1981b, Purification and characterization of an endoglucanase (1,4-β-D-glucan glucanohydrolase) from *C. thermocellum, Biochem. J.* **199**:341–350.

Ng, T. K., and Zeikus J. G., 1982, Differential metabolism of cellobiose and glucose by *C. thermocellum* and *Clostridium thermohydrosulfuricum, J. Bacteriol.* **150**:1391–1399.

Ng, T. K., and Zeikus, J. G., 1988, Endoglucanase from *Clostridium thermocellum, Methods Enzymol.* **160**:351–355.

Ng, T. K., Weimer, P. J., and Zeikus, J. G., 1977, Cellulolytic and physiological properties of *Clostridium thermocellum, Arch. Microbiol.* **114**:1–7.

Ng, T. K., Ben-Bassat, A., and Zeikus, J. G., 1981, Ethanol production by thermophilic bacteria: Fermentation of cellulosic substrates by co-cultures of *Clostridium thermocellum* and *Clostridium thermohydrosulfuricum, Appl. Environ. Microbiol.* **41**:1337–1343.

Norman, B. E., 1982, A novel debranching enzyme for application in the glucose syrup industry, *Starch/Starke* **34**:340–346.

Ohba, R., and Ueda, S., 1973, Purification, crystallization, and some properties of intracellular pullulanase from *Aerobacter aerogenes, Agric. Biol. Chem.* **37**:2821–2826.

Ohba, R. and Ueda, S., 1975, Some properties of crystalline extra- and intra-cellular pullulanase from *Aerobacter aerogenes, Agric. Biol. Chem.* **37**:2821–2826.

Oppenheim, F. and Franzblau, C., 1978, A modified procedure for the purification of Clostridial collagenase, *Prep. Biochem.* **8**:387–407.

Peterkosky, B., 1982, Bacterial collagenase, in: *Methods in Enzymology,* Vol. 82 (L. W. Cunningham and D. W. Frederiksen, ed.), Academic Press, New York, pp. 453–471.

Petre, J., Longin, R., and Millet, J., 1981, Purification and properties of an endo-β-1,4-glucanase from *C. thermocellum, Biochimie* **63**:629–639.

Petre, D., Millet, J., Longin, R., Beguin, P., Girard, H., and Aubert, J.-P., 1986, Purification and properties of the endoglucanase C of *Clostridium thermocellum* produced in *Escherichia coli, Biochimie* **68**:687–695.

Plant, A. R., Patel, B. K. C., Morgan, H. W., and Daniel, R. W., 1987a, Starch degradation by thermophilic anaerobic bacteria, *Syst. Appl. Microbiol.* **9**:158–162.

Plant, A. R., Clemens, R. M., Daniel, R. M., and Morgan, H. W., 1987b, Purification and preliminary characterization of an extracellular pullulanase from *Thermoanaerobium* Tok6-B1, *Appl. Microbiol. Biotechnol.* **26**:427–433.

Pugsley, A. P., Chapon, C., and Schwartz, M., 1986, Extracellular pullulanase of *Klebsiella pneumoniae* is a lipoprotein, *J. Bacteriol.* **166**:1083–1088.

Romaniec, M. P. M., Clarke, N. G., and Hazlewood, G. P., 1987, Molecular cloning of *Clostridium thermocellum* DNA and the expression of further novel endo β-glucanases genes in *Escherichia coli, J. Gen. Microbiol.* **133**:1297–1307.

Saha, B. C., Shen, G.-J., and Zeikus, J. G., 1987, Behaviour of a novel thermostable β-amylase on raw starch, *Enzyme Microb. Technol.* **9**:598–601.

Saha, B. C., Mathupala, S. P., and Zeikus, J. G., 1988, Purification and characterization of a highly thermostable, novel pullulanase from *Clostridium thermohydrosulfuricum, Biochem. J.* **247**:343–348.

Schink, B. and Zeikus, J. G., 1983, Characterization of pectinolytic enzymes of *Clostridium thermosulfurogenes, FEMS Microbiol. Lett.* **17**:295–298.

Schwarz, W. H., Schimming, K. S., and Staudenbauer, W. L., 1987, High-level expression of *Clostridium thermocellum* cellulase genes in *Escherichia coli, Appl. Microbiol. Biotechnol.* **27:**50–57.

Scott, D. and Hedrick, L. R., 1952, The amylase of *Clostridium acetobutylicum, J. Bacteriol.* **63:** 795–803.

Seifter, S. and Harper, E., 1970, Collagenases, in: *Methods of Enzymology* (G. E. Perlmann and L. Lorand, eds.), Academic Press, New York, pp. 613–635.

Seifter, S., and Gallop, P. M., 1962, Collagenase from *Clostridium histolyticum*, in: *Methods in Enzymology*, Vol. 5 (S. P. Colowick and N. O. Kaplan, eds.), Academic Press, New York, pp. 659–665.

Seifter, S. and Harper, E., 1971, The collagenases, in: *The Enzymes*, Vol. 3 (P. D. Boyer, ed.), Academic Press, New York, pp. 649–697.

Seifter, S., Gallop, P. M., Klein, L., and Meilman, E., 1959, Collagen. II. Properties of purified collagenase and its inhibition, *J. Biol.Chem.* **234:**285–293.

Sheiman, M. I., Macmillan, J. D., Miller, L., and Chase, T., 1976, Coordinated action of pectinesterase and polygalacturonate lyase complex of *Clostridium multifermentans, Eur. J. Biochem.* **64:**565–572.

Shen, G.-J., Saha, B. C., Lee, Y. Y., Bhatnagar, L., and Zeikus, J. G., 1988, Purification and characterization of a novel thermostable β-amylase from *Clostridium thermosulfurogenes, Biochem. J.* **254:**835–840.

Simon, H., Bader, J., Guntner, H., Neumann, S., and Thanos, J., 1985, Chiral compounds synthesized by biocatalytic reductions, *Agnew. Chem. Int. Ed. Engl.* **24:**539–553.

Sleat, R., Mah, R. A., and Robinson, R., 1984, Isolation and characterization of an anaerobic, cellulolytic bacterium, *Clostridium cellulovorans* sp. nov., *Appl. Environ. Microbiol.* **48:**88–93.

Strauch, L., and Grassmann, W., 1966, Purification of collagenase. III. Isolation of enzymically homogeneous collagenase, *Z. Physiol. Chem.* **344:**140–158.

Takasaki, Y., 1976, Purification and enzymatic properties of β-amylase and pullulanase from *Bacillus cereus* var. mycoides, *Agric.Biol. Chem.* **48:**1523–1530.

Tanaka, H., and Stadtman, T. C., 1979, Selenium-dependent clostridial glycine reductase: Purification and characterization of the two membrane-associated protein components, *J. Biol. Chem.* **254:**447–452.

Tanaka, T., Shimomura, Y., Himejima, M., Taniguchi, M., and Oi, S., 1986, Characterization of xylan-utilizing anaerobes from mesophilic and thermophilic methane sludge and their xylan degrading enzymes, *Agric. Biol. Chem.* **50:**2185–2192.

Tanaka, T., Shimomura, Y., Taniguchi, M., and Oi, S., 1987a, Raw starch-utilizing anaerobe from mesophilic methane sludge and its production of amylase, *Agric. Biol. Chem.* **51:**591–592.

Tanaka, T., Ishimoto, E., Shimomura, Y., Taniguchi, M., and Oi, S., 1987b, Purification and some properties of raw starch-binding amylase of *Clostridium butyricum* T-7 isolated from mesophilic methane sludge, *Agric. Biol. Chem.* **51:**309–405.

Taniguchi, H., Jae, C. M., Yoshigi, N., and Maruyama, Y., 1983, Purification of *Bacillus circulans* F2 amylase and its general properties. *Agric. Biol.Chem.* **47:**511–519.

Taya, M., Suzuki, Y., and Kobayashi, T., 1984, A thermophilic anaerobe (*Clostridium* species) utilizing various biomass-derived carbohydrates, *J. Ferment. Technol.* **62:**229–236.

Ueda, S., and Ohba, R., 1972, Purification, crystallization and some properties of extracellular pullulanase from *Aerobacter aerogenes, Agric. Biol. Chem.* **36:**2381–2391.

Ueda, S., and Saha, B. C., 1981, Raw starch digestion by microbial amylase, *Utiliz. Res.* **1:**9–17.

Viljoen, J. A., Fred, E. G., and Peterson, W. H., 1926, The fermentation of cellulose by thermophilic bacteria, *J. Agric. Sci.* **16:**1–17.

Walker, G. T., 1968, Metabolism of the reserve polysaccharide of *Streptococcus mitis:* some properties of a pullulanase, *Biochem. J.* **108:**33–40.

Wang, D. I. C., Bioic, I., Fang, H. -Y., and Wang, S. -D., 1979, Direct microbiological conversion of cellulosic biomass to ethanol, in: *Proceeding of the Third Annual Biomass Energy Systems Conference*, NTIS, Springfield, VA, p. 61.

Whelan, W. J., 1971, Enzymic explorations of the structures of starch and glycogen. The fourth CIBA medal lecture, *Biochem. J.* **122:**609–622.

Whelan, W. J., and Nasr, H., 1951, The amylase of *Clostridium butyricum, Biochem. J.* **48:**416–422.

Wu, J. H. D., and Demain, A. L., 1985, Purification of the extracellular cellulase system of *C. thermocellum, Am. Soc. Microbiol.*, Ann. Meet. Abstr., p. 248.

Wuensch, E., 1963, Synthesis of proline-peptides. II. collagenase substrates, *Z. Physiol. Chem.* **332:**295–299.

Wuensch, E., and Heidrich, H.-G., 1963, Determination of collagenase, *Z. Physiol. Chem.* **333:** 149–151.

Yagisawa, S., Morita, F., Nagai, Y., Noda, H., and Ogura, Y., 1965, Kinetic studies on the action of collagenase, *J. Biochem.* **58:**407–416.

Yoshida, E., and Noda, H., 1965, Isolation and characterization of collagenases I and II from *Clostridium histolyticum, Biochem. Biophys. Acta* **105:**562–574.

Zeikus, J. G., and Saha, B. C., 1989, U.S. Patent 4,814,267.

Toxigenic Clostridia   8

CLIFFORD C. SHONE and PETER HAMBLETON

1. INTRODUCTION: CLOSTRIDIA AND HUMAN DISEASE

The members of the genus *Clostridium* elaborate a wide range of exoproteins, many of which may function as virulence factors since many strains within the genus are pathogenic to both man and animals. Since the spores of clostridia are widely distributed in the environment, contaminating soil, dust, feces, insects, raw foodstuffs, and even cooked foods, it is not surprising that these organisms are frequently found in association with diseases arising primarily from contamination of wounds and foodstuffs.

For detailed accounts of the association of clostridia with diseases of man and animals, the reader is referred to other, more comprehensive reviews (MacLennan, 1962; Willis, 1969; Boriello, 1985a), but a summary of the more recognized associations of clostridia with diseases is given in Table I. The most important species in relation to human disease are *C. botulinum*, *C. tetani*, *C. perfringens*, *C. difficile*, and *C. septicum*; the significance of exotoxins produced by these species in the pathogenesis of disease has now been well characterized in a number of cases.

The neurotoxins produced by *C. botulinum* and *C. tetani* form the most potent group of microbial toxins known. The toxins act by inhibiting the release of neurotransmitters from presynaptic nerve terminals (Wellhoner, 1982; Shone, 1987). Botulinum toxin primarily affects peripheral cholinergic synapses inducing a flaccid paralysis whereas tetanus toxin (tetanospasmin) acts more specifically on the central nervous system blocking inhibitory mechanisms, thereby giving rise to the typical spastic paralysis. Although the symptoms induced by the toxins appear dramatically different, the toxins have similarities in their structures and modes of action (Sugiyama, 1980). Thus, both appear to exert their effects by interference with some intracellular process associated with calcium-mediated secretion of neurotransmitting substances.

CLIFFORD C. SHONE and PETER HAMBLETON ● Division of Biologics, Public Health Laboratory Service Centre for Applied Microbiology and Research, Porton Down, Salisbury, Wiltshire SP4 0JG, England.

Table I. Clostridial Diseases of Man and Animals

Species	Disease	Ref.
C. botulinum	Botulism: food-borne, wound, infant	Christie, 1980; Arnon, 1981; Lewis, 1981
C. tetani	Tetanus	Christie, 1980
C. perfringens	Gas gangrene	MacLennan, 1962
(*C. welchii*)	Food poisoning	Hobbs, 1979
	Enteritis necroticans	Hobbs, 1979; Walker, 1985
	Meningitis	
	Animal enterotoxemia	Boriello and Carman, 1985
C. difficile	Pseudomembranous enterocolitis	Boriello and Larson, 1985
C. butyricum	Gas gangrene	Pedersen *et al.*, 1976
	Neonatal necrotizing entero-colitis	Howard *et al.*, 1977; Phillips *et al.*, 1985
C. septicum	Gas gangrene	MacLennan, 1962; Phillips *et al.*, 1985
	Neutropenic enterocolitis	Editorial, 1987
	"Braxy" in sheep	Boriello and Carman, 1985
C. spiroforma	Animal enterotoxemias	Boriello and Carman, 1985
C. sordelli (*C. bifermentans*)	Gas gangrene	MacLennan, 1985; Phillips *et al.*, 1985
	Animal enterotoxemias	Boriello and Carman, 1985
C. novyi	Gas gangrene	MacLennan, 1962; Phillips *et al.*, 1985
	Necrotic hepatitis (sheep, cattle)	Boriello and Carman, 1985
C. histolyticum *C. terticum* *C. sporogenes* *C. fallax* *C. oedematiens*	Gas gangrene	MacLennan, 1962; Phillips *et al.*, 1985
C. chauvoei	Gas gangrene (animals)	MacLennan, 1962
C. colinum	Avian ulcerative colitis	Boriello and Carman, 1985

Fortunately, both botulism and tetanus are relatively rare in Western countries; the widespread use of immunization, coupled with high standards of hygiene, limits the incidence of tetanus, while the effectiveness of food preservation and handling procedures serves to control the incidence of food-borne botulism. Nevertheless, the potency of the toxins and the dramatic nature of the symptoms they induce serve to maintain botulism and tetanus among the most highly emotive of microbially induced diseases of man.

Whereas *C. botulinum* and *C. tetanus* produce single toxins with highly specific modes of action and, clearly defined pathological effects, strains of *C. perfringens* (*C. welchii*) produce a diversity of exoproteins having patho-

**Table II. Classification of *C. perfringens* Types
on Basis of Production of Major Lethal Antigens**

Type toxins	Disease associations
A α	Poisoning
B α, β, ε	Lamb dysentery, sheep enteritis
C α, β	Enteritis necroticans (man), "struck" in sheep, enteritis in piglets, sheep, and calves
D α, ε	Enterotoxemia in sheep, goats, cattle
E α, ι	Pathogenicity doubtful, enteritis in sheep, cattle

logical activity. Some of the exoproteins have been clearly identified as virulence factors whereas others, despite having identifiable biological activity, may have little or no significance in the pathogenesis of disease. *Clostridium perfringens* has been associated with a wide variety of pathological conditions (Table I) including gas gangrene and necrotic enteritis in man and various animal enterotoxemias. Pathogenicity is related to the production of one or more lethal toxins and the classification of *C. perfringens* into five types (A–E) is based on the nature of the major lethal toxins produced (Table II). Type A strains of *C. perfringens* may also produce a potent enterotoxin during sporulation that is a major cause of food-poisoning outbreaks (Hobbs, 1979; Stringer, 1985).

Clostridium difficile is not normally found in the gut flora of healthy persons but asymptomatic colonization of the intestinal tract may occur, particularly following antibiotic therapeutic regimes, which may substantially unbalance the normal intestinal flora. Pseudomembranous colitis (PMC) is the most extreme form of antibiotic-associated colitis and appears to result from colonization of the intestinal tract by *C. difficile*. The colonizing bacteria produce two potent toxins that cause PMC. One, type A, is an enterotoxin and the other, type B, has cytotoxic activity (Lyerly *et al.*, 1988); of these, it is the former which appears to be responsible for the pathogenic sequelae of PMC.

The involvement of other clostridial species in human disease would appear to be mainly with gangrene and toxemia arising from contamination of wounds, failure of asepsis in surgery (MacLennan, 1962; Willis, 1969), or enterocolitis (Editorial, 1987). Although many clostridial species have been associated with human histotoxic infections, *C. perfringens*, *C. novyi*, and *C. septicum* are probably the most important (MacLennan, 1962; Editorial, 1987). In all cases exotoxins have been implicated in the pathogenesis of disease, although the interactions between invading parasite and host would appear to be complex.

2. MODES OF ACTION OF TOXINS

2.1. *C. botulinum*

2.1.1. Botulinum Neurotoxins

Various types of *C. botulinum* produce seven antigenically different neurotoxins (types A–G), which act primarily on the neuromuscular junction blocking the release of the neurotransmitter acetylcholine (Sugiyama, 1980; Simpson, 1981; Shone, 1987). Action of any of these toxins on the nervous system results in widespread muscular weakness, impaired vision, and, ultimately, death due to respiratory failure. These symptoms are characteristic of the syndrome botulism, a rare but frequently fatal condition which affects both man and animals. By far the most common cause of botulism in humans is the ingestion of preformed toxin in contaminated foodstuffs. So potent are the botulinum toxins as neuroparalytic agents that ingestion of as little as 1–2 μg of toxin may prove fatal. Fortunately, the effectiveness of modern food-preserving processes makes outbreaks of botulism uncommon: in the 29-year period 1950–1979, 215 outbreaks of botulism were recorded in the United States (Feldman *et al.*, 1981). Cases of food-borne botulism in humans appear to be confined to four toxin types (types A, B, E, and F) of the seven so far identified. Of these, the vast majority of outbreaks are caused by types A, B, and E neurotoxins, with type F being responsible for only a handful of cases.

Other types of human botulism are caused by direct infection with the bacterium. Wound botulism is an extremely rare form of the disease in which *C. botulinum* infects and produces toxin at the site of a wound. Infant botulism, so called because only infants up to 35 weeks old are susceptible, appears to be caused by colonization of the intestinal tract by *C. botulinum* (Sugiyama, 1980; Arnon, 1981). Toxin produced by the organism in the gut gives rise to muscular paralysis and may be a cause of sudden infant death. With increasing awareness of the disease, more and more cases are now being reported. Since it was first diagnosed in 1976 (Midura and Arnon, 1976), several hundred cases have been reported in the United States.

The botulinum toxins are secreted from the bacterium in the form of protein complexes. Toxin complexes from *C. botulinum* types C, E, F, and hemagglutinin-negative strains of type D generally consist of the neurotoxin moiety (approx. 150,000 mol. wt.) in association with one other similar-sized nontoxic protein. These complexes vary in molecular weight between 230,000 and 350,000 depending on the toxin type. Toxin complexes from *C. botulinum* type A, B, and hemagglutanin-positive strains of type D are larger (450,000–500,000 mol. wt.) and contain a third protein constituent which has hemagglutinin activity. The botulinum toxin complexes are considerably more toxic than the purified neurotoxins when assessed by the oral route (Ohishi, 1984). The nontoxic proteins associated with the neurotoxin appear to protect the latter from the hostile environment of the gut, thus enhancing its absorption into the lymph and blood systems (Sugii *et al.*, 1977). In contrast, when toxicity is assessed by the parenteral route, the specific toxicities of the purified 150,000 mol. wt. neurotoxins are greater than their respective complexes.

With the exception of type G toxin, all the botulinum neurotoxins have been purified to near or complete homogeneity. The toxins are similar in molecular weight (140,000–170,000) and in their most active forms comprise two asymmetrical subunits (Fig. 1); a heavy chain (85,000–105,000 mol. wt.) and a light chain (50,000–59,000 mol. wt.) linked together by at least one disulfide bridge (DasGupta and Sugiyama, 1972). The neurotoxins are synthesized as single polypeptides which, depending on the botulinum type, may be cleaved subsequently by proteases to produce the dichain form of the neurotoxin. Of the three neurotoxin types most intensively studied (types A, B, and E), the light chain appears to constitute the amino terminus end of the polypeptide chain. Fragments of the *C. botulinum* type A neurotoxin gene were recently cloned into *E. coli* and the complete amino acid sequence of the neurotoxin is now known (N. P. Minton, personal communication). Not all neurotoxins are secreted from the bacterium in their most active forms. Type E neurotoxin is secreted in its single-chain form and that of type B is secreted as a mixture of single and dichain toxin forms (DasGupta, 1981). Treatment of the single-chain type E neurotoxin by trypsin converts the toxin to the dichain form and increases the specific toxicity by a factor of 80- to 100-fold. There is a considerable amount of evidence that the activation of the neurotoxins, whether it occurs *in vivo* or *in vitro*, is not simply a cleavage of the polypeptide chain at a single site to give the dichain neurotoxin form, but a more complex process involving several sites of proteolysis (Sugiyama, 1980; Shone, 1987).

The botulinum toxins are extremely potent neuroparalytic agents with specific toxicities ranging from 2×10^7 to 2×10^8 mouse 50% lethal doses per mg of protein, which act presynaptically by blocking the release of the neurotransmitter acetylcholine at the neuromuscular junction. At least

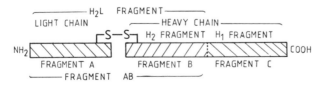

Figure 1. Overall structure of the clostridial neurotoxins showing currently used notation for the fragments of the botulinum neurotoxins (above the diagram) and for tetanus toxin (below the diagram).

three stages are involved in the toxins' inhibitory action: (1) an initial step in which the toxin binds to acceptor molecules present on the presynaptic nerve surface; (2) an internalization stage in which the toxin crosses the plasmalemma and enters the nerve terminus; and (3) one or more steps in which the acetylcholine release mechanism is disabled (Simpson, 1981).

The acceptor molecules to which the botulinum neurotoxins bind on the presynaptic nerve surface have not been characterized. The matter is complicated by the fact that the neurotoxins do not all bind to the same acceptor molecules. Using rat brain synaptosomes as a nerve model, type A neurotoxin does not compete with type B toxin for acceptors on the synaptosomal membrane surface. Type A and E neurotoxins appear to share the same acceptor molecules (Kozaki, 1979), as do type C_1 and D toxins (Murayama et al., 1984), but the relationship between the type C_1 and D neurotoxin acceptors and those of type A, B, and E toxins has yet to be established. Neurotoxins A, B, C_1, and D all appear to bind to more than one type of acceptor on rat brain synaptosomal membranes (Williams et al., 1983; Agui et al., 1985; Evans et al., 1987). In each case high-affinity binding (dissociation constant 0.08–0.6 nM) to a small pool of acceptors was observed. Neurotoxins A, B, and E appear to bind to these acceptors by an active (binding) site region located on the COOH terminal half of the heavy subunit (Shone et al., 1985; Kozaki et al., 1986, 1987). Treatment of rat brain synaptosomal membranes with either neuraminidase or proteases reduced the binding of neurotoxins A and B, which suggests that sialic acid residues and proteins may be components of the neurotoxin acceptors. The involvement of sialic acid residues in the acceptor molecules for several of the neurotoxins has prompted suggestions that a ganglioside or ganglioside-like molecule may act as the neurotoxin acceptors. These speculations have been supported by the finding that the toxicity of several of the botulinum neurotoxins is reduced by incubation with trisialogangliosides.

Most studies on the botulinum acceptor have been performed on either rat or mouse brain synaptosomes, and it is not known whether or not acceptor molecules on these tissues bear any resemblance to those of the peripheral nervous system. The precise nature of the botulinum neurotoxin acceptors thus remains unclear.

Once bound to its acceptor on the presynaptic nerve surface, the neurotoxin is internalized into the nerve ending by an energy-dependent process. Studies on type A and B neurotoxins using mouse hemidiaphragm preparations showed the internalization process to reach a maximum after approximately 90 min at 22°C (Dolly et al., 1984; Black and Dolly, 1986). This process most probably resembles that of receptor-mediated endocytosis in which molecules are encapsulated in clathrin-coated endosomes whose internal environment becomes acidic as they progress through the cytosol. Several studies have shown that the heavy chain of the neurotoxins is capable of forming channels in lipid membranes at low pH which may be relevant to the internalization process (Hoch *et al.*, 1985; Shone *et al.*, 1987). The fate of the toxin once internalized is unknown. It is presently not clear if the toxin remains within a membrane or whether or not the whole toxin or a fragment is delivered to the cytosol.

After internalization, the action of the neurotoxin is completed by one or more steps in which the mechanism of transmitter release is inhibited. Under normal conditions, an impulse arriving at the nerve ending triggers the influx of calcium ions, which in turn promotes the exocytosis of acetylcholine from vesicles at "active zones" on the plasmalemma (Fig. 2). The botulinum toxins do not appear to affect the nerve impulse-mediated influx of calcium; rather they appear to act on an event which occurs after calcium influx. Several possible mechanisms have been proposed: (1) the neurotoxins may act directly on a component of the acetylcholine release mechanism (Simpson, 1981; Knight, 1986), possibly a vesicle protein or a component of the cytoskeleton; (2) the toxins may act to stimulate the rate of calcium disposal so that the calcium level never reaches the required level for transmitter release to occur (Molgo and Thesleff, 1984); and (3) neurotransmitter release may require the entry of calcium ions into an internal milieu. It is possible that the toxin blocks this latter process (Sanchez-Prieto *et al.*, 1987).

There are presently insufficient data to favor any one of the above mechanisms. There is also growing evidence that not all the botulinum neurotoxins block transmitter release by a common mechanism. Thus the

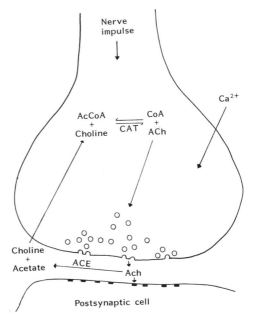

Figure 2. Diagrammatic representation of the neuromuscular junction showing selected events. Acetylcholine (ACh) is synthesized from choline and acetyl coenzyme A by the enzyme choline acetyltransferase (CAT) and loaded into synaptic vesicles. The arrival of a nerve impulse at the synapse evokes the influx of calcium ions into the nerve terminus which triggers the transmitter release mechanism resulting in the fusion of ACh-filled vesicles with the plasmalemma. The released ACh crosses the synaptic cleft and interacts with receptors (■) on the muscle end-plate cell (postsynaptic cell) giving rise to an end-plate potential. ACh remaining in the synaptic cleft is rapidly hydrolyzed by the enzyme acetylcholinesterase (ACE).

mechanism of action of the botulinum neurotoxins at the molecular level is still a complete mystery. In view of the very high specific toxicity of the neurotoxins and the longevity of their paralytic effects, it seems likely that the botulinum toxins exert their toxicity by enzymic activity (Simpson, 1981; Shone, 1987). Recent reports that types C_1 and D neurotoxins possess ADP-ribosylating activity (Matsuoka *et al.*, 1987; Ohashi and Narumiya, 1987) appear to be unfounded (Rosener *et al.*, 1987) and as yet no such enzymic activity has been demonstrated in any of the botulinum neurotoxins.

2.1.2. Botulinum Type C_2 Enterotoxin

Although not involved in human disease, botulinum type C_2 toxin is of interest since its mechanism of action was recently elucidated (Simpson, 1984; Aktories *et al.*, 1986; Vandekerckhove *et al.*, 1988). Type C_2 toxin is a minor toxin component produced by several strains of both *C. botulinum* type C and D. Structurally the toxin is very similar to the botulinum neurotoxins, with the exception that the interchain disulfide bridge is missing. Botulinum type C_2 toxin has no neurotoxic activity and acts on a broader range of tissues, increasing vascular permeability. Symptoms of C_2 toxin poisoning are hypotension, hemorrhaging, and collection of fluid in the lungs. Both subunits are required for toxicity but these do not have to be associated. It is now known that the toxin exerts its toxic effect by ADP-ribosylating a variety of protein substrates, including actin. This enzymic activity is contained within the light subunit of the toxin.

2.2. Tetanus Neurotoxin

Spasm of voluntary muscles with convulsions commencing with trismus are typical symptoms associated with *C. tetani* infection (Bizzini, 1979; Wellhoner, 1982; Bleck, 1986). The incidence of the disease has been significantly lowered by vaccination, but still each year several hundred thousand cases occur worldwide. The infection usually occurs at the site of a wound where, under anaerobic conditions, the organism grows producing the neurotoxin. Tetanus neurotoxin enters the peripheral nervous system and is transported intra-axonally to its site of action, the spinal cord. Here the toxin only affects the polysynaptic reflexes in which regulatory interneurons (Renshaw cells) are involved. Under normal conditions, the Renshaw cell interneurons moderate the amplitude of the action potential in the motoneuron by releasing the neurotransmitter glycine. Tetanus neurotoxin acts by blocking the release of glycine from the Renshaw cells, thus eliminating their moderating effect on the synapse. With the inhibitory effect of the Renshaw cells removed, the polysynaptic reflex activity is greatly enhanced, ultimately resulting in the characteristic muscular spasm.

Many similarities exist between tetanus toxin and the botulinum neurotoxins, both in their molecular structures and in their modes of action. Structurally, tetanus toxin has an architecture comparable to the botulinum toxins, with similar molecular size and subunit structure. The toxin is synthesized as a single polypeptide chain (150,700 mol. wt.), which is subsequently cleaved, as it is secreted from the bacterium, to yield the dichain toxin form consisting of two asymmetrical subunits (98,300 and 52,288 mol. wt.) linked by a disulfide bond (Fig. 1). The tetanus toxin gene has been cloned in *E. coli* and the complete amino acid sequence of the neurotoxin is now known (Eisel *et al.*, 1986; Fairweather and Lyness, 1986). Comparison of the amino acid sequence of tetanus toxin with that of botulinum type A reveals a considerable degree of homology (approximately 30%), which reflects their overall structural similarities and may also be an indication of similarities in their modes of action.

The manner in which tetanus toxin inhibits neurotransmitter release is also similar to that observed in the botulinum toxins. At least three steps are involved: binding, translocation, and paralysis (Dreyer and John, 1981). Gangliosides have been implicated as the acceptor component for tetanus toxin on the presynaptic nerve surface. At pH 6.0, disialo- and trisialogangliosides (G_{Dlb} and G_{Tlb}, respectively) bind to tetanus neurotoxin with high affinity (dissociation constant 6–10 nM; Rogers and Snyder, 1981) and neutralize its neurotoxic action. The active site region on the neurotoxin involved in the interaction with gangliosides appears to be located on a 47,000 mol. wt. fragment (fragment C) from the COOH terminal region of the heavy chain, since the purified fragment retains the high-affinity binding to the gangliosides (Morris *et al.*, 1980).

Recent data, however, cast doubt on the role of gangliosides as the primary acceptor for the neurotoxin. At pH 7.4, the binding of tetanus toxin to rat brain synaptosomes is very different from that observed at pH 6.0 (Pierce *et al.*, 1986). At the higher pH, at least two pools of acceptor molecules are evident, which include a small pool of high-affinity sites sensitive to proteases. Further studies are evidently required to determine the exact nature of the tetanus toxin acceptor.

As is the case for botulinum toxin, the molecular details of the mechanism by which tetanus toxin blocks the release of the neurotransmitter glycine are not known. Tetanus toxin appears to act on an essential process controlling exocytosis after the entry of calcium ions. The active site region responsible for the paralytic step appears to be located on a fragment of the molecule consisting of the light subunit linked to the NH_2 terminal half of the heavy chain since this fragment, when intracellularly injected to chromaffin cells, effectively blocks exocytosis (Penner *et al.*, 1986). There is now a considerable amount of evidence to suggest that both tetanus and botulinum neurotoxins are effective in blocking the release of a broad range of neurotransmitters from a number of different tissue types, and that their internal site(s) of action is on some fundamental process(es) common to all calcium-mediated secretory mechanisms. Double-poisoning experiments on nerve-muscle preparations using botulinum and tetanus toxins suggest a common site of action for tetanus toxin and botulinum type B neurotoxin, both of which are distinct from that of botulinum type A toxin (Gansel et al., 1987).

It now seems certain that once the complex mechanisms of action of these clostridial neurotoxins have been deciphered, a better understanding of the calcium-mediated release process will be obtained.

2.3. *C. perfringens* (*C. welchii*)

Since the end of the last century it has been known that certain pathological conditions of man and animals were associated with *C. perfringens* infection. During the 1914–1918 war *C. welchii* was recognized as being the most important causal organism of anaerobic myonecrosis (gas gangrene) in man. At around the same period, the organism was also associated with outbreaks of mild diarrhea and cramps (see Hobbs, 1979) that resemble the *C. perfringens* food poisoning recognized today.

Strains of *C. perfringens* are classified into five groups (types A–E) on the basis of their production of lethal toxins (Sterne and Warrack, 1964; Willis, 1969; Lee and Cherniak, 1974) and the relationship between strain type and major lethal antigen(s) (α, β, ϵ, and ι) is shown in Table II. In addition to these major lethal antigens, a number of other exoproteins are produced that have defined biological activities and which may be involved in the pathogenesis of disease (Table III). The pathological conditions caused by *C. perfringens* are varied but can be broadly classified into necrotic-type diseases and enterotoxemias. While the association of *C. perfringens* with these diseases is clear, in many instances the precise involvement of the various lethal and other toxins is not fully understood.

Of the five types of *C. perfringens*, only A and C appear to be significant causes of disease in man (Table III). Type A strains cause gas gangrene (Willis, 1969; Ispolatovskaya, 1971) as well as necrotic enteritis and mild diarrhea (Boriello and Carman, 1985). Although type A strains produce the α toxin as their major lethal antigen, the pathogenesis of gangrene may involve another toxic component (MacLennan, 1962; Willis, 1969; Ispolatovskaya, 1971). Indeed, the development of *C. perfringens* gas gangrene may be determined more by the susceptibility of the host tissue to colonization than by microbial virulence factors (Willis, 1969).

The most common enteric condition caused by Type A strains of *C. perfringens* is a relatively mild diarrhea with abdominal cramps caused by a potent exterotoxin produced in the bowel by sporulating organisms rather than by ingestion of preformed toxin (see reviews by Hobbs, 1979 and Stringer, 1985).

Table III. Exoproteins Produced by *C. perfringens*[a]

Toxin types	Biological activity	Production by *C. perfringens*				
		A	B	C	D	E
α	Lethal, necrotizing phospholipase C	++[b]	+	+	+	+
β	Lethal, necrotizing	−	++	++	−	−
ε	Trypsin activated, lethal, necrotizing	−	++	−	++	−
ι	Trypsin activated, lethal, necrotizing	−	−	−	−	++
γ	Lethal	+	+	−	−	
δ	Hemolysin	−	+	+	−	−
η	Lethal(?)	+	−	−	−	−
θ	Hemolysin	+	++	++	++	++
κ	Collagenase	++	+	+	+	+
λ	Proteinase	−	++	−	+	+
μ	Hyaluronidase	+	+	+	+	+
ν	Deoxyribonuclease	+	+	+	+	+

[a]Derived from Hauschild (1971), Hobbs (1979), Boriello and Carman (1985).
[b]++ Most strains. + Some strains. − Not produced.

Type C strains *C. perfringens* are responsible for particularly severe forms of food poisoning that manifest as a necrotizing hemorrhagic jejunitis (enteritis necroticans). A condition called Darmbrand described in Germany in 1947 was associated with consumption of type C-contaminated canned meat (Hobbs, 1979; Walker, 1985). The symptoms were severe cramps, vomiting, nausea, severe diarrhea, and necrotic inflammation of the small intestine with a high mortality rate. A similar condition—"pig-bel"—is a major cause of death in young adults in Papua New Guinea and results from eating contaminated pig meat in ritual feats (Walker, 1985).

The involvement of *C. perfringens* in diseases of animals is well described by Boriello and Carman (1985). Type A enterotoxemia, caused by α-toxigenic strains, appears to be the most common animal disease, with some necrotic enteritis. Type B-associated enterotoxemias appear to be restricted geographically to Europe, South Africa, and the Middle East. Type C organisms cause a disease—"struck"—in weaned sheep, primarily in the Romney Marsh area of England, particularly in late winter and early spring; its pathogenicity resembles that of pig-bel in humans (Walker, 1985). Type D enterotoxaemia is important in sheep. Type E strains are not generally regarded as highly pathogenic but type E enterotoxemia in cattle has been described (see Boriello and Carman, 1985).

2.3.1. Lethal and Necrotizing Toxins of *C. perfringens*

In addition to the four major toxins used to biotype *C. perfringens*, a further eight soluble proteins have been identified with biological activities that may justify their being considered as toxins of some significance to the pathogenesis of disease. The characterization of these various toxins is incomplete, but the main features in our current understanding of their activities, structures, and modes of action are summarized below.

2.3.1a. α Toxin. This is recognized as being a phospholipase C having a molecular weight of about 30,000 and an isoelectric point of 5.7 (Mollby and Wadstrom, 1973). Although originally considered to be a major virulence factor (Evans, 1945), the toxin itself does not induce the observed local and systemic charges of clostridial myonecrosis (Ispolatovskaya, 1971). The absence of polymorphonuclear leucocytes (PMNL) from areas of necrosis suggested that the toxin might act as a leucocytolysin (Ispolatovskaya, 1971), but although purified toxin has been shown to induce perturbation of the plasma membrane of PMNL, such changes appear not to radically affect PMNL function (Stevens *et al.*, 1987).

2.3.1b. β Toxin. Early studies identified β toxin as being lethal and necrotizing, capable of inducing changes in blood pressure in animals and morphological changes in guinea pigs (Hauschild, 1971). More specific studies on *C. perfringens* type C-induced enteritis necroticans suggest that the toxin causes necrotic changes in intestinal villi (Walker, 1985) and may affect villous mobility (Parnas, 1976). The toxin has been characterized as a single chain polypeptide of about 30,000 mol. wt. (Sakurai and Duncan, 1978), which is heat labile and trypsin-sensitive. It is also inactivated by oxidizing agents and contains thiol groups essential for biological activity (Sakurai *et al.*, 1980).

2.3.1c. ε Toxin. This is produced as an inactive prototoxin with a molecular weight of about 33,000 (Habeeb, 1975; Worthington and Mulders, 1977). The toxin is readily activated by proteases which cleave off a basic fragment of about 1500 mol. wt., causing a shift in pI from 8.02 into fractions of pI 5.36 and 5.74, respectively. The active toxin is lethal to mice but its mode of action remains obscure. It has been established, however, that lethal activity is associated with the single tryptophan residue of the polypeptide (Sakurai and Nagahama, 1985).

2.3.1d. i Toxin. Initially identified as a lethal, necrotizing factor produced by *C. perfringens* type E, little was known of this toxin other than that it was activated by trypsin (Hauschild, 1971). It is now recognized (Stiles and Wilkins, 1986; Simpson *et al.*, 1987) as comprising two separate polypeptide chains that act synergistically to cause death in mice. The so-called light chain is an enzyme that ADP-ribosylates certain amino acids. The heavy chain does not affect the enzyme activity of the light chain but may act as a binding component, which directs the enzyme to susceptible cells. The intracellular site of action of the toxin is not known but it has been suggested that ι toxin and *C. botulinum* type C_2 toxin may act similarly (Stiles and Wilkins, 1986; Simpson *et al.*, 1987).

2.3.1e. θ Toxin. Purified θ-hemolysin preparations appear to be heterogeneous, consisting of four components (Smyth, 1975; Stevens *et al.*, 1987) with molecular weights of 59,000–62,000, and pI values of 6.1, 6.3, 6.5, and 6.8, respectively. The nature of the observed heterogeneity is unclear; it may be artefactual or reflect degrees of partial toxoiding; the latter might explain the observed range of hemolytic activities of the four components (Smyth, 1975). Purified θ toxin possesses potent leukocytolytic activity (Stevens *et al.*, 1987) with effects on PMNL migration, chemotaxis, and chemiluminescence being described, together with induction of morphological changes.

2.3.1f. δ Toxin. This toxin was characterized in crude culture filtrates by an apparently restricted spectrum of hemolytic activity, limited to even-toed ungulates (Sterne and Warrack, 1984). Little was known of its structure until Alouf and Jolivet-Reynaud (1981) succeeded in purifying the protein to homogeneity. The toxin is a single-chain basic polypeptide of 42,000 mol. wt. It appears to act in a multistep process in which binding to G_{M2} ganglioside, or a similar receptor molecule, precedes membrane damage and subsequent cytolysis (Jolivet-Reynaud *et al.*, 1982). Erythrocytes from different species demonstrates different capabilities to bind, and hence different sensitivities, to the δ toxin. The toxin also appears to be selectively toxic for certain rabbit leukocyte populations (Jolivet-Reynaud *et al.*, 1982).

2.3.1g. Other Toxins. Characterization of the other identified toxic components of *C. perfringens* remains incomplete. Several have been identified as having enzymic activity. The κ toxin is a collagenase that causes muscle disintegration, the λ toxin is a nonspecific proteinase that does not attack collagen, and the μ toxin is a hyaluronidase that appears to function as a spreading factor rather than causing direct pathological effects (Hauschild, 1971). The nature of the γ and η toxins apparently remains obscure; the ν toxin is a deoxyribonuclease (Sterne and Warrack, 1964; Hobbs, 1979).

2.3.2. *C. perfringens* Type A Enterotoxin

Food poisoning due to *C. perfringens* type A is a major food-borne illness in the United Kingdom (PHLS, 1986) and the United States (Shandera *et al.*, 1983). The disease is characterized by severe diarrhea and lower abdominal cramp pains with onset 8–24 hr after consuming contaminated food. The duration is generally short and symptoms usually abate within 24 hr. It is rarely fatal, except among elderly and/or debilitated persons. The causative factor, the enterotoxin, is produced only by sporulating organisms (Duncan *et al.*, 1972). Although sporulation may take place in contaminated food, the illness generally results from ingestion of food heavily contaminated with vegetative organisms. These sporulate readily in the intestine, resulting in production of high levels of toxin, which causes massive outpouring of fluid into the bowel lumen, resulting in diarrhea in a manner analogous to the toxins of *Vibrio cholera* and *Bacillus cereus*.

A toxin preparation of high purity has been obtained (Granum and Whitaker, 1980; Reynolds *et al.*, 1986). It is a polypeptide of about 34,000 mol. wt. and a pI of 4.3, stable within the pH range 5–12 but relatively heat-labile, though inactivated rapidly at 60°C (Reynolds, 1987); the complete amino acid sequence has been determined (Richardson and Granum, 1985). The mode of action of the toxin remains obscure but biological activity seems to involve a multistep process (McClane and McDonel, 1981) in which toxin initially binds to susceptible epithelial cells, particularly in the tips of intestinal villi. In subsequent steps, which may be calcium-dependent (McClane and McDonel, 1981; Matsuda *et al.*, 1986), host cell permeability is affected, creating an ion imbalance that results in a net influx of ions and water with the formation of membrane "blebs" and eventual cell death.

The precise biological activity of the toxin is still unclear, but Granum (1982) showed that a 16,000 mol. wt. fragment was capable of inhibiting protein synthesis in a cell-free system; a 15,000 mol. wt. fragment isolated by Horiguchi *et al.*, (1987) may be associated with the binding of toxin to susceptible cells.

2.4. *C. difficile* Toxins

Pseudomembranous colitis (PMC) is a frequently fatal diarrheal disease characterized by severe inflammation of the colonic mucosa with the appearance of yellow plaques or pseudomembranes composed of mucus, fibrin, and various inflammatory cells. Although PMC has been recognized since 1893, it was not until the 1950s and 1960s that the incidence of the disease increased dramatically as a complication of antibiotic therapy (Bartlett, 1979, 1983; Bolton and Thomas, 1986; McFarland and Stamm, 1986;

Lyerly *et al.*, 1988). Patients treated with broad-spectrum antibiotics such as clindamycin and lincomycin, which are effective against anaerobes, seemed particularly susceptible to the development of PMC. Antibiotic-associated PMC was first attributed to *Staphylococcus aureus* infection, but it is now known that infection by *C. difficile* is the prime cause. Most outbreaks of PMC occur in hospitals among elderly patients undergoing antibiotic treatment. Antibiotics appear to disrupt the gut flora allowing infection by spores of *C. difficile,* which then colonize the intestinal tract. Pseudomembranous colitis does not arise directly from the presence of the bacterium in the gut but from the effects of two exotoxins produced by the organism: an enterotoxin (toxin A) and a potent cytotoxin (toxin B).

Structurally toxins A and B appear to be similar. Polyacrylamide gel electrophoresis of the toxins under denaturing conditions suggests that both toxins consist of a single high molecular weight polypeptide chain (Sullivan *et al.*, 1982). The molecular weight of each toxin has yet to be determined accurately, and reported values vary considerably, but it is now generally agreed that the subunit size of each toxin is between 250,000 and 300,000 mol. wt. Under native conditions each toxin appears to exist as a high molecular weight complex between 350,000 and 500,000, presumably composed of subunit dimers. The type A toxin gene has recently cloned and partially sequenced. The gene coding for type A toxin was found to be over 6 kilobases, which is in good agreement with its large subunit size (Lyerly *et al.*, 1988).

Type A toxin is a powerful enterotoxin which attacks the brush border membranes of the gut mucosa causing extensive tissue damage and fluid loss. The toxin is lethal to mice with a specific toxicity of 2×10^5 mouse lethal doses per mg of protein (intraperitoneal route) and has also been reported to have a slight cytotoxic activity. The enterotoxic activity of toxin A appears to be mediated by acceptors present on the brush border membranes. Studies on hamster brush border membranes showed the toxin binding to be temperature-dependent; strangely, considerably more toxin was found to bind at 4°C than at 37°C (Krivan *et al.*, 1986). Similar acceptors for type A toxin have been identified on erythrocyte membranes and studies on these have shown them to contain the trisaccharide Gal α 1-3Gal β 1-4GlcNAc (Krivan *et al.*, 1986). At high toxin concentrations, toxin A has hemagglutinin activity and this activity is retained by a toxin fragment expressed by a 2.1-kilobase gene fragment. The latter fragment, while showing no toxicity, appears to contain the acceptor-binding region of the intact toxin. The amino acid sequence of the fragment shows it to contain several repeated sequences of a block of about 20 amino acids, and it has been suggested that toxin may be capable of binding to more than one acceptor molecule (Lyerly *et al.*, 1988). If this is the case, then acceptor density on the brush border membrane may be an important factor in toxin binding and could explain the temperature-sensitive nature of the binding process. As has been found for so many bacterial toxins, the toxin fragment involved in acceptor binding is located at the COOH terminus of the molecule. The sequence of events by which type A toxin exerts its potent tissue-damaging enterotoxic activity, once it has bound its acceptor, has yet to be determined. It is possible that the toxin acts directly on the outer brush border cell membranes, but in view of findings with other bacterial toxins, a mechanism in which the toxin is internalized to act enzymically on an internal target molecule seems more likely.

Less is known about the cytotoxic *C. difficile* type B toxin. Toxin B is active against a wide variety of animal cells in culture at concentrations of less than 1 pg/ml (Banno *et al.*, 1984) and its toxicity in mice (approx. 2 × 10^5 mouse lethal doses per mg, intraperitoneal route) is approximately the same as that observed for toxin A (Sullivan *et al.*, 1982). The mechanism of action of type B toxin is presently undetermined: the toxin has been compared to diphtheria toxin, with reports of ADP-ribosylating activity (Florin and Thelestam, 1986), and appears to have some effect on protein synthesis, but it is not clear whether or not this is the primary action of the cytotoxin since it appears to affect a number of cellular mechanisms. Although most of the symptoms of PMC can be attributed to type A toxin, it seems likely that type B toxin also plays a role since hamsters need to be vaccinated with both toxins in order to be fully protected from the disease (Fernie *et al.*, 1983).

2.5. *C. sordelli* (*C. bifermentans*)

This organism is associated with gas gangrene in man (MacLennan, 1962) and enteritis and entertoxemia in animals (Boriello and Carman, 1985). It elaborates a variety of putative virulence factors including a lethal or edema-producing toxin (LT), hemorrhagic toxin (HT), phospholipase, protease, hemolysins, DNase (Arseculeratne *et al.*, 1969), and cytotoxin (Nakamura *et al.*, 1983). The neutralization of *C. difficile* cytotoxin activity by *C. sordelli* antitoxins (Chang *et al.*, 1978) suggested a serological commonality between toxic component(s) of the two species.

Purified LT is a protein, possibly a glycoprotein, of 240,000–250,000 mol. wt., with a pI of 4.55 (Popoff, 1987). It is lethal to mice, cytotoxic to Vero cells, erythemetous and edematous in guinea pigs, and induces fluid accumulation in isolated intestinal loops. Tryptophan and methionine residues appear to be important for lethal activity. Lethal toxin-specific antiserum neutralized LT biological activities and also the lethal and cytotoxic activities of *C. difficile* B toxin (Popoff, 1987). The latter, however, was not recognized by anti-LT in immunoblotting, indicating limitations to the serological commonality of the two toxins.

2.6. *C. septicum*

This organism was a common cause of gangrene during the 1914–1918 war and its virulence was ascribed to its toxins. The α toxin causes spreading hemorrhagic necrosis and is cardiotoxic. DNAase (β toxin), hyaluronidase (γ toxin), and gelatinase may also have significant *in vivo* activity (Willis, 1969). The spectrum of human infection has since changed markedly, with patients with neoplasia, diabetes, or atherosclerosis now being at risk of acquiring infection via the gut (Editorial, 1987). Patients with neutropenic enterocolitis seem predisposed to clostridial infections. *Clostridium septicum* is the most common infectious agent (Boriello, 1985b), but the involvement of other species such as *C. perfringens*, *C. sordelli*, *C. spheroides*, and *C. paraperfringens* have been reported (Editorial, 1987; Newbold *et al.*, 1987). The exotoxins produced by the invading clostridia can give rise to acute gangrene of the bowel and toxemia. Neutropenic enterocolitis with *C. septicum* infection has a high fatality rate and requires rapid surgery since the condition is unlikely to respond to antibiotic therapy alone (Editorial, 1987).

3. PROPHYLACTIC AGENTS

3.1. Vaccines against Tetanus Toxin

Despite the availability of tetanus vaccines for over 60 years, the disease is still a significant public health problem, even in the most affluent of Western countries. In the United States, close to 100 cases of tetanus are reported each year and surveys indicate that approximately 10% of the population are inadequately protected against the disease due either to infrequency or total lack of immunization (Bleck, 1986). At present, commercially available tetanus vaccines consist of partially purified (approx. 50%) preparations of the neurotoxin which have been chemically inactivated with formaldehyde to give a toxoid. The toxoid is administered in an absorbed form, frequently combined with diphtheria and pertussis vaccines, for the immunization of children. A course of three doses of the absorbed toxoid is sufficient to confer effective protection against the disease and subsequent booster doses every 10 years are sufficient to sustain the immune response. Many unnecessary cases of tetanus occur when individuals neglect these booster doses and their tetanus antibody levels fall too low to offer effective protection.

The present tetanus vaccines, if administered at regular intervals, are extremely efficient at conveying protection against the disease with failure rates of less than 1 in 10 million. Commercial vaccines, however, contain proteins other than the tetanus neurotoxin which serve only to dilute the immune response and, as such, could be improved further by using highly purified neurotoxin preparations. In addition, the tetanus toxin gene was recently cloned in *E. coli* (Eisel *et al.*, 1986; Fairweather *et al.*, 1986), which may open up the possibility of alternative vaccines in the form of nontoxic fragments of the tetanus neurotoxin substituting for the formaldehyde-inactivated toxin. Recently, a nontoxic cloned fragment of the neurotoxin, representing the COOH terminal half of the heavy subunit (fragment C), was reported to induce neutralizing antibodies in mice (Fairweather *et al.*, 1987). Studies on the immunogenic properties of various neurotoxin fragments have shown that fragment AB (Fig. 1) produces a protective immune response in mice as efficient as the intact toxin molecule (Matsuda *et al.*, 1983). These fragments, if they could be produced cheaply by genetic manipulation, might be worthy candidates for future tetanus vaccines.

3.2. Vaccines against the *C. botulinum* Neurotoxins

The effectiveness of modern food-preserving processes in Western countries has made outbreaks of botulism resulting from the consumption of contaminated foodstuffs extremely rare, and consequently there is little justification on health or economic grounds for immunizing the general population against botulism. However, *C. botulinum* is frequently used as a test organism by microbiologists in the food industry and its neurotoxins are used by a growing number of neurobiochemists studying nerve action. In order to protect such laboratory workers from these extremely potent toxins, human vaccines have been developed against the botulinum toxins.

The formulation of the botulinum vaccines has changed little since their initial production in the early 1950s: partially purified preparations of the neurotoxins are first toxoided by formaldehyde treatment and then adsorbed onto precipitated aluminum salts. Polyvalent vaccines for human immunization (ABCDE and ABEF) were formulated in the late 1950s and toxoids prepared by these guidelines are still in current use. The main drawback with these early vaccines is the high proportion (60–90%) of contaminating proteins in the toxoid preparations, which, in addition to reducing considerably the desired immune response, give rise to a great deal of batch-to-batch variation. At the present time, stocks of botulinum vaccines are becoming low and a second generation vaccine is now required, not only to immunize laboratory workers, but also to produce antisera for passive immunotherapy of the rare, but often fatal, outbreaks of botulism. An obvious first step toward the production of an improved botulinum vaccine would be to toxoid highly purified preparations of the neurotoxins in place of the partially purified toxins presently used. With the exception of type G, all the botulinum neurotoxins have been purified to near-homogeneity: rapid affinity methods have been developed for purification of types A and B neurotoxins (Shone *et al.*, 1985; Evans *et al.*,

1987) and high-pressure liquid chromatography techniques have been employed in the purification of several of the other neurotoxin types (Schmitt *et al.*, 1986). While providing a much improved vaccine, the use of purified neurotoxins is not without its drawbacks. The toxins have to be produced under high-containment conditions and, to prevent possible reversion of toxoids to their active state, low concentrations of formaldehyde are required in the final product.

The immunogenic potential of the subunits of the botulinum neurotoxins has been investigated by several laboratories. In general, the separated toxin subunits produce a poor immune response compared to the intact toxin and appear to have little potential as vaccines. Unpublished data from our laboratory have shown that a toxin fragment consisting of the light chain linked to the NH_2 terminal half of the heavy chain (analogous to fragment AB of tetanus toxin) produces an immune response in guinea pigs comparable with that of the intact toxin and, since this fragment appears to lack the active site region responsible for binding the neurotoxin to the presynaptic nerve surface, it is nontoxic. Such a fragment, if cloned and expressed in a suitable host, might provide an excellent candidate for a future vaccine.

4. CLOSTRIDIAL ANTITOXINS

4.1. Gas Gangrene

Serotherapy was formerly used routinely as an adjunct to surgery and antibiotic therapy (MacLennan, 1962) and a polyvalent serum containing *C. perfringens, C. novyi*, and *C. septicum* antitoxins was recommended by the Medical Research Council (UK). Passive immunization was effective in preventing experimental gas gangrene caused by *C. novyi* in high-velocity missile wounds (Boyd *et al.*, 1972). However, the problem of adverse reactions associated with the use of equine-derived antitoxins limits the value of this form of treatment, and the combination of surgery and antibiotics remains the treatment of choice.

4.2. Tetanus

Since tetanus toxin is internalized by nerve cells, antitoxin cannot dislodge the toxin or reverse its effects. Therefore once clinical symptoms of tetanus have been presented, antitoxin can only serve to neutralize circulating toxin and prevent further toxin binding. While there is no overwhelming evidence for its efficacy, antitoxin may be used prophylactically in patients with severe wounding, especially where there is tissue trauma or contamination by soil or dust. In addition to its doubtful efficacy, a further objection to serotherapy is that equine antitoxins can give rise to very serious side reactions. The use of human tetanus immunoglobulin (HTI), derived from donors actively immunized with tetanus toxoid, offers several advantages, including the lack of serum sickness side effects and anaphylaxis, as well as a prolonged half-life (Christie, 1980). The recommended procedure for treatment of nonimmunized patients having possibly contaminated deep wounds is to give intramuscular injections of tetanus toxoid and 250 units of HTI (or equine antitoxin if HTI is not available) at different sites coupled with antibiotic therapy (Christie, 1980). Although it might be considered that the two components (toxoid and antitoxin) might interfere with one another, this appears not to happen and the desired combination of immediate and future immunity seems to be achieved.

4.3. Botulism

As with tetanus, by the time clinical symptoms of botulism are present, toxin has bound and been internalized by a significant number of nerve synapses, inducing a blockade of acetylcholine release. Although treatment with antitoxin at this stage will serve to neutralize any unbound toxin, it will not reverse or bring relief from existing symptoms. Furthermore, the use of heterologous equine antitoxin only introduces the risk of severe side effects, as described above.

The use of human immune plasma or immunoglobulin preparations (Lewis, 1981), which eliminates the problems of serum sickness and affords a longer half-life, avoids some of the side reactions associated with the use of equine sera but it is no more effective in treating established food-borne botulism. The relatively long half-life of human antibodies would, however, be advantageous in the treatment of toxic infections where infecting *C. botulism* organisms produce toxin over a period of time.

In infant botulism, the bowel is colonized by toxigenic *C. botulinum* organisms which may be present in the gut for several weeks (Arnon, 1981). Immunoglobulin therapy could provide adequate toxin-neutralizing protection during and beyond the period of gut colonization. A similar approach should also be beneficial in the treatment of wound botulism. The preparation of high-titer immune plasma and immunoglobulin derived from donors immunized with pentavalent botulism toxoid were described by Lewis (1981). He suggested that an alternative approach might be to use hybridoma technology to prepare specific monoclonal antibodies with all the perceived advantages for continued production.

5. BOTULINUM NEUROTOXIN TYPE A AS THERAPEUTIC AGENT

Considering the dramatic and emotive nature of many clostridial diseases, it is perhaps surprising that the most potent of microbial toxins is now finding use as a therapeutic drug as an alternative to surgical manipulation of certain aberrant muscular functions (Hambleton *et al.*, 1987). Unlike most pharmaceutical products, the mode and site of action of type A botulinum toxin was relatively well understood prior to its clinical application. Indeed, it was the knowledge of the specificity of the toxin which led Scott (1980) to introduce the technique of injecting *C. botulinum* type A toxin into extraocular muscles to correct squint. This innovative therapy has now been extended by Scott and many others to induce temporary flaccid paralysis in other conditions associated with muscles undergoing involuntary spasmodic contraction, e.g., focal dystonias.

5.1. Strabismus

For treatment of squint, electromyographic recording techniques are used to ensure precise location of the required muscle prior to injection of toxin. The effects of toxin are localized by injecting only small volumes (100 μl) and quantities (approx. 6.25×10^{-6} μg; Lee *et al.*, 1988). The type A toxin induces a dose-dependent flaccid paralysis in the injected muscle but the effect is abolished when motor fusion is reestablished, so that repeat injections are necessary. Such schedules do, however, achieve satisfactory, stable improvements in a number of patients (Lee *et al.*, 1988) and can be considered a safe alternative to surgery in some cases.

5.2. Blepharospasm

The use of botulinum toxin to cause selective weakening of overactive muscles has been applied to other conditions, in particular, benign essential blepharospasm. This condition is a focal dystonia characterized by repeated involuntary closure of the eyelids, primarily due to spasmodic contractions of the orbicularis muscle. The condition, which may arise from organic dysfunction of the basal ganglia, is often severe enough to cause functional blindness.

Botulinum toxin therapy has now been established as the treatment of choice, in preference over the surgical alternatives of stripping of the orbicularis muscle or bilateral partial avulsion of the facial nerve. Several thousand patients have now received regular treatment with total bilateral dosages of 5–20 ng (Elston, 1988; Patrinely *et al.*, 1988) and have generally been free of untoward side effects. As with squint, the treatment is not a cure and, with the reestablishment of motor control, repeat injections (at a lower dose) may be required as dictated by the response of the individual patient to the induced paralysis.

5.3. Other Applications

Other indications for the use of botulinum toxin therapy of ocular muscle dysfunction include postretinal detachment surgery, orbital trauma, VIth nerve palsy, and nystagmus (Lee *et al.*, 1985). It has also been used for the management of lateral rectus paresis (Scott and Kraft, 1985) and strabismus at the end stage of endocrine orbital myopathy (Scott, 1984). In many instances, treatment requires injection under elec-tromyogram (EMG) control.

Symptoms of hemifacial spasm (Elston, 1988), Meige syndrome (Mauriello, 1985), facial synkinesis (Elston and Ross-Russell, 1985), spas-modic torticollis (Tsui *et al.*, 1986), and spasmodic dysphonia (Miller *et al.*, 1987) have all been relieved by injecting the type A toxin. The treatment may also be extended to other conditions arising from focal spasmodic dystonias, e.g., writer's cramp (J. Elston, personal communication).

5.4. Side Effects of Toxin Therapy

Generally speaking, side effects from botulinum toxin therapy are minor, self-limited, and localized (Patrinely *et al.*, 1988), and can be mini-mized by careful placement of injections or possibly by localized injections of homologous botulinum antitoxin (Scott, 1988). The long-term side ef-fects of the treatment, as well as the maximum allowable dose, have not yet been determined. Concern has been expressed about the possible chronic stimulation of the immune system by repeated injections of toxin (Patrinely *et al.*, 1988) since the presence of circulating antibody could neutralize the efficiency of the treatment (Scott *et al.*, 1973). However, in our experience and that of Gonnering (1988), antibody cannot be detected in the sera of persons who have had repeated injections of toxin.

5.5. Preparation of Botulinum Toxin for Therapeutic Use

To be acceptable for clinical use, botulinum toxin, in common with other pharmaceutical agents, must be produced to appropriate standards of quality, safety, and efficacy according to current good manufacturing practice techniques (Melling *et al.*, 1988). Type A toxin for clinical use is manufactured at CAMR and supplied as purified complex toxin freeze-dried with a carrier protein (human serum albumin) and an inert (lactose) bulking agent (Hambleton *et al.*, 1987). The simplicity of the procedures for treatment of focal dystonias allows this to be carried out in the patient's own locality rather than in specialist centers as for strabismus, where EMG monitoring is required, and the relatively low amounts of toxin involved do not require those administering the toxin to receive toxoid immunization.

REFERENCES

Agui, T., Syuto, B., Oguma, K., Iida, H., and Kubo, S., 1985, Binding of *Clostridium botulinum* type C to rat brain synaptosomes, *J. Biochem.* **94**:521–527.

Aktories, K., Barmann, M., Ohishi, I., Tsuyama, S., Jakobs, K. and Habermann, E., 1986, Botulinum C2 toxin ADP-ribosylates actin, *Nature* **322**:390–392.

Alouf, J. E. and Jolivet-Reynaud, C. 1981, Purification and characterization of *Clostridium perfringens* delta toxin, *Infect. Immun.* **31**:536–546.

Arnon, S., 1981, Infant botulism, pathogenesis, clinical aspects and relation to crib death, in: *Biomedical Aspects of Botulism* (G. E. Lewis Jr., ed.), Academic Press, New York, pp. 331–345.

Arseculeratne, S. N., Pannabokke, R. G., and Wijesundera, S., 1969, The toxins responsible for the lesions of *Clostridium sordelli* gas gangrene, *J. Med. Microbiol.* **2**:37–53.

Banno, Y. T., Kobayashi, T., Kono, H., Watanabe, K., Ueno, K., and Nozawa, Y., 1984, Biochemical characterization and biologic action of two toxins (D1 and D2) from *Clostridium difficile*, *Rev. Infect. Dis.* **6**:S11–S20.

Bartlett, J. G., 1979, Antibiotic associated pseudo-membranous colitis, *Rev. Infect. Diseases* **1**: 530–539.

Bartlett, J. G., 1983, Pseudomembranous colitis in: *Human* intestinal microflora in health and disease (D. J. Hentges, ed.), Academic Press, New York, pp. 447–479.

Bizzini, B., 1979, Tetanus toxin, *Microbiol. Rev.* **43**:224–240.

Black, J. D., and Dolly, J. O., 1986, Interaction of ^{125}I-labeled botulinum neurotoxins with nerve terminals, *J. Cell. Biol.* **103**:521–534.

Bleck, T. P., 1986, Pharmacology of tetanus, *Clin. Neuropharmacol.*, **9**:103–120.

Bolton, R. P., and Thomas, D. F. M., 1986, Pseudomembranous colitis in children and adults, *Br. J. Hosp. Med.* **35**:37–41.

Boriello, S. P., 1985a, *Clostridia in Gastrointestinal disease*, CRC, Boca Raton.

Boriello, S. P., 1985b, Newly described clostridial diseases of the gastrointestinal tract: *Clostridium perfringens* enterotoxin-associated diarrhoea and neutropenic enterocolitis due to *Clostridium septicum*, in: *Clostridia in Gastrointestinal Disease* (S. P. Boriello, ed.), CRC, Boca Raton pp. 224–229.

Boriello, S. P. and Carman, R. J., 1985, Clostridial diseases of the gastrointestinal tract in animals, in: *Clostridia in Gastrointestinal Disease* (S. P. Boriello, ed.), CRC, Boca Raton, pp. 195–221.

Boriello, S. P., and Larson, H. E., 1985, Pseudomembranous and antibiotic-associated colitis, in: *Clostridia in Gastrointestinal Disease* (S. P. Boriello, ed.), CRC, Boca Raton, pp. 145–164.

Boyd, N. A., Walker, P. D., and Thomson, R. O., 1972, The prevention of experimental *Clostridium novyi* gas gangrene in high velocity missile wounds by passive immunization, *J. Med. Microbiol.* **5**:459–465.

Chang, T. W., Gorbach, S. C., and Bartlett, J. B., 1978, Neutralization of *Clostridium difficile* toxin by *Clostridium sordelli* antitoxin, *Infect. Immun.* **22**:418–422.

Christie, A. B., 1980, *Infectious diseases: epidemiology and clinical practice* 3rd ed. Churchill Livingstone, New York.

DasGupta, B. R., 1981, Structure-function relation of botulinum toxins, in: *Biomedical Aspects of Botulism* (G. E. Lewis Jr., ed.), Academic Press, New York, pp. 1–19.

DasGupta, B. R., and Sugiyama, H., 1972, A common subunit structure in *Clostridium botulinum* type A, B and E toxins, *Biochem. Biophys. Res. Commun.* **48**:108–112.

Dolly, J. O., Black, J. D., Williams, R. S., and Melling, J., 1984, Acceptors for botulinum neurotoxin reside on motor nerve terminals and mediate its internalization, *Nature* **307**: 457–460.

Dreyer, A. S. F., and John, C., 1981, At least three sequential steps are involved in tetanus toxin-induced block of neuromuscular transmission, *Naun. Schmiedeberg's Arch. Pharmacol.* **317**:326–330.

Duncan, C. L., Strong, D. H., and Sebald, M., 1972, Sporulation and enterotoxin production by a mutant of *Clostridium perfringens*, *J. Bacteriol.* **110**:378–391.

Editorial, 1987, *Clostridium septicum* and neutropenic enterocolitis, *Lancet* **2**:608.

Eisel, U., Jarausch, W., Goretzki, K., Henschen, A., Engels, J., Weller, U., Hudel, M., Habermann, E., and Niemann, H., 1986, Tetanus toxin: Primary structure, expression in E. coli, and homology with botulinum toxins, *EMBO J.* **5**:2495–2502.

Elston, J. S., 1988, Botulinum toxin therapy for involuntary eye movement, *Eye* **2**:12–15.

Elston, J. S., and Ross-Russell, R. W., 1985, Effect of treatment with botulinum toxin on neurogenic blepharospasm, *Br. med. J.* **240**:1857–1859.

Evans, D. M., Williams, R. S., Shone, C. C., Hambleton, P., Melling, J., and Dolly, J. O., 1987, Botulinum type B, its purification, radioiodination and interaction with rat brain synaptosomes, *Eur. J. Biochem.* **154**:409–416.

Evans, D. G. B., 1945, The treatment with anti-toxin of experimental gas gangrene produced in guinea pigs by (a) *Clostridium welchii*, (b) *Clostridium oedematiens* and (c) *Clostridium septicum*, *Br. J. Exp. Pathol.* **26**:104–111.

Fairweather, N. F. and Lyness, V. A., 1986, The complete nucleotide sequence of tetanus toxin, *Nucl. Acids. Res.* **14**:7809–7812.

Fairweather, N. F., Lyness, V. A., and Maskell, D. J., 1987, Immunization of mice against tetanus and fragments of tetanus toxin synthesized in E. coli, *Infect. Immun.* **55**:2541–2545.

Feldman, R. A., Morris, J. G. and Pollard, R. A., 1981, Epidemiologic characteristics of botulism in the United States, in: *Biomedical Aspects of Botulism* (G. E. Lewis Jr., ed.), Academic Press, New York, pp. 271–284.

Fernie, D. S., Thomson, R. O., Batty, I., and Walker, P. D., 1983, Active and passive immunissation to protect against antibiotic associated caecitis in hamsters, *Dev. Biol.* **53**:325–332.

Florin, I., and Thelestam, M., 1986, ADP-ribosylation in cultured cells treated with *Clostridium difficile* toxin B, *Biochem. Biophys. Res. Commun.* **139**:64–70.

Gansel, M., Penner, R., and Dreyer, F., 1987, Distinct sites of action of clostridial neurotoxins revealed by double-poisoning of mouse motor nerve terminals, *Pfugers Arch.* **409**:533–539.

Gonnering, R. S., 1988, Negative antibody response to long-term treatment of facial spasm with botulinum toxin, *Am. J. Ophthalmol.* **105**:313–315.

Granum, P. E., 1982, Inhibition of protein synthesis by a tryptic polypeptide of *Clostridium perfringens* type A enterotoxin, *Biochim. Biophys. Acta* **708**:6–11.

Granum, P. E., and Whitaker, J. R., 1980, Improved methods for purification of entertoxin from *Clostridium perfringens* type A, *Appl. Env. Microbiol.* **39**:1120–1122.

Habeeb, A. F. S. A., 1975, Studies on ε-prototoxin of *Clostridium perfringens* type D. Physicochemical and chemical properties of ε-prototoxin, *Biochim. Biophys. Acta.* **412**:62–69.

Hambleton, P., Shone, C. C., and Melling, J., 1987, Botulinum Toxin-Structure, action and clinical use, in: *Neurotoxins and Their Pharmacological Application* (P. Jenner, ed.), New York, Raven Press, pp. 233–260.

Hauschild, A. M. W., 1971, *Clostridium perfringens* toxin types B, C, D and E, in: *Microbial Toxins,* Vol. IIa (S. Kadis, T. C. Montie and S. J. Ajl, eds.), Academic Press, New York, pp. 159–188.

Hobbs, B. C., 1979, *Clostridium perfringens* gastroenteritis, in: *Foodborne infections and Intoxication* (H. Riemann and F. L. Bryan, eds.), Academic Press, London, pp. 131–171.

Hoch, D. H., Romero-Mira, M., Ehrlich, B. E., Finkelstein, A., DasGupta, B. R., and Simpson, L. L., 1985, Channels formed by botulinum, tetanus and diphtheria toxins in planar lipid bilayers: Relevance to translocation of proteins across membranes, *Proc. Natl. Acad. Sci. USA* **82**:1692–1696.

Horiguchi, Y., Akai, T., and Sakaguchi, G., 1987, Isolation and function of a *Clostridium perfringens* enterotoxin fragment, *Infect. Immun.* **55**:2912–2915.

Howard, F. M., Flynn, D. M., Bradley, J. M., Noone, P., and Szawatkowski, M., 1977, Outbreak of necrotising enterocolitis caused by *Clostridium butyricum*. *Lancet* **2**:1099–1102.

Ispolatovskaya, M. V., 1971, Type A *Clostridium perfringens* toxin, in: *Microbiol. Toxins,* Vol. IIa (S. Kadis, T. C. Montie, and S. J. Ajl, eds.), Academic Press, New York, pp. 109–158.

Jolivet-Reynaud, C., Cavaillan, J. M., and Alouf, J. E., 1982, Selective cytotoxicity of *Clostridium perfringens* delta toxin on rabbit leucocytes, *Infect. Immun.* **38**:860–864.

Knight, D. E., 1986, Botulinum toxin types A, B and D inhibit catecholamine secretion from bovine adrenal medullary cells, *FEBS Lett.* **207**:222–226.

Kozaki, S., 1979, Interaction of type A, B and E derivative toxins with synaptosomes of rat brain, *Nauny. Schmeideberg's Arch. Pharmacol.* **308**:67–70.

Kozaki, S., Kamata, Y., Takafumi, N., Ogasawara, J., and Sakaguchi, G., 1986, The use of monoclonal antibodies to analyse the structure of *Clostridium botulinum* type E derivative toxin, *Infect. Immun.* **52**:786–791.

Kozaki, S., Ogasawora, J., Shimote, Y., Kamata, Y., and Sakaguchi, G., 1987, Antigenic Structure of *Clostridium botulinum* type B neurotoxin and its interaction with gangliosides, cerebrosides and free fatty acids, *Infect. Immun.* **55**:3051–3056.

Krivan, H. C., Clark, G. F., Smith, D. F., and Wilkins, T. D., 1986, Cell surface binding of *Clostridium difficile* enterotoxin: evidence for a glycoprotein containing the sequence Gal α 1-3gal β 1-4GlcNAc, *Infect. Immun.* **53**:573–581.

Lee, J., Elston, J., Vickers, S., Powell, C., Ketley, J., and Hogg, C., 1988, Botulinum toxin therapy for squints, *Eye* **2**:24–28.

Lee, L. and Cherniak, R., 1974, Capsular polysaccharide of *Clostridium perfringens Infect. Immun.* **9**:318–322.

Lewis, G. E., Jr., 1981, Approaches to the prophylaxis immunotherapy and chemotherapy of botulism, in: *Biomedical Aspects of Botulism* (G. E. Lewis Jr., ed.), Academic Press, New York, pp. 261–270.

Lyerly, D. M., Krivan, H. C., and Wilkins, T. D., 1988, *Clostridium difficile:* its diseases and toxins, *Clin. Microbiol. Rev.* **1**:1–18.

MacLennan, J. D., 1962, The histotoxic clostridial infection of man, *Bacteriology* **26**:177–174.

Matsuda, M., Makinga, G., and Hirai, T., 1983, Studies on the antibody composition and neutralizing activity of tetanus antitoxin sera from various species of animals in relation to the antigenic substructure of the tetanus toxin molecule, *Biken. J.* **26**:133–143.

Matsuda, M., Ozutsumi, K., Iwashi, M., and Sugimoto, N., 1986, Primary action of *Clostridium perfringens* type A enterotoxin on Hela and Vero cells in the absence of extracellular calcium: rapid and characteristic changes in membrane permeability, *Biochem. Biophys. Res. Commun.* **141**:704–710.

Matsuoka, I., Syuto, B., Kurihara, K., and Kubo, S., 1987, ADP-ribosylation of specific membrane proteins in pheochromocytoma and primary cultured brain cells by botulinum neurotoxins type C and D, *FEBS Lett.* **216**:295–299.

Mauriello, J. A., 1985, Blepharospasm, Meige syndrome and hemifacial spasm: Treatment with botulinum toxin, *Neurology* **35**:1499–1500.

McClane, B. A., and McDonel, J. C., 1981, Protective effects of osmotic stabilisers on morphological and permeability alterations induced in Vero cells by *Clostridium perfringens* enterotoxin, *Biochim. Biophys. Acta* **641**:401–409.

McFarland, L. V., and Stamm, W. E., 1986, Review of *Clostridium difficile* associated diseases, *Am. J. Infect. Control* **14**:99–109.

Melling, J., Hambleton, P., Shone, C. C., 1988, *Clostridium botulinum* toxins: Nature and preparation for clinical use, *Eye* **2**:16–23.

Midura, T. F., and Arnon, S., 1976, Infant botulism, identification of *Clostridium botulinum* and its toxin in faeces, *Lancet* **2**:934–936.

Miller, R. M., Woodson, G. E., and Jankovic, J., 1987, Botulinum toxin injection of the vocal fold for spasmodic dysphonia, *Arch. Otolary Tngol Head Neck Surg.* **113**:603–605.

Molgo, J., and Thesleff, S., 1984, Studies on the mode of action of botulinum toxin type A at the frog neuromuscular junction, *Brain Res.* **279**:309–316.

Mollby, R. and Wadstrom, T., 1973, Purification of phospholipase C (alpha toxin) from *Clostridium perfringens, Biochim. Biophys. Acta.* **321**:569–584.

Morris, N. P., Consiglio, E., Hohn, L. D., Habig, W. H., Hardegree, M. C., and Helting, T. B.,

1980, Interaction of fragments B and C of tetanus toxin with neural and thyroid membranes and with gangliosides, *J. Biol. Chem.* **255**:6071–6076.

Murayama, S., Syoto, B., Oguma, K., Iida, H., and Kubo, S., 1984, Comparison of *Clostridium botulinum* toxin types D and C_1 in molecular property, antigenicity and binding ability to rat brain synaptosomes, *Eur. J. Biochem.* **142**:487–492.

Nakamura, S., Tanabe, N., Yamakawa, K., and Nishida, S., 1983, Cytotoxin products by *Clostridium sordelli* strains, *Microbiol. Immunol.* **27**:695–502.

Newbold, K. M., Lord, M. G., and Baglin, T. P., 1987, Role of clostridial organisms in neutropenic enterocolitis, *J. Clin. Pathol.* **40**:471.

Ohashi, Y., and Narumiya, S., 1987, ADP-ribosylation of a M_r 21,000 membrane protein by type D botulinum toxin, *J. Biol. Chem.* **262**:1430–1433.

Ohishi, I., 1984, Oral toxicities of *Clostridium botulinum* type A and B toxins from different strains, *Infect. Immun.* **43**:487–490.

Parnas, J., 1976, The effects of *Clostridium perfringens* beta toxin (type C) on the mobility of intestinal segments *in vitro*, *Abstr. Zentralblat. Bacteriol. Hyg.* **234**:243–246.

Patrinely, J. R., Whiting, A. S. and Anderson, R. L., 1988, Local side effects of botulinum toxin infections, *Adv. Neurol.* **49**:493–500.

Pedersen, P. V., Hansen, F. H., Halveg, A. B., Christiansen, E. D., Justesten, T., and Hogh, P., 1976, Necrotising enterocolitis in the newborn. Is it gas gangrene of the bowel?, *Lancet* **2**:715–716.

Penner, R., Neher, E., and Dreyer, F., 1986, Intracellularly injected tetanus toxin inhibits exocytosis in bovine adrenal chromaffin cells, *Nature* **324**:76–78.

Phillips, K. D., Brazier, J. S., Levett, P. N., and Willis, A. T., 1985, Clostridia, in: *Isolation and Identification of Microorganisms of Medical and Veterinary Importance* (C. H. Collins and J. M. Grange, eds.), Academic Press, London, pp. 215–236.

PHLS. CDSC., 1986, Foodborne disease surveillance in England and Wales 1984, *Commun. Dis. Rep.* **34**:3–6.

Pierce, E. J., Davidson, M. D., Parton, R. G., Habig, W. H., and Critchley, D. R., 1986, Characterization of tetanus toxin binding to rat brain membranes, *Biochem. J.* **236**:845–852.

Popoff, M. R., 1987, Purification and Characterization of *Clostridium sordelli* lethal toxin and cross-reactivity with *Clostridium difficile* cytotoxin, *Infect. Immun.* **55**:35–43.

Reynolds, D., 1987, The large scale purification of *Clostridium pertringens* type A enterotoxin and *Staphylococcus aureus* enterotoxin A M.Phil thesis, Council for National Academic Awards, UK.

Reynolds, D., Tranter, H. S., and Hambleton, P., 1986, Scaled-up production and purification of *Clostridium perfringens* type A enterotoxin, *J. Appl. Bacteriol.* **60**:517–525.

Richardson, M., and Granum, P. E., 1985, The amino acid sequence of the enterotoxin from *Clostridium perfringens* type A, *FEBS Lett.* **182**:479–486.

Rogers, T. B., and Snyder, S. H., 1981, High affinity binding of tetanus toxin to mammalian brain membranes, *J. Biol. Chem.* **256**:2402–2407.

Rosener, S., Chatwal, G. S. and Aktories, K., 1987, Botulinum ADP-ribosyltransferase C3 but not neurotoxins C1 and D ADP-ribosylates low molecular mass GTP-binding proteins, *FEBS Lett.* **224**:38–42.

Sakurai, J., and Duncan, C. L., 1978, Some properties of Beta-toxin produced by *Clostridium perfringens* type C, *Infect. Immun.* **21**:678–680.

Sakurai, J., Gujii, Y., and Matsuura, M., 1980, Effect of oxidizing and sulfhydryl group reagents on beta toxin from *Clostridium perfringens* type C, *Microbiol. Immunol.* **24**:595–601.

Sakurai, J., and Nagahama, M., 1985, Role of one tryptophan residue in the lethal activity of *Clostridium perfrigens* epsilon toxins, *Biochem. Biophys. Res. Commun.* **128**:760–766.

Sanchez-Prieto, J., Shira, T. S., Evans, D., Ashton, A., Dolly, J. O., and Nicholls, D. G., 1987,

Botulinum toxin A blocks glutamate exocytosiis from guinea-pig cerebral cortical synaptosomes, *Eur. J. Biochem.* **165:**675–681.

Schmitt, J. J., and Siegel, L. S., 1986, Purification of type E botulinum neurotoxin by high performance ion exchange chromatography, *Anal. Biochem.* **156:**213–219.

Scott, A. B., 1980, Botulinum toxin injection into extraocular muscles as an alternative to strabismus surgery, *Ophthalmology* **87:**1044–1049.

Scott, A. B., 1984, Injection treatment of endocrine orbital myopathy, *Doc. Ophthalmol.* **58:** 141–145.

Scott, A. B., 1988, Antitoxin reduces botulinum side effects, *Eye* **2:**29–32.

Scott, A. B., and Kraft, S. P., 1985, Botulinum toxin injection in the management of lateral rectus paresis, *Ophthalmol.* **92:**676–683.

Scott, A. B., Rosenbaum, A. C., and Collins, C. C., 1973, Pharmacological weakening of extraocular muscles, *Invest. Ophthalmol.* **12:**924–927.

Shandera, W. X., Tacket, C. O., and Blake, P. A., 1983, Food poisoning due to *Clostridium perfringens* in the United States, *J. Infect. Dis.* **147:**167–170.

Shone, C. C., 1987, *Clostridium botulinum* neurotoxins, their structures and modes of action, in: *Natural Toxicants in Foods* (D. Watson, ed.), Ellis Horwood Ltd., Chichester, pp. 11–57.

Shone, C. C., Hambleton, P., and Melling, J., 1985, Inactivation of *Clostridium botulinum* type A neurotoxin by trypsin and purification of two tryptic fragments, *Eur. J. Biochem.* **151:**75–85.

Shone, C. C., Hambleton, P., and Melling, J., 1987, A 50-kDa fragment from the NH$_2$-terminus of the heavy subunit of *Clostridium botulinum* type A neurotoxin forms channels in lipid vesicles, *Eur. J. Biochem.* **167:**175–180.

Simpson, L. L., 1981, The origin, structure and pharmacological activity of botulinum toxin, *Pharmacol. Rev.* **33:**155–188.

Simpson, L. L., 1984, Molecular basis for the pharmacological action of *Clostridium botulinum* type C$_2$ toxin, *J. Pharmacol. Exp. Ther.* **230:**665–669.

Simpson, L. L., Stiles, B. G., Zepeda, H. H., and Wilkins, T. D., 1987, Molecular Basis for the Pathological Actions of *Clostridium perfringens* Iota toxin, *Infect. Immun.* **55:**118–122.

Smyth, C. J., 1975, The identification and purification of multiple forms of θ-haemolysin (θ toxin) of *Clostridium perfringens* type A., *J. Gen. Microbiol.* **87:**219–238.

Sterne, M., and Warrack, G. M., 1984, The types of *Clostridium perfringens, J. Pathol. Bacteriol.* **88:**279–283.

Stevens, D. L., Mitten, J., and Henry, C., 1987, Effects of and toxins from *Clostridium perfringens* on human polymorphonuclear leukocytes, *J. Infect. Dis.* **156:**324–333.

Stiles, B. G., and Wilkins, T. D., 1986, *Clostridium perfringens* iota toxin: Synergism between two proteins, *Toxicon.* **24:**767–773.

Stringer, M. F., 1985, *Clostridium perfringens* type A food poisoning, in: *Clostridia in Gastrointestinal disease* (S. P. Boriello, ed.), CRC, Boca Raton, pp. 117–143.

Sugii, S., Ohishi, I., and Sakaguchi, G., 1977, Intestinal absorption of botulinum toxins of different molecular sizes in rats, *Infect. Immun.* **17:**910–914.

Sugiyama, H., 1980, *Clostridium botulinum* neurotoxins, *Microbiol. Rev.* **44:**419–448.

Sullivan, N. M., Pellett, M. S., and Wilkins, T. D., 1982, Purification and characterisation of toxins A and B of *Clostridium difficile*, *Infect. Immun.* **35:**1032–1040.

Tsui, J. K. C., Eisen, A., Stoessl, A. J., Calne, S., and Calne, D. B., 1986, Double-Blind study of botulinum toxin in spasmodic torticollis. *Lancet* **2:**245–247.

Vandekerckhove, J., Schering, B., Barmann, M., and Aktories, K., 1988, Botulinum C2 toxin ADP-ribosylates cytoplasmic β/γ-actin in arginine 177, *J. Biol.Chem.* **263:**696–700.

Walker, P. D., 1985, Pig-Bel, in: *Clostridia in gastrointestinal disease*, (S. P. Boriello, ed.), CRC, Boca Raton, pp. 93–115.

Wellhoner, H.-H., 1982, Tetanus neurotoxin, *Rev. Physiol. Biochem. Pharmacol.* **93:**1–68.

Williams, R. S., Tse, C.-K.., Dolly, J. O., Hambleton, P., and Melling, J., 1983, Radioiodination of botulinum neurotoxin type A and retention of biological activity and its binding to rat brain synaptosomes, *Eur. J. Biochem.* **131**:437–445.

Willis, A. T., 1969, *Clostridia of Wound Infection*, Butterworths, London.

Worthington, R. W., and Mulders, M. S. G., 1977, Physical changes in the epsilon prototoxin molecule of *Clostridium perfringens* during enzymatic activation, **18**:549–551.

Index